Advances in Intelligent Systems and Computing

Volume 1204

The series "Advances in Intelligent Systems and Computing" contains publications on theory, applications, and design methods of Intelligent Systems and Intelligent Computing. Virtually all disciplines such as engineering, natural sciences, computer and information science, ICT, economics, business, e-commerce, environment, healthcare, life science are covered. The list of topics spans all the areas of modern intelligent systems and computing such as: computational intelligence, soft computing including neural networks, fuzzy systems, evolutionary computing and the fusion of these paradigms, social intelligence, ambient intelligence, computational neuroscience, artificial life, virtual worlds and society, cognitive science and systems, Perception and Vision, DNA and immune based systems, self-organizing and adaptive systems, e-Learning and teaching, human-centered and human-centric computing, recommender systems, intelligent control, robotics and mechatronics including human-machine teaming, knowledge-based paradigms, learning paradigms, machine ethics, intelligent data analysis, knowledge management, intelligent agents, intelligent decision making and support, intelligent network security, trust management, interactive entertainment, Web intelligence and multimedia.

The publications within "Advances in Intelligent Systems and Computing" are primarily proceedings of important conferences, symposia and congresses. They cover significant recent developments in the field, both of a foundational and applicable character. An important characteristic feature of the series is the short publication time and world-wide distribution. This permits a rapid and broad dissemination of research results.

**** Indexing: The books of this series are submitted to ISI Proceedings, EI-Compendex, DBLP, SCOPUS, Google Scholar and Springerlink ****

More information about this series at http://www.springer.com/series/11156

Pedro M. Arezes · Ronald L. Boring
Editors

Advances in Safety Management and Human Performance

Proceedings of the AHFE 2020 Virtual Conferences on Safety Management and Human Factors, and Human Error, Reliability, Resilience, and Performance, July 16–20, 2020, USA

Springer

Editors
Pedro M. Arezes
DPS, School of Engineering
University of Minho
Guimaraes, Portugal

Ronald L. Boring
Idaho National Laboratory
Idaho Falls, ID, USA

ISSN 2194-5357 ISSN 2194-5365 (electronic)
Advances in Intelligent Systems and Computing
ISBN 978-3-030-50945-3 ISBN 978-3-030-50946-0 (eBook)
https://doi.org/10.1007/978-3-030-50946-0

This Springer imprint is published by the registered company Springer Nature Switzerland AG
The registered company address is: Gewerbestrasse 11, 6330 Cham, Switzerland

Advances in Human Factors and Ergonomics 2020

AHFE 2020 Series Editors

Tareq Z. Ahram, Florida, USA
Waldemar Karwowski, Florida, USA

11th International Conference on Applied Human Factors and Ergonomics and the Affiliated Conferences

Proceedings of the AHFE 2020 Virtual Conferences on Safety Management and Human Error, Reliability, Resilience, and Performance, July 16–20, 2020, USA

Advances in Neuroergonomics and Cognitive Engineering	Hasan Ayaz and Umer Asgher
Advances in Industrial Design	Giuseppe Di Bucchianico, Cliff Sungsoo Shin, Scott Shim, Shuichi Fukuda, Gianni Montagna and Cristina Carvalho
Advances in Ergonomics in Design	Francisco Rebelo and Marcelo Soares
Advances in Safety Management and Human Performance	Pedro M. Arezes and Ronald L. Boring
Advances in Human Factors and Ergonomics in Healthcare and Medical Devices	Jay Kalra and Nancy J. Lightner
Advances in Simulation and Digital Human Modeling	Daniel N Cassenti, Sofia Scataglini, Sudhakar L. Rajulu and Julia L. Wright
Advances in Human Factors and Systems Interaction	Isabel L. Nunes
Advances in the Human Side of Service Engineering	Jim Spohrer and Christine Leitner
Advances in Human Factors, Business Management and Leadership	Jussi Ilari Kantola, Salman Nazir and Vesa Salminen
Advances in Human Factors in Robots, Drones and Unmanned Systems	Matteo Zallio
Advances in Human Factors in Cybersecurity	Isabella Corradini, Enrico Nardelli and Tareq Ahram

(continued)

(continued)

Advances in Human Factors in Training, Education, and Learning Sciences	Salman Nazir, Tareq Ahram and Waldemar Karwowski
Advances in Human Aspects of Transportation	Neville Stanton
Advances in Artificial Intelligence, Software and Systems Engineering	Tareq Ahram
Advances in Human Factors in Architecture, Sustainable Urban Planning and Infrastructure	Jerzy Charytonowicz
Advances in Physical, Social & Occupational Ergonomics	Waldemar Karwowski, Ravindra S. Goonetilleke, Shuping Xiong, Richard H.M. Goossens and Atsuo Murata
Advances in Manufacturing, Production Management and Process Control	Beata Mrugalska, Stefan Trzcielinski, Waldemar Karwowski, Massimo Di Nicolantonio and Emilio Rossi
Advances in Usability, User Experience, Wearable and Assistive Technology	Tareq Ahram and Christianne Falcão
Advances in Creativity, Innovation, Entrepreneurship and Communication of Design	Evangelos Markopoulos, Ravindra S. Goonetilleke, Amic G. Ho and Yan Luximon

Preface

This volume combines the proceedings of two affiliated conferences of the 2020 Applied Human Factors and Ergonomics conference: the 7th International Conference on Safety Management and Human Factors, chaired by Pedro Arezes of University of Minho, Portugal, and the 4th International Conference on Human Error, Reliability, Resilience, and Performance, chaired by Ron Boring of Idaho National Laboratory, USA.

Safety Management and Risk Prevention is a common thread throughout every workplace, yet keeping employee safety and health knowledge current is a continual challenge for all employers. The discipline of Safety Management and Human Factors is a cross-disciplinary area concerned with protecting the safety, health and welfare of people engaged in work or employment and in society at large. The book offers a platform to showcase research and for the exchange of information in safety management and human factors. Mastering Safety Management and Human Factors concepts is fundamental to the creation of both products and systems that people are able to use and for work systems design, avoiding stresses and minimizing the risk for accidents.

This book focuses on the advances in the safety management and its relationship with human factors, which are a critical aspect in the design of any human-centered technological system. The ideas and practical solutions described in the book are the outcome of dedicated research by academics and practitioners aiming to advance theory and practice in this dynamic and all-encompassing discipline.

The International Conference on Human Error, Reliability, Resilience, and Performance (HERRP) is unlike other risk conferences, which have tended to be centered largely on probabilistic risk of hardware systems. HERRP has a decidedly human factors' angle. The research presented explores human error from a human factors perspective, not solely a risk modeling perspective.

The purpose of the HERRP conference is to bring together researchers and practitioners from different fields who broadly share the study of human error. The HERRP conference is intended to serve as an umbrella for human error topics by providing an annual forum for otherwise disjoint research efforts. As such, the conference is intended to complement but not replace existing specialized forums on particular facets of human error. The HERRP conference is distinctly interdisciplinary, encouraging the submission of papers in focused technical domains that would benefit from interaction with a wide human factors' audience. Additionally, the HERRP conference aims to provide a yearly, high-quality, archival collection of papers that may be readily accessed by the current and future research and practitioner community.

Six sections are presented in this book:

Section 1 Safety Management
Section 2 Assessing Risks
Section 3 Emergent Issues in Safety Management
Section 4 Resilience and Recovery
Section 5 Cognitive Factors in Human Performance
Section 6 New Findings and Methods in Human Performance

Sections 1-3 cover topics related to safety management, while Sections 4-6 cover topics related to human error, reliability, resilience and performance. Thematically, the two conferences complement each other. The former focuses on prevention and management of risk, while the latter discusses causes of human error.

To err is human, and human error is consistently implicated as a significant factor in safety incidents and accidents. Yet, as pervasive and important as human error is, the study of human error has been fragmented into many different fields. In fact, in many of these fields, the term "human error" is considered negative, and terms such as human variability and human failure are preferred. Across differences in terminology and approach, the common link remains an interest in how, why and when humans make incorrect decisions or commit incorrect actions. Human error often has significant consequences, and a variety of approaches have emerged to identify, prevent or mitigate it. These different approaches find a unified home in this volume covering a wide spectrum of safety and risk topics.

Each section contains research papers that have been reviewed by members of the International Editorial Board. Our sincere thanks and appreciation to the board members as listed below:

We wish to thank the authors for their exceptional contributions and to Scientific Advisory Board for encouraging strong submissions:

Safety Management and Human Factors

S. Costa, Portugal
J. Domingues, Portugal
A. Drummond, Ireland
L. Franz, Brazil
F. Guldenmund, The Netherlands
C. Jacinto, Portugal
L. Kocůrková, Czech Republic
T. Larsson, Sweden
M. Martínez-Aires, Spain
R. Melo, Portugal
M. Menozzi, Switzerland
A. Miguel, Portugal
B. Mrugalska, Poland
D. Nathanael, Greece
S. Nazir, Norway/Italy
M. Neves, Portugal
I. Nunes, Portugal
M. Pillay, Australia
R. Pope, USA
M. Rodrigues, Portugal
J. Rubio-Romero, Spain
J. Santos Baptista, Portugal
T. Saurin, Brazil
M. Shahriari, Turkey
S. Silva, Portugal
M. Silva Borges, Brazil
P. Sivaprakash, India
P. Swuste, The Netherlands
G. Szabo, Hungary
W. Van Wassenhove, France

Human Error, Reliability, Resilience and Performance

H. Blackman, USA
Y. Chang, USA
S. Filho, Brazil
D. Gertman, USA
K. Groth, USA
X. He, Sweden
Y. Kim, Korea
B. Kirwan, France
K. Laumann, Norway
Z. Li, China
P. Liu, China

R. McDonald, Norway
R. McLeod, UK
M. Merad, France
N. Meshkati, USA
A. Obenius-Mowitz, Sweden
J. Park, Korea
M. Pillay, Australia
A. Salway, Canada
C. Smidts, USA
O. Straeter, Germany
P. Trbovich, Canada
M. Weinger, USA
A. Whaley, USA
D. Yacht, USA

July 2020

Pedro Arezes
Ronald Laurids Boring

Contents

Emergent Issues in Safety Management

Resilience and Recovery

Cognitive Factors in Human Performance

Safety Management

Utilization of Machine Learning in Analyzing Post-incident State of Occupational Injuries in Agro-Manufacturing Industries

Fatemeh Davoudi Kakhki[1(✉)], Steven A. Freeman[2], and Gretchen A. Mosher[2]

[1] Department of Technology, San Jose State University, San Jose, CA 95192, USA
fatemeh.davoudi@sjsu.edu

[2] Department of Agricultural and Biosystems Engineering, Iowa State University, Ames, IA 50011, USA
{sfreeman, gamosher}@iastate.edu

Abstract. Due to the high frequency and costs of occupational incidents in agro-manufacturing operations, as well as substantial impact of occupational injuries on labor-market outcomes, predicting the post-incident state of an injury and identifying its contributory factors is vital to protect workers, improve workplace safety, and reduce overall costs of injuries. This study evaluates the performance of machine learning algorithms in classifying post-incident outcomes of occupational injuries in agro-manufacturing operations. Injury factors extracted from 14,000 workers' compensation claims recorded between 2008 and 2016 in the Midwest region of the United States were used to develop machine learning models. The models predicted incident outcomes based on injury details with an overall accuracy rate of 78%, and high accuracy rate for medical post-incident state (97–98%). The results emphasize the significance of quantitative analysis of empirical injury data in safety science, and contributes to enhanced understanding of occupational incidents root causes using predictive modeling along with safety experts' perspective.

Keywords: Occupational safety · Occupational incident analysis · Machine learning · Decision trees · Safety management

1 Introduction

Occupational incidents have direct impact on labor-market outcomes, including health of the injured workers, income reduction, and job loss [1]. Moreover, workplace incidents include medical and indemnity costs as well as indirect costs such as equipment damage and repair, incident investigation time, a slowdown for production schedules, lower worker motivation to return to work, training new personnel for replacement of the injured ones, and an increase in insurance premiums for the year following the incidents [2–4]. Occupational incidents are among the leading challenges in most industries [5]. Despite ongoing improvement in technology, coordinated prevention measures, higher education of the workforce, and improved training [6],

P. M. Arezes and R. L. Boring (Eds.): AHFE 2020, AISC 1204, pp. 3–9, 2020.
https://doi.org/10.1007/978-3-030-50946-0_1

agriculturally-related industries are routinely among the most hazardous work environments [7]. Agro-manufacturing industries include operations involved in manufacturing and distribution of farm supplies, production operations on the farm, and the storage, processing, and distribution of farm commodities [8].

The term "post-incident state" is used to describe the health status of an injured person when a non-fatal occupational injury has occurred, in the post-incident period when the worker returns to work, either immediately with zero days away from work (medical state) or after a disability period (disability state) [5, 9]. On average, the consequences and cost of disability state of an occupational incident is higher than the medical outcome [10, 11], and in agribusiness occupational incidents, as well [6, 12, 13]. Due to the high frequency and inflated costs of occupational incidents in agribusiness industrial operations [14], as well as substantial impact of occupational injuries on labor-market outcomes, predicting the post-incident state of an injury and identifying its contributory factors is vital to protect workers, improve workplace safety, and reduce overall costs of injuries [4, 15–17].

The combination of factors such as location, individual, environmental, equipment, and occupational activities are among responsible elements for the occurrence of occupational incidents [18]. Considering the substantial human capital and financial losses from occupational injuries, researchers have continually sought solutions to improve the accuracy of predicting the likelihood of future incidents severity and outcomes [4, 19]. Among various sources for injury data, workers' compensation claims provide useful details of occupational incidents such as injury type, cause, and nature, injured body part(s), injury narratives, and demographics of the injured workers [20]. Due to the significance of occupational injury management from engineering and economic points of view in industry, occupational injury analysis is vital for identifying prevalent injury patterns to design proper preventive measures that reduce the likelihood of future incidents [7, 12, 21, 22].

The primary aim of ML modeling is to detect, interpret, and predict qualitative and quantitative patterns in data, which leads to gaining applied insights and knowledge [23]. Detailed classification and prediction of occupational injuries plays a significant role in reducing and preventing incidents at workplace [24]. Among various ML methods, decision trees (DT) have proven powerful tools for classification and prediction tasks in injury analysis [4, 12, 25–27].

This study utilized two algorithms for developing decision tress: classification and regression trees (CART), and chi-square automatic interaction detection (CHAID) method. The usefulness of CART and CHAID models in analyzing occupational incidents and predicting outcomes of injuries and workplace due to straightforward visualization, meaningful rule generation, and high classification and prediction accuracy rate [23, 28, 29]. Mistikoglu et al. [23] used CHAID with 66.8% accuracy in constructions industry, and identified fall distance, cause of injury, safety training as the most significant predictors of an occupational incident outcome. Shirali et al. [26] did an analysis of the outcomes of occupational incidents in steel industry using CART (81.78% accuracy) and CHAID (80.73% accuracy), and showed that workers' age, level of education, and place of incident were the most important predictors of an incident classified as minor, severe, or fatal. Sarkar et al. [30] applied CART in predicting slip-trip-fall occupational incidents with an accuracy rate of 77.69% using injury narrative data.

Despite the wide use of machine learning (ML) methods in classifying and predictive modeling of future events, their application in occupational incident analysis in agro-manufacturing industries is relatively new. The objective of this paper is to discuss an application of selected ML methods in analyzing and predicting the post-incident state of an occupational incident in agro-manufacturing operations, and use the result to generate interpretable safety decision rules explaining the factors behind the occurrence of such incidents.

2 Materials and Methods

We used 13,867 workers' compensation claims that were filed between 2008 and 2016 in the Midwest region of the United Stated. All of the occupational incidents used in this study occurred in agro-manufacturing industries. The ML modeling was completed using SPSS IBM 26. The descriptive analysis of 13,867 incident records showed that grain elevator operations had the highest frequency of occupational incidents (50%), followed by refined fuel, feed mills for livestock operation, food distributors, fertilizer blending/distribution, poultry hatchery/grower/processor, and grain milling operations (18%, 11%, 8%, 8%, 3%, and 2%, respectively).

The other variables in the study were injured body group, type of incident (main cause of incident), nature of injury, and the injured occupations. The injured body part group refers to the main classification of an injury afflicting a specific part of the body such as the upper extremities body group which includes elbow(s), lower arm, finger (s), shoulder(s), etc. Cause of injury group has ten categories, each including some specific cause. For example, the struck or injured by group includes specific causes of hand tool or machine in use, falling or flying objects, object being lifted or handled, etc. The nature of injury group has three levels such as the specific injuries category that includes a detailed nature of the injury, such as laceration, fracture, amputation, concussion, etc. The numerical variables in the data were age (average = 43 years old) and experience of the injured workers (average = 6 years). The 13,867 claims include a column titles as "type of injury" which includes the labels of medical only, temporary total or temporary partial disability, and minor permanent partial or total disability. To perform the binary classification analysis, all those injury types that included any disability form were renames as "disability", and those medical only claims kept the same label. The new dichotomous variable was inserted in a new column with the title of "post-incident state". After such classification, 77% of the data were labelled as *medical*, while 23% were labelled as *disability*.

Two classifiers (1) CHAID, and (2) CART were applied for classifying and predicting the post-incident state of an occupational incident in agro-manufacturing operations. The response variable is binary, having two levels of medical and disability as injury outcome. The main predictors (independent variables) of post-incident state included injury cause group, injury nature, injured body part group, and workers' age, experience and occupation. The data was divided into training set (70%), and test set (30%), using stratified resampling. Regarding ML model performance, the overall model accuracy is gained through dividing the total number of cases predicted correctly as a labeled class over the total number of data points in the study.

3 Results

The models built using the train data were applied on the test data to evaluate their performance. Table 1 shows the model performance on the test data. Both classifiers showed high overall prediction performance, over 78%. All ML models could predict a medical post-incident state with accuracy over 97%. However, the prediction accuracy rate for disability post-incident state is low in all ML models (11.6–16.8%). This was expected due to the considerably lower frequency of disability cases in both train and test data.

Table 1. ML model performance on test data (%).

ML methods	Medical accuracy	Disability accuracy	Overall accuracy
CHAID	97.1	16.8	78.6
CART	98.3	11.6	78.2

Analyzing the DTs graphs provided useful information about the most significant predictors of post-incident states. Injury nature and cause were the most statistically significant determinants of injury severity outcomes. Considering the incidents outcome, injuries in lower extremities with nature of fracture caused by fall, slip, trip, lifting or handling has the highest chance of leading into disability status (77.7%), while the same injuries with nature of laceration and puncture caused by fall, slip, or trip were mostly medical (88.7%).

Furthermore, looking at the effect of age in determining the injury severity outcomes, higher ages have an increased chance of an injury turning into disability. For injuries with nature of contusion, strain, concussion, carpal tunnel syndrome, dislocation, and sprain, as age group changes from 28 years old or younger, to ages between 28 and 43, to higher than 43 years old, the chances of disability post-incident vs medical increases from 16% to 33.7%.

Considering the injured body parts, injuries in neck, and lower extremities have a high chance of disability outcome (70.8%) compared to injuries in trunk, head, and upper extremities (48.8%). However, these probabilities significantly decrease for injuries caused by cut, puncture, striking against or stepping on that had a nature of laceration, inflammation, or puncture.

4 Conclusion

This study identified injured body parts, nature of the injury, the age, years of experience, and occupations of the injured worker as statistically significant predictors of post-incident states of an occupational injury in agribusiness operations, based on workers' compensation data recorded during this period. The results emphasize the significance of quantitative analysis of empirical injury data in safety science, and contributes to enhanced understanding of occupational incidents root causes using predictive modeling along with safety experts' perspective. The analytical approach and results of this study provide insightful understanding of factors influencing medical and disability injuries and has applications in augmenting the experience of safety professionals in agro-manufacturing operations to develop effective countermeasures for prevention of occupational incidents and improve safety intervention efforts.

References

1. Boden, L.I., O'Leary, P.K., Applebaum, K.M., Tripodis, Y.: The impact of non-fatal workplace injuries and illnesses on mortality. Am. J. Ind. Med. (2016). https://doi.org/10.1002/ajim.22632
2. Tsoukalas, V.D., Fragiadakis, N.G.: Prediction of occupational risk in the shipbuilding industry using multivariable linear regression and genetic algorithm analysis. Saf. Sci. **83**, 12–22 (2016). https://doi.org/10.1016/j.ssci.2015.11.010
3. Gavious, A., Mizrahi, S., Shani, Y., Minchuk, Y.: The costs of industrial accidents for the organization: developing methods and tools for evaluation and cost-benefit analysis of investment in safety. J. Loss Prev. Process Ind. (2009). https://doi.org/10.1016/j.jlp.2009.02.008
4. Kakhki, F.D., Freeman, S.A., Mosher, G.A.: Evaluating machine learning performance in predicting injury severity in agribusiness industries. Saf. Sci. (2019). https://doi.org/10.1016/j.ssci.2019.04.026
5. Altunkaynak, B.: A statistical study of occupational accidents in the manufacturing industry in Turkey. Int. J. Ind. Ergon. **66**, 101–109 (2018). https://doi.org/10.1016/j.ergon.2018.02.012
6. Robert, K., Elisabeth, Q., Josef, B.: Analysis of occupational accidents with agricultural machinery in the period 2008–2010 in Austria. Saf. Sci. (2015). https://doi.org/10.1016/j.ssci.2014.10.004
7. Field, W.E., Heber, D.J., Riedel, S.M., Wettschurack, S.W., Roberts, M.J., Grafft, L.M.J.: Worker hazards associated with the use of grain vacuum systems. J. Agric. Saf. Health (2014). https://doi.org/10.13031/jash.20.9989
8. Zylbersztajn, D.: Agribusiness systems analysis: origin, evolution and research perspectives. Rev. Adm. (2017). https://doi.org/10.1016/j.rausp.2016.10.004
9. Kakhki, F.D., Freeman, S.A., Mosher, G.A.: Use of logistic regression to identify factors influencing the post-incident state of occupational injuries in agribusiness operations. Appl. Sci. (2019). https://doi.org/10.3390/app9173449
10. Sears, J.M., Blanar, L., Bowman, S.M.: Predicting work-related disability and medical cost outcomes: a comparison of injury severity scoring methods. Injury (2014). https://doi.org/10.1016/j.injury.2012.12.024

11. Zakiei, A., Kiani, N., Morovati, F., Komasi, S.: Classification of various types of disability and determining their predictive causes in western Iran. Clin. Epidemiol. Glob. Health (2018). https://doi.org/10.1016/j.cegh.2018.11.003
12. Davoudi Kakhki, F., Freeman, S.A., Mosher, G.A.: Use of neural networks to identify safety prevention priorities in agro-manufacturing operations within commercial grain elevators. Appl. Sci. **9**(21), 4690 (2019). https://doi.org/10.3390/app9214690
13. Ramaswamy, S.K., Mosher, G.A.: Using workers' compensation claims data to characterize occupational injuries in the biofuels industry. Saf. Sci. (2018). https://doi.org/10.1016/j.ssci.2017.12.014
14. Kakhki, F.D., Freeman, S.A., Mosher, G.A.: Segmentation of severe occupational incidents in agribusiness industries using latent class clustering. Appl. Sci. (2019). https://doi.org/10.3390/app9183641
15. Meyers, A.R., Al-Tarawneh, I.S., Wurzelbacher, S.J., Bushnell, P.T., Lampl, M.P., Bell, J. L., Bertke, S.J., Robins, D.C., Tseng, C.Y., Wei, C., Raudabaugh, J.A., Schnorr, T.M.: Applying machine learning to workers' compensation data to identify industry-specific ergonomic and safety prevention priorities: Ohio, 2001 to 2011. J. Occup. Environ. Med. (2018). https://doi.org/10.1097/JOM.0000000000001162
16. Tremblay, A., Badri, A.: A novel tool for evaluating occupational health and safety performance in small and medium-sized enterprises: the case of the Quebec forestry/pulp and paper industry. Saf. Sci. (2018). https://doi.org/10.1016/j.ssci.2017.09.017
17. Leigh, J.P.: Economic burden of occupational injury and illness in the United States. Milbank Q. (2011). https://doi.org/10.1111/j.1468-0009.2011.00648.x
18. Khanzode, V.V., Maiti, J., Ray, P.K.: Occupational injury and accident research: a comprehensive review (2012). https://doi.org/10.1016/j.ssci.2011.12.015
19. Lord, D., Mannering, F.: The statistical analysis of crash-frequency data: a review and assessment of methodological alternatives. Transp. Res. Part A Policy Pract. (2010). https://doi.org/10.1016/j.tra.2010.02.001
20. Wurzelbacher, S.J., Al-Tarawneh, I.S., Meyers, A.R., Bushnell, P.T., Lampl, M.P., Robins, D.C., Tseng, C.Y., Wei, C., Bertke, S.J., Raudabaugh, J.A., Haviland, T.M., Schnorr, T.M.: Development of methods for using workers' compensation data for surveillance and prevention of occupational injuries among state-insured private employers in Ohio. Am. J. Ind. Med. (2016). https://doi.org/10.1002/ajim.22653
21. Jacinto, C., Canoa, M., Soares, C.G.: Workplace and organisational factors in accident analysis within the Food Industry. Saf. Sci. (2009). https://doi.org/10.1016/j.ssci.2008.08.002
22. Bevilacqua, M., Ciarapica, F.E., Giacchetta, G.: Industrial and occupational ergonomics in the petrochemical process industry: a regression trees approach. Accid. Anal. Prev. (2008). https://doi.org/10.1016/j.aap.2008.03.012
23. Mistikoglu, G., Gerek, I.H., Erdis, E., Usmen, P.E.M., Cakan, H., Kazan, E.E.: Decision tree analysis of construction fall accidents involving roofers. Expert Syst. Appl. (2015). https://doi.org/10.1016/j.eswa.2014.10.009
24. Kang, K., Ryu, H.: Predicting types of occupational accidents at construction sites in Korea using random forest model. Saf. Sci. (2019). https://doi.org/10.1016/j.ssci.2019.06.034
25. Koyuncugil, A.S., Ozgulbas, N.: Financial early warning system model and data mining application for risk detection. Expert Syst. Appl. (2012). https://doi.org/10.1016/j.eswa.2011.12.021
26. Shirali, G.A., Noroozi, M.V., Malehi, A.S.: Predicting the outcome of occupational accidents by CART and CHAID methods at a steel factory in Iran. J. Public health Res. (2018). https://doi.org/10.4081/jphr.2018.1361

27. Zheng, Z., Lu, P., Lantz, B.: Commercial truck crash injury severity analysis using gradient boosting data mining model. J. Saf. Res. (2018). https://doi.org/10.1016/j.jsr.2018.03.002
28. De Oña, J., López, G., Abellán, J.: Extracting decision rules from police accident reports through decision trees. Accid. Anal. Prev. (2013). https://doi.org/10.1016/j.aap.2012.09.006
29. Kashani, A.T., Shariat Mohaymany, A., Ranjbari, A.: Analysis of factors associated with traffic injury severity on rural roads in Iran. J. Inj. Violence Res. (2012). https://doi.org/10.5249/jivr.v4i1.67
30. Sarkar, S., Raj, R., Vinay, S., Maiti, J., Pratihar, D.K.: An optimization-based decision tree approach for predicting slip-trip-fall accidents at work. Saf. Sci. (2019). https://doi.org/10.1016/j.ssci.2019.05.009

Safety Decision-Making in Academia

Anastasia Kalugina[1] and Thierry Meyer[1,2](✉)

[1] Group of Physical and Chemical Safety, Institute of Chemical Sciences and Engineering, Ecole Polytechnique Fédérale de Lausanne, Station 6, 1015 Lausanne, Switzerland
{anastasia.kalugina, thierry.meyer}@epfl.ch
[2] Safety Competence Center, Ecole Polytechnique Fédérale de Lausanne, Station 6, 1015 Lausanne, Switzerland

Abstract. This paper illustrates the challenges of safety management in Academia, which are difficult to overcome using conventional methods of risk management. The positive aspects and disadvantages of the risk management tools specifically designed for the laboratory collection are addressed. We propose a different approach to hazard detection that can ease the burden for Academia safety experts of tedious time-consuming risk assessment. Based on the observations of Academia's safety management, we are attempting to build a semi-quantitative risk analysis model that will conform to the concept of safety-II. This paper suggests an approach that allows to integrate this model into human oriented performance driven decision making.

Keywords: Safety-II · Risk management · Safety climate · Academia

1 Introduction

Modern society emphasizes safety requirements and reduces the rate of inappropriate performance. The growing need to feel safe or at least be able to predict when and where a situation becomes dangerous has led to the development of numerous risk assessment techniques [1]. The first generation of safety management systems were developed on the poor reliability of the machines. The main objective was to detect software malfunctions and engineering errors. Despite the development of engineering level and machine reliability, accidents proved that risk analysis methods such as Hazard Operability study (HAZOP), Failure Mode and Effects Analysis (FMEA), Fault and Event trees are no longer sufficient to provide a desired level of safety. Various industrial accidents of the 1970s and 1980s [2, 3] focused on possible human errors and their reduction or complete elimination in the development of human performance assessment. It soon proved insufficient to consider pure human factors without an organizational context [4]. Nowadays there are several attempts to implement new approaches, which will work with intractable systems [5]. It does not mean old approaches are no longer useful, but any situation with a strong socio-technical presence requires a new approach to safety management.

Safety in Academia has never been a topic of discussion until early 90's. In view of the last incidents that occurred in research laboratories, society realized that education

P. M. Arezes and R. L. Boring (Eds.): AHFE 2020, AISC 1204, pp. 10–17, 2020.
https://doi.org/10.1007/978-3-030-50946-0_2

is definitely not as safe as they would like to think [6]. Academic and experimental laboratories are a typical example of intractable systems that work despite being extremely decentralized compared to the industry. For Academia to succeed in its primary goal of scientific discovery and research, the performance variability of the system's participants is an important prerequisite. According to the Cambridge dictionary, safe means not to be not in danger or unlikely to be harmed. The majority of industrial accidents are considered highly unlikely [7] which shall mean that all currently functioning systems are safe. However, companies continue to invest in safety in order to achieve the goal of zero accidents [8]. Statistical information on incidents and near misses is hardly available in Academia, moreover quantification of the accident rate cannot serve as a measure of change in safety.

Conventional strategies based on hazard detection and risk assessment must be coupled with the inclusion of various factors defining the working and social environment to ensure optimal protection in Academia. It means not only compliance with and imposition of safety measures, but also a last-minute adaptation or gradual improvement of existing working conditions. Work-as-imagine model shall be replaced by work-as-done concept. It includes in field assessments and other analytical research that can shed light on security gaps.

2 Conventional Safety Management Systems

According to the Occupational Safety and Health Administration (OSHA) Safety Management System (SMS) is "a proactive, collaborative process for finding and fixing workplace hazards before employees become injured or ill" [9]. Any SMS includes at least two main components: Safety Risk Management (SRM, including hazards identification, risk assessment, selection and implementation of mitigation measures) [10] and Safety Assurance (SA) [11] which are supported by Safety Policy (SP) and Constant Safety Promotion (CSP) [12]. Since any SMS is very complex and requires a detailed description of all its components, we concentrate in this paper on its key components. While the last two components (SP and CSP) can rely on the common strategies for both tractable and intractable systems, the system's organizational features are highly dependent on the first two components.

2.1 Hazard Identification

The hazard identification process vastly depends on the objectives of the risk assessment. The first goal can be to describe the situation in the laboratory as a whole and recognize all potential hazards without any relation to the processes. Another objective is the assessment within the processes. The first case takes place when there is no preliminary information on the laboratory. Usually, however, information is already available on hazardous materials used and stored in laboratories. In fact, it is almost impossible to make a risk analysis of all hazards related to all processes due to the wide variety of different activities in research laboratories.

In the second case, the purpose of hazard identification is to collect all pertaining information on the risk assessment of the hazardous processes carried out in laboratory

[13]. There are various methods of hazard classification, allowing systematic and eased identification. According to OSHA, chemicals can be classified based on their health and physical hazards [14]. This standard focuses solely on chemicals, although in a research laboratory it cannot be applied alone. Canadian center for occupational health and safety distinguishes 7 categories of hazards: chemicals, ergonomic, health, physical, psychosocial, safety, and workplace. All these categories have their own subcategories [15]. Another approach for classification is "energy-based" [16] which is useful for understanding causation but has huge applicational limitation in multi hazardous environment [17]. According to Macdonald [18] hazards can be classified based on the context and conditions. This method is very versatile and, however, for the research laboratories it is too cumbersome. The ACHiL methodology [19] for hazard identification is partly based on the GHS categories and specifically designed for laboratory, which can be applied in an academic setting during the risk assessment.

2.2 Risk Analysis and Treatment

Three types of risk analysis are available: qualitative, semi-quantitative and quantitative [20]. Qualitative approaches that are commonly used in various industries such as HAZOP [21] or Job Safety Analysis (JSA) [22] do not allow different risks to be compared, focus solely on deviations or require the development of a full job portfolio. Failure Mode, Effects & Criticality Analysis (FMECA) [23] is one of the most popular quantitative methods used in the industry. It is very detailed and can be used with automated and fixed processes, such as nuclear engineering [24] and pharmaceutical industry [25]. However, it is time and resource-consuming and can hardly be applied on a daily basis in research laboratories where processes are constantly changing [26]. Another family of quantitative methods such as Fault Tree Analysis (FTA) [27] or Event Tree Analysis (ETA) [23] can be used even if there is a lack of statistical data, using Fuzzy logic [28] or Bayesian networks [29]. These methods, however, are still too resources-consuming to be used daily in research laboratories.

Recently, two approaches have been developed specifically for laboratory risk assessment. Lab-HIRA is a tool explicitly designed to address chemical hazards, present during chemical synthesis [30]. It calculates an Overall Hazards Index (OH) using the following formula:

$$\begin{aligned}\mathrm{OH} = \sum \Big(\sum \mathrm{IV}_{\text{Dis. Prop.}} + \sum \mathrm{IV}_{\text{Dis. Reac. Cond.}} + \sum \mathrm{IV}_{\text{Index Value Spec. Chem.l Hazard}} \\ + \mathrm{I(V)}_{\text{Name reaction}} + \mathrm{I(V)}_{\text{Type reaction}}\Big)\end{aligned} \quad (1)$$

The method is useful for initial risk assessment by laboratory staff, but not for risk management, and it does not consider the presence of another type of hazard. LARA methodology [31] calculates a Lab Criticality Index for each particular hazard within the process, using several variables such as: severity, probability, detectability and various worsening factors, such as ergonomics or existence of other hazards. This tool proposes safety measures for risk reduction and also considers the feasibility of such measures. Nonetheless, due to a high number of similar hazards, even this tool, which was specifically designed for an Academic environment, has its drawbacks, such as the cumbersomeness of the risk analysis process. Moreover, it still maintains the same view of the Human Factors as a potential worsening factor.

As a rule, each environment management establishes the level of unacceptable risk. It can be done using either the "as low as reasonably practicable" (ALARP) [32], "Globalement Au Moins Aussi Bon" (GAMAB) [33] or "Minimum Endogenous Mortality" (MEM principles) [34] all of which are based on the idea of financial reasonability of risk reduction. The effect of the measures on the hazard and their potential for risk reduction are compared with the primary risk assessment after evaluation of the risk and its position in the acceptability zone. When speaking about risks with low probability, however, any risk assessment would fail into the "regulator paradox" and the majority of expensive measures would be considered not rational [35].

3 Risk Assessment for Resilient Safety

Traditional methods of risk analysis focus on functionalities of system elements, possible deviations from expected results/outcomes [36]. Focusing on functions is hardly possible in academia, and even less predicting deviations accurately.

3.1 Hazard Identification

Any risk management's purpose is to prevent any possible undesirable outcomes, and/or reduce their probability or severity [37]. Different internal characteristics of hazards can sometimes have the same target and the same exposure route (e.g. skin toxicity and skin corrosion). During the analysis of the work of safety experts, it was observed that risk assessment of a process involving chemicals includes repetitive assessments while working with GHS hazard categories. Meanwhile, experts tend to use the same type of protective measures to protect from hazards. On the basis of these observations, another approach for hazard identification is proposed. In order to improve the time-efficiency of a risk assessment, the experts will first determine what is the main target of a hazard which could be health, environment or materials. In particular for the health target, expert shall determine possible ways of exposure. Afterwards, based on the results of a preliminary evaluation on the extent of possible harm, paths of exposure shall be protected (Fig. 1).

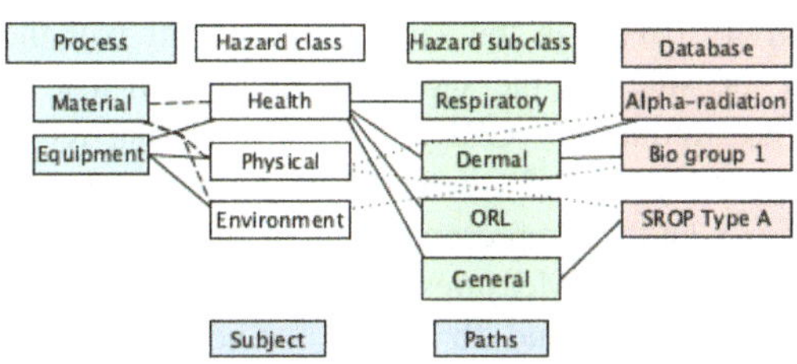

Fig. 1. Matching between GHS hazard classification and subject-based classification.

This classification greatly reduces the number of repetitive steps and can help to represent versatile and unique solutions for future decision making (DM). This classification not only reduces the analysis time of experts but can be used by anyone without expert knowledge on planning or conducting an experiment.

3.2 Risk Analysis

The risk index is usually calculated to quantify and compare different risks. Conventional methods either do not consider the human influence on the performance or consider such as worsening desired performance and use strategies of elimination and minimization of such influences. Nevertheless, in an environment with high human involvement in everyday activities, such as Academia, assessment of general working setup is more suitable. In order to understand how to better treat risk in a particular environment, it is proposed to make an evaluation of the Safety Climate (SC) in the laboratory (Fig. 2). After the calculation of the risk index, the effects of different measures on the initial conditions are evaluated and compared. To consider specific characteristics of the environment, feasibility (e.g. process/environment compatibility and user acceptability) is assessed.

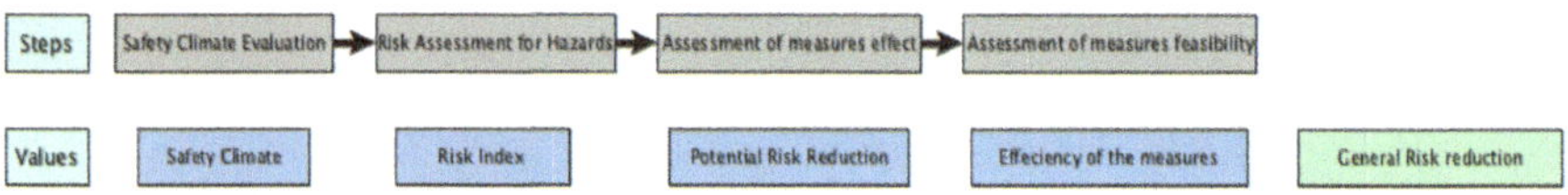

Fig. 2. Risk analysis process scheme with corresponding output parameters.

In order for the method to achieve maximum flexibility and efficiency of the analyses carried out for different conditions, three groups of parameters can be distinguished:

- Hazard/risk-related parameters (severity, probability, detectability, etc.)
- Measure-related parameters (costs, acceptability (AC)) and simplicity of the use (S), compatibility with process (CP)/environment (CE), risk reduction (RR))
- Safety climate parameter.

The reliability of any safety measure will depend heavily, not only on its own characteristics, but also on the quality of measure-human interaction and the extent to which this measure is sensitive to such interactions [38]. The reliability of measures with different sensitivity (SHF) in the same environment will therefore be different, while the reliability measures of the same sensitivity will also vary in the different settings. General Risk Reduction (GRR) is the actual risk reduction impact of the measures that can be calculated (2), considering human-measure-environment interactions. Coefficients 0.8 and 0.2 represent an input of possible failures either human- (HR) or machine-related (TR). Measures-related parameters are evaluated using the scale from 1 to 5, thus 10 in 3a and 3b is used for normalization.

$$GRR = (0.8 * HR + 0.2 * TR) * RR \tag{2}$$

$$\text{where } TR = (CP + CE)/10 \tag{3a}$$

$$HR = SC * (SHF^{(AC+S)/10}) \tag{3b}$$

Therefore, GRR is a combination of technical and human reliability of the measure compounded by potential risk reduction. One of the most difficult parameters to estimate is SC. A Safety Climate survey was conducted among various universities (more than 2500 responses) to collect information on the main factors contributing to SC in the laboratory. Correlations between major factors obtained from data processing can be used in the further development of SC estimation methodologies. It is expected that, using data obtained from survey results, experts making safety visits to the laboratory and conducting risk assessment will be able to estimate the current safety level of the visited laboratory.

3.3 Decision Making

The last step of the DM process is the actual selection of decision alternatives. Conventional safety DM methods can be used to select between two alternatives. The first option is to make an investment in order to prevent something from happening. The second alternative is not to make any investments. After all, any decision is based on selection of acceptable probability for different levels of severity of negative consequences. Nevertheless, this approach does not consider whether any of the proposed solutions will actually be effective in terms of performance. Using safety-II approach decision makers will select between two other alternatives:

- To make an investment to improve performance
- Not to make an investment, due to existing high level of existing performance.

Thus, in the first case, RR will be the factor determining the final decision, while in the second case it is GRR (with having regard to acceptable limits of HR and TR). It also allows decision makers to select safety alternatives which will be targeting improvement of performance of already existing one.

4 Conclusions

Many methods of risk management have been developed through decades of safety management studies. Meanwhile, requirements for the safety and working conditions have also changed. While many industries have chosen the way to eliminate human factors, human involvement in many hazardous activities is vital for Academia. This article suggests a different approach for risk assessment in order to provide a desirable level of safety for an academic environment and to ensure its high scientific efficiency.

References

1. Hollnagel, E.: Safety-I and safety-II. Ashgate Publishing Limited, Farnham (2014)
2. Centemeri, L.: The Seveso disaster legacy. In: Nature and History in Modern Italy, pp. 251–273. Ohio University Press & Swallow Press (2010)
3. IAEA. https://inis.iaea.org/collection/NCLCollectionStore/_Public/13/677/13677904.pdf
4. Khan, A.H., Hasan, S., Sarkar, M.A.R.: Analysis of possible causes of Fukushima disaster. Int. J. Nucl. Quantum Eng. **12**(2), 53–58 (2018)
5. Eurocontrol. https://www.skybrary.aero/bookshelf/books/2437.pdf
6. https://cen.acs.org/articles/91/i18/Importance-Teaching-Safety.html. Accessed 08 Dec 2019
7. Bilir, S., Gürcanli, G.: A method for determination of accident probability in construction industry. Teknik Dergi **29**(4), 8537–8561 (2018)
8. https://oshwiki.eu/wiki/Zero_accident_vision. Accessed 08 Dec 2019
9. https://www.osha.gov/dsg/hospitals/mgmt_tools_resources.html. Accessed 08 Dec 2019
10. Ludwig, J., Bastin, G., Chewings, V., Eager, R., Liedloff, A.: Leakiness: a new index for monitoring the health of arid and semiarid landscapes using remotely sensed vegetation cover and elevation data. Ecol. Ind. **7**, 442–454 (2007)
11. https://www.itf-oecd.org/sites/default/files/why-sms.pdf (Retrieved: 08/12/2019)
12. https://www.faa.gov/about/initiatives/sms. Accessed 08 Dec 2019
13. http://www.hse.gov.uk/Research. Accessed 08 Dec 2019
14. https://www.osha.gov/Publications/OSHA3844.pdf. Accessed 08 Dec 2019
15. https://www.ccohs.ca/topics/hazards. Accessed 08 Dec 2019
16. Viner, D.: Accident analysis and risk control. VHMS, Fairfield (1994)
17. The core body of knowledge for generalist OHS professionals. Safety Institute of Australia, Tullamarine (2012)
18. Macdonald, W.: A hierarchy of risk control measures for prevention of work-related musculoskeletal disorders. In: International Ergonomics Conference on Humanising Work and the Work Environment, Guwahati, India (2005)
19. Marendaz, J., Suard, J., Meyer, T.: A systematic tool for assessment and classification of hazards in laboratories (ACHiL). Saf. Sci. **53**, 168–176 (2013)
20. Valis, D., Koucky, M.: Selected overview of risk assessment techniques. Problemy ekspoatacji, 19–23 (2009)
21. Hashemi-Tilehnoee, M., Pazirandeh, A., Tashakor, S.: HAZOP-study on heavy water research reactor primary cooling system. Ann. Nucl. Energy **37**, 428–433 (2010)
22. Rozenfeld, O., Sacks, R., Rosenfeld, Y., Baum, H.: Construction job safety analysis. Saf. Sci. **48**, 491–498 (2010)
23. Ericson, C.: Hazard Analysis Techniques for System Safety. Wiley-Interscience, Hoboken (2010)
24. Guimarães, A., Lapa, C.: Fuzzy inference to risk assessment on nuclear engineering systems. Appl. Soft Comput. **7**, 17–28 (2007)
25. Díaz, R., Fernánde, G., Muzzio, C.: Practical application of quality risk management to the filling process of betamethasone injections. Pharm. Eng. **31**, 84–89 (2011)
26. Dai, W., Maropoulos, P., Cheung, W., Tang, X.: Decision-making in product quality based on failure knowledge. Int. J. Prod. Lifecycle Manag. **5**, 143 (2011)
27. Park, A., Lee, S.: Fault tree analysis on handwashing for hygiene management. Food Control **20**, 223–229 (2009)
28. You, X., Tonon, F.: Event-tree analysis with imprecise probabilities. Risk Anal. **32**, 330–344 (2011)

29. Givehchi, S., Heidari, A.: Bayes networks and fault tree analysis application in reliability estimation (case study: automatic water sprinkler system). Environ. Energy Econ. Res. **2**(4), 325–341 (2018)
30. Leggett, D.: Lab-HIRA: hazard identification and risk analysis for the chemical research laboratory: part 1. Preliminary hazard evaluation. J. Chem. Health Saf. **19**, 9–24 (2012)
31. Ouédraogo, A., Groso, A., Meyer, T.: Risk analysis in research environment – part I: modeling lab criticity index using improved risk priority number. Saf. Sci. **49**, 778–784 (2011)
32. Kletz, T.A.: Looking beyond ALARP: overcoming its limitations. Process Saf. Environ. Prot. **83**, 81–84 (2005)
33. Kron, H.H.: On the evaluation of risk acceptance principles. http://citeseerx.ist.psu.edu/viewdoc/download?doi=10.1.1.455.4506&rep=rep1&type=pdf
34. Wigger, P.: Experience with safety integrity level (SIL) allocation in railway applications (2001)
35. Gosman, S.: Justifying safety: the paradox of rationality. 90 Temp. L. Rev. 155 (2018). SSRN. https://ssrn.com/abstract=2956022
36. Leimeister, M., Kolios, A.: A review of reliability-based methods for risk analysis and their application in the offshore wind industry. Renew. Sustain. Energy Rev. **91**, 1065–1076 (2018)
37. Herrmann, J.: Engineering Decision Making and Risk Management. Wiley, New York (2015)
38. Eggemeier, F.: Properties of workload assessment techniques. Adv. Psychol. **52**, 41–62 (1988)

The Role of Workers' Representative and OHS Performance: An Interpretative Framework

Paolo Trucco(✉), Rossella Onofrio, and Raffaella Cagliano

School of Management, Politecnico di Milano, Via Lambruschini. 4b, 20156 Milan, Italy
{Paolo.Trucco,Rossella.Onofrio, Raffaella.Cagliano}@Polimi.it

Abstract. In the context of OH&S, the positive role of workers' representation, involvement and consultation for better OH&S management arrangements is largely discussed in literature; however, the focus is typically on personal traits. The objective of this paper is to expand the scope of previous studies by exploring the relationship between OH&S organizational maturity and the role of workers' representatives, and consequently the effect on OH&S performance. To this end, a structured questionnaire was administered to a large-scale sample of Italian representatives. Results suggest four types of participative models of organization/representative at different levels of maturity and show clear correlation with OH&S performance.

Keywords: Occupational Health and Safety (OHS) · Workers' representative · Performance · Framework

1 Introduction

The shift from prescriptive to process-based regulation took place in the late 1980s in Europe and other developed countries such as Canada, Australia and New Zealand. It requires employers to manage OH&S in a systematic, informed and participative way [1, 2]. Correspondingly, the Safety Worker Representative (SWR) has become a key player in advanced OH&S management systems.

This study starts from the consolidated national and international evidence on the role of the SWR in fostering effective OH&S management in organizations. The objective is to expand existing knowledge by exploring the relationship between OH&S management maturity and the role of the SWR, and consequently their contribution to OH&S performance. In doing so, the study tackles the criticalities that SWRs are still facing in playing their role. While the OH&S management system is conceptualized in terms of the organizational set-up, processes and safety culture, the role of SWR is conceptualized according to the three dimensions of consultation, representation and involvement [3], taking into consideration also the main preconditions.

P. M. Arezes and R. L. Boring (Eds.): AHFE 2020, AISC 1204, pp. 18–25, 2020.
https://doi.org/10.1007/978-3-030-50946-0_3

2 Theoretical Framework

Several national and international studies have scrutinized the role of the Safety Worker Representative (SWR) within OH&S management systems. In particular, literature provides evidence that the attentive involvement of SWRs and a cooperative approach correlates positively and significantly with the effective implementation of preventive OH&S management [3, 4]. In order to clarify the characteristics of cooperative approaches, the literature distinguishes between two states, when workers act as passive recipients of information about the practice of OH&S management, and when they are able to actively influence the managers' directions. The latter approach is aligned with advancing cooperative dialogue between workers and managers. However, the cooperative dialogue can be developed either by "direct participation" of workers and/or by "collective representation". Regarding the latter, the SWR acts as the collective representative of workers by consulting with managers through participation means [5]. Thus, the role of SWRs relies essentially on worker representation, consultation and involvement to achieve systematic management of OH&S. The majority of literature on this topic, applied a standardized questionnaire within a representative sample of firms/representatives to explore the role of representatives within the OH&S management system, typically in relation to the regulatory system in place in each country. Results shown the positive role of workers' representation, involvement and consultation for better OH&S management arrangements [3, 5–9]. In addition, previous studies show the criticalities that SWRs face in performing their role due to numerous factors. The benefits and criticalities are typically identified within various business sizes, sectors, territories, and organizational models. In addition, in order to comprehensively understand the contribution of SWRs to the achievement of enhanced OH&S performance, several studies underline the importance of considering the level of maturity of the corporate OH&S management system and its safety culture. In particular, different models of organizations' maturity have been identified with respect to the level of completeness and awareness of the safety culture in various countries [10]. Numerous researches, at national and international scale, have investigated the close relationship between adopting OH&S management practices, safety culture and safety performance of organizations [11]. Though, literature that jointly investigate the maturity of SWR's role with the maturity of organization's OH&S management systems is scanty. Furthermore, due to the apparent role of local regulations and cultures, results of national studies cannot be fully generalized to other nations and contexts. In contrast with many other nations, in Italy there has not been so far any specific investigation of the role of SWRs and their influence on OH&S management maturity and performance.

The present study aims at covering this gap by using extant literature to frame a research framework that synthesizes the relevant variables and the expected relationships between them (Fig. 1). In particular, we aim at answering the following research questions:

RQ1. What are the participatory models of Safety Worker Representatives at different maturity levels of OH&S management?

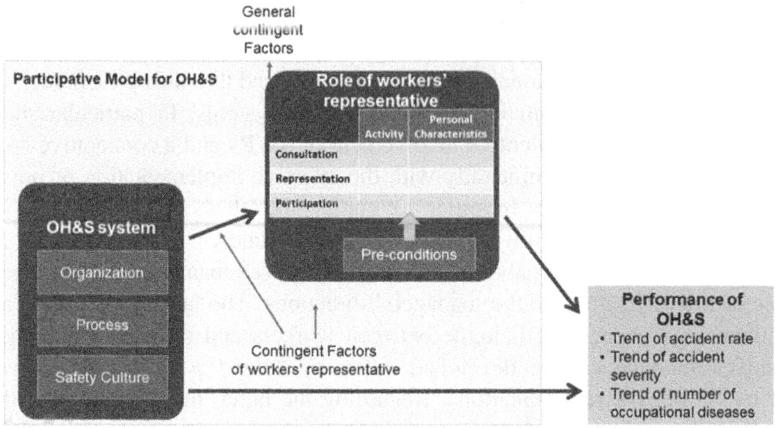

Fig. 1. Theoretical framework.

RQ2. Does the participatory model of Safety Worker Representatives influences OH&S performance?
RQ3. What contingent factors moderate the relationship between the participatory model and OH&S performance?

3 Methodology

A survey questionnaire was specifically designed taking inspiration from previous studies [10–14]. An extensive survey was conducted across Italy, covering almost all sectors, and addressing a large-scale sample of the Italian population of SWRs. Statistical sampling and on-line questionnaire administration were implemented.

The questionnaire was divided into sections aimed at deepening the specific themes of interest: 1) The socio-demographic and working characteristics of SWRs; 2) The company context; 3) Occupational hazards, health and safety conditions; 4) Safety culture and OH&S management system; 5) SWR access to information; 6) SWR training; 7) SWR consultation, involvement and representation; 8) The status and individual characteristics of the SWR; 9) Interactions with the other actors of the OH&S management system; 10) Industrial relations (e.g. the role of trade unions) and their relationship with SWRs. In line with prior studies, two different scale measurement systems have been applied: categorical scales (nominal or ordinal) with an average of 4 or 5 response categories (to understand descriptive elements or adoption of certain practices); Likert scale (from 1 to 5) to identify the maturity level of the relevant attributes.

The unit of analysis is the Safety Worker Representative (SWR). The reference population includes all the SWRs of Italian organizations with ≥ 10 employees; since the actual size of the population is not known, it was indirectly estimated on the population of private and public companies (from Italian National Institute of Statistics) by applying the requirements of the national Legislative Decree 81/2008, which sets the minimum requirements: one SWR in organizations with $\leq$200 employees; three SWRs for 201 $\geq$ employees $\geq$ 1000; six SWRs for organizations with $\geq$1000 employees. The sample size calculation was made to ensure its representativeness, and in particular to ensure a confidence interval of 95% and a margin of error of ±3%. The final sample was finally segmented by sector, company size and macro-regions. A response rate of 25% was assumed. Overall, to achieve 1,167 responses, 2,428 SWRs in private companies and 2,240 in public companies were contacted. After data collection and cleaning (responses with <70% of answered questions were discarded), the final sample includes 2039 respondents (with an average response rate of 43,6%), with the following coverage: Private (66%); Public (17%); Companies controlled by the public administration (13%). Sample composition by sector is: Services (44%); Industry_Non manufacturing (37%); Industry_Manufacturing (17%); Agriculture (2%).

Data analysis was focused on the combination of two dimensions: 1) the maturity level of the company's OH&S management system, and 2) the maturity level of SWR's role. The OH&S management system construct includes: OH&S policy, planning (risk assessment, requirements, goals and programs), implementation (resources, responsibilities, training, documentation, procedures and control), Monitoring and verification (performance measurement, accident analysis, audit, recording and reporting, corrective measures). The mean of the answers to all variables of all the constructs (all expressed on a scale from 1 to 5) is taken as the maturity level of Organization OH&S management system. Companies with an index $\leq$ 2.5 are considered with a low maturity level, high otherwise. The maturity of the SWR was operationalized by means of five constructs: Information sharing, Training, Consultation, Representation, and Participation. Some parameters were re-scaled in order to have homogenous measurement scales between all variables. The mean of the answers to all variables of all the constructs (all expressed on a scale from 1 to 5 after homogenous rescaling) is taken as the maturity level of the SWR's role. Companies with an index $\leq$ 2.5 are considered with a low maturity level of SWR's role, high otherwise.

4 Findings

Cluster analysis was carried out at two levels: first, for grouping companies against the maturity level of their OH&S management system (high and low); second, for grouping companies against the maturity level of the SWR's role (high and low). It led to the identification of four types of "OH&S participative models" (Table 1); when the OH&S maturity is low (*bureaucratic*), the participative model can be either:

- *divergent* (when the maturity of SWR's role is high – 102 respondents: 5% of total sample). The SWR is apparently involved in management processes (high participation), but within a bureaucratic and unstructured OH&S management system. Recognition of the role is therefore frustrated in its ability to contribute to the prevention system.

Table 1. Four profiles of participative models for OH&S (scale from 1 to 5).

Maturity of the role of workers' representatives	Low maturity of OH&S	High maturity of OH&S
High	DIVERGENT: • Bureaucratic OH&S (Medium Level of maturity = 2,5) • Frustrated role of workers' representative (Medium Level of maturity = 3,3)	VIRTUOUS: • Proactive OH&S (Medium Level of maturity = 3,9) • Proactive role of workers' representative (Medium Level of maturity = 3,6)
Low	LOCKED: •Bureaucratic OH&S (Medium Level of maturity = 2,2) • Formal Role of workers' representative (Medium Level of maturity = 2,3)	INCOMPLETE: • Proactive OH&S (Medium Level of maturity = 3,6) • Marginal role of worker representative (Medium Level of maturity = 2,6)

- or *locked* (when the maturity level of SWR's role is low – 889 respondents: 44% of total sample). The SWR plays a marginal role (low participation and consultation), where only formal aspects are guaranteed (information, training and representation).

In a similar vein, when the OH&S maturity level is high (*proactive*), the participative model can be either:

- *virtuous* (when the maturity level of SWR's role is high – 608 respondents: 30% of total sample). OH&S management system is proactive, recognizes the contribution of workers and benefits from the role of SWRs in a virtuous relationship (proactive information sharing and consultation).
- or *incomplete* (when the maturity level of SWR's role is low – 440 respondents: 21% of total sample). The OH&S management system is characterized by continuous (proactive) improvement yet does not include an active contribution from the SWRs, which is familiar with OH&S programs, but plays a marginal role.

Figure 2 shows in more detail the differences between the four participative models.

Participative models with higher level of maturity show better planning, monitoring and verification. The virtuous model has also a higher value associated to implementation. As for the maturity level of the SWR's role, there is a clear differentiation between factors: in the Locked and Incomplete profiles, SWRs are granted the training and information needed, but the consultation and participation roles are marginalized; on the other hand, the Virtuous and Divergent profiles differentiate on the level of consultation activities.

The analysis of OH&S performance within the four participative models (Fig. 3) shows a strong positive correlation between maturity and performance. The trends of the accident rate, accident severity and the number of occupational diseases were used for the purpose. We checked for the correlation between the participative model and the overall performance, using the arithmetic mean of a linear scale that reflects the order of

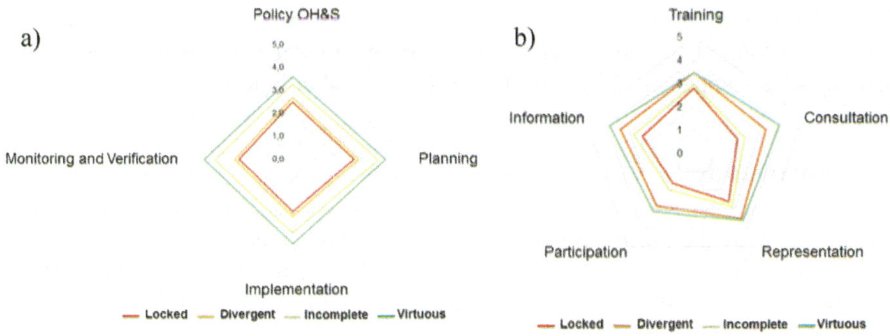

Fig. 2. Maturity of the OH&S management system phases (a) and of the role of workers' representative (b) in the four participative models.

preference of the trend: decreasing (5); stable (3); increasing (1). Clusters with high OH&S maturity (Virtuous and Incomplete) are characterized by a higher number of companies with high OH&S performance.

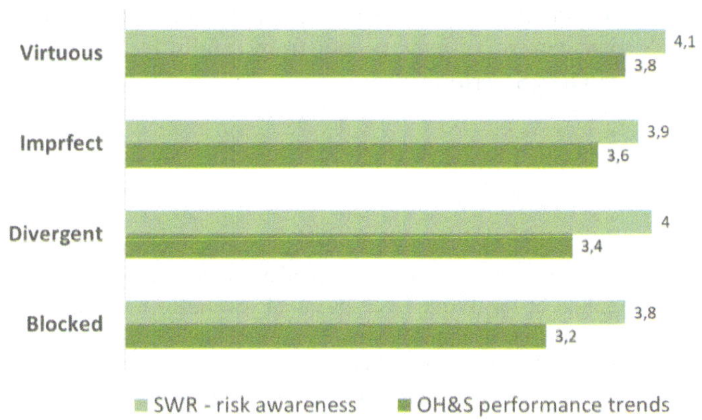

Fig. 3. Correlation between OH&S maturity and performance (significance: OH&S performance, p = 0,005; risk awareness, p = 0,000).

In small companies, the Locked model represents nearly half of the sample (49%). A Virtuous model is in place in about 30% of medium and large companies. The Incomplete model is more frequent in large companies (23%). Furthermore, about half (54%) of large companies are characterized by high OH&S management maturity.

The Locked model is widespread in the Public Administration (53%), where only 22% of the companies have reached a Virtuous participative model. Overall, about one third of public and private companies (31%) present Virtuous models. In the private sector, however, the number of companies with low degree of maturity models is higher: 42% of companies with Locked model, against the 38% in companies controlled by the public administration. This difference can be explained by a relatively higher number of SMEs in the private sector, where participatory models are less mature. Looking at the private sector, the Virtuous model is the most widespread in multinational companies (37% versus 27% in national companies).

5 Conclusions

Thirty years after the European Directive 89/391, by which the new pre-emptive model was launched, many difficulties remain in recognizing an active role of SWRs in different businesses and sectors. In the majority of companies, SWRs' contribution is still hindered by a number of factors that limit an active and fully participative role; this is mainly because of less mature OH&S management systems where processes still do not grant adequate information, consultation and participation. Nevertheless, cases of Virtuous participative models are present across sectors, types (private vs public) and sizes (large vs SME). The survey showed that companies with higher maturity on both OH&S management and SWRs' role have probability of better OH&S performance.

Results contribute to OH&S management literature by several means. First, by proposing a novel conceptual framework for addressing the role of SWR within OH&S management systems. Second, the paper offers new insights from a robust large-scale survey with no prior examples. From a managerial perspective, the paper suggests that effective OH&S management systems benefit from a fully recognized role to SWRs leading to enhanced OH&S performance.

References

1. Walters, D., Dalton, A.J.P., Gee, D.: Worker representation on health and safety in Europe. European Trade Union Technical Bureau for Health and Safety (1993)
2. Vogel, L.: Prevention at the Workplace: an initial review of how the 1989 Community Framework Directive is being implemented. European Trade Union Technical Bureau for Health and Safety (ETUC), Brussels (1993)
3. Walters, D.: The role of worker representation and consultation in managing health and safety in the construction industry, pp. 1–48. ILO (2010)
4. Shannon, H.S., Robson, L.S., Sale, J.E.: Creating safer and healthier workplaces: role of organizational factors and job characteristics. Am. J. Ind. Med. **40**(3), 319–334 (2001)
5. Walters, D.: Worker representation and health and safety in small enterprises in Europe. Ind. Relat. J. **35**, 169–186 (2004)
6. Biggins, D., Holland, T.: The training and effectiveness of health and safety representatives. In: Eddington, I. (ed.) Towards Health and Safety at Work. Technical Papers of the Asia Pacific Conference on Occupational Health and Safety (1995)
7. Biggins, D., Phillips, M.: A survey of health and safety representatives in Queensland. Part 1: activities, issues, information sources. J. Occup. Health Saf. – Aust. N. Z. **7**(3), 195–202 (1991)
8. Biggins, D.R., Phillips, M., O'Sullivan, P.: Benefits of worker participation in health and safety. Labour Ind.: J. Soc. Econ. Relat. Work **4**(1), 138–159 (1991)
9. Gaines, J., Biggins, D.: A survey of health and safety representatives in the Northern Territory. J. Occup. Health Saf. Aust. N. Z. 8, 421 (1992)

10. Cooper, M.D., Phillips, R.A.: Exploratory analysis of the safety climate and safety behavior relationship. J. Saf. Res. **35**(5), 497–512 (2004)
11. Fernández-Muñiz, B., Montes-Peón, J.M., Vázquez-Ordás, C.J.: Relation between occupational safety management and firm performance. Saf. Sci. **47**(7), 980–991 (2009)
12. Goncalves Filho, A.P., Andrade, J.C.S., de Oliveira Marinho, M.M.: A safety culture maturity model for petrochemical companies in Brazil. Saf. Sci. **48**(5), 615–624 (2010)
13. Ostrom, L., Wilhelmsen, C., Kaplan, B.: Assessing safety culture. Nucl. Saf. **34**(2), 163–172 (1993)
14. Zohar, D.: Safety climate in industrial organizations: theoretical and applied implications. J. Appl. Psychol. **65**(1), 96 (1980)

Process Operator Students' Outlook on Safety

Susanna Mattila(✉), Sanna Nenonen, Noora Nenonen, and Sari Tappura

Tampere University, 33014 Tampere, Finland
{susanna.mattila,sanna.nenonen,noora.nenonen, sari.tappura}@tuni.fi

Abstract. This paper discusses the outlook on and understanding of safety among process operator students participating in vocational education and training. The data were collected through an e-survey directed to the process operator students of five Finnish vocational institutions in fall 2019, and 122 students answered the questionnaire. According to the results, the students had a mainly positive outlook on safety. For example, more than half of the respondents were of the opinion that accidents can be prevented. The majority of the students thought safety is an important part of their future work. On-the-job learning and working in industry before the vocational studies seems to relate to students' outlook on safety.

Keywords: On-the-job learning · Vocational student · Safety climate · Safety attitude · Vocational education and training · Process industry

1 Introduction

Potentially dangerous materials and extreme conditions are typical in the process industry and create a remarkable accident risk [1]. In line with the increasing complexity of chemical processes, this calls for good process safety competence from industrial operators [2]. In the safety-critical process industry, companies require a serious attitude toward safety and the ability to prioritize safety over production from students carrying out on-the job-learning. Moreover, process operator students' safety values and attitudes are essential components of their process safety competence [3].

Young workers have greater risk of occupational injuries than older workers [4]. Vocational education and training (VET) is increasingly implemented in workplaces as the VET reform in Finland emphasizes the role of on-the-job learning [5]. VET students' safety competence is based on the qualification requirements of the specific vocational field (e.g., process industry). Students are provided with a basic understanding of safety by attending the safety activities of the vocational institution [6, 7]. However, this knowledge needs to be further strengthened at the workplace to take company-specific safety requirements into account [8]. In conclusion, students may not gain adequate knowledge about safety during studies [9], and these skills evolve only through work experience [10].

This paper discusses the outlook on and understanding of safety among process operator students participating in VET. The aim is to study whether students have the

P. M. Arezes and R. L. Boring (Eds.): AHFE 2020, AISC 1204, pp. 26–32, 2020.
https://doi.org/10.1007/978-3-030-50946-0_4

safety attitudes and abilities desired by process industry companies [3]. The focus is on the importance of safety in general, in studies, and during on-the-job learning. The specific interest is on the differences in insights between students who have not yet participated in on-the-job learning and those who have. This study is part of a research project studying process operator students' safety competences and collaboration between vocational institutes and process industry companies [3, 7, 8].

2 Materials and Methods

The data were collected through an anonymous e-survey directed to the process operator students of five vocational institutions in fall 2019. At some of the institutions, students answered the questionnaire during a lesson, while other students answered the questionnaire during their free time. A total of 122 students answered the questionnaire, but five questionnaires were removed because all of the responses were neutral. Age of the respondents varied between 15 and 50, with a mean age of 23. A majority of the students were first or second-year students. Most of the students had not yet done on-the-job learning, and most did not have previous work experience. Table 1 shows the number of respondents in different categories.

Table 1. Demographic data of respondents.

Demographic category	Item	Percentage
Study year (n = 117)	First	51
	Second	32
	Third	17
On-the-job learning (n = 117)	Yes	33
	No	67
Work experience (n = 117)	Not at all	67
	Less than 6 months	17
	6 months or more	16

The questionnaire was based on previous studies and surveys [9, 11–14] regarding safety perceptions among employees and students. The questionnaire consisted of 30 Likert-scale items covering perceptions about accidents, safety instructions, communication, the role of safety, safety competence and training, and safety performance. The data were summarized by descriptive statistics and analyzed with statistical tests. The Mann-Whitney test was used for the effect of on-the-job learning and Kruskal-Wallis for the effect of work experience on safety attitudes. Effect of study year on safety attitudes was not tested because first-year students had not done on-the-job learning so that and the study year would correlate.

3 Results

3.1 Outlook on Safety in General

The frequencies and mean scores of the items in which the means were under 1.5 (disagree) or over 4.5 (agree) are shown in Table 2. More than 30% of respondents neither agreed nor disagreed with the items Safe working is hardly recognized during the studies at vocational institution, Safe working is hardly recognized during on-the-job learning, Safety matters should be paid more attention to during the studies at vocational institution, Safety matters should be paid more attention to during on-the-job learning, and I would like to get more safety training before on-the-job learning.

Table 2. Frequencies and mean of responses for items with mean <1.5 or >4.5.

Item	n	1%	2%	3%	4%	5%	Mean
I always try to learn from accidents	117	2	1	3	15	80	4.69
During the studies at vocational institution, I always follow the safety instructions given to me	116	3	1	4	23	69	4.55
During on-the-job learning, I always follow the safety instructions given to me	117	2	2	3	13	81	4.70
During on-the-job learning, I always inform superior or workplace guidance personnel of the hazards I notice	115	1	0	4	21	74	4.67
Safety matters are not an essential part of the work of a process operator	117	80	9	6	2	4	1.43
If I am reminded about unsafe behavior, I correct my behavior immediately	117	0	0	8	17	75	4.68

1 = disagree, 2 = partly disagree, 3 = neither agree nor disagree, 4 = partly agree, 5 = agree

3.2 Effect of on-the-Job Learning and Work Experience on Outlook

On-the-job learning relates to a student's outlook on safety. Compared to those who had not attended on-the-job learning, the majority of those who had done on-the-job learning thought that accidents can be prevented and stated that they know how to act in an emergency situation. At the same time, thinking that safety regulations make work more difficult was more common among those who had done on-the-job learning compared to those who had not. Most of those who had not learned on the job would like to get more safety training before on-the-job learning compared to those who had already learned on the job. The frequencies and mean scores by group for the items in which differences between groups were statistically significant ($p < 0.05$) are shown in Fig. 1.

Working in industry before beginning studies also related to student outlooks on safety, where 74% of those without work experience and of those who had at least six months of work experience in industry before studies said that they always follow safety instructions during studies at the vocational institution. Of the respondents who had under six months of work experience, only 42% said that they always follow safety instructions at the vocational institution. According to responses, 87% of those who had no work experience always follow safety instructions during on-the-job learning, compared to 65% of those who had under six months of work experience and 74% of those who had at least six months of work experience.

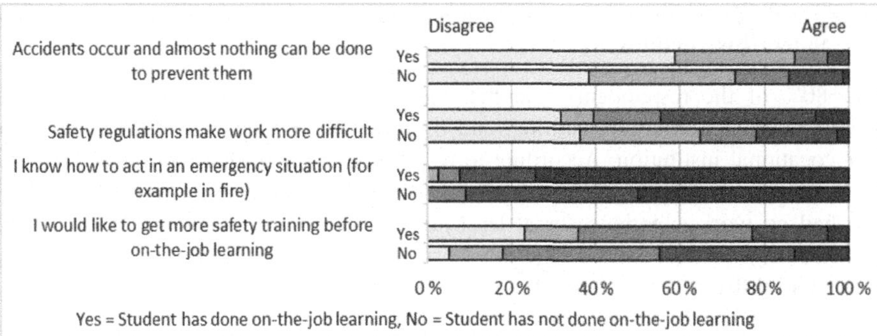

Fig. 1. Frequencies of responses between students who had done on-the-job learning and those who had not for the items with a statistically significant difference.

There were similar differences when considering use of protective equipment and safety as part of daily activities at the vocational institution. Most of those with at least six months' work experience agreed that safety matters should get more attention during studies at vocational institution, compared to those who had no work experience and those with less than six months' work experience before studies. The frequencies and mean scores by group for statistically significant items are shown in Fig. 2 ($p < 0.05$).

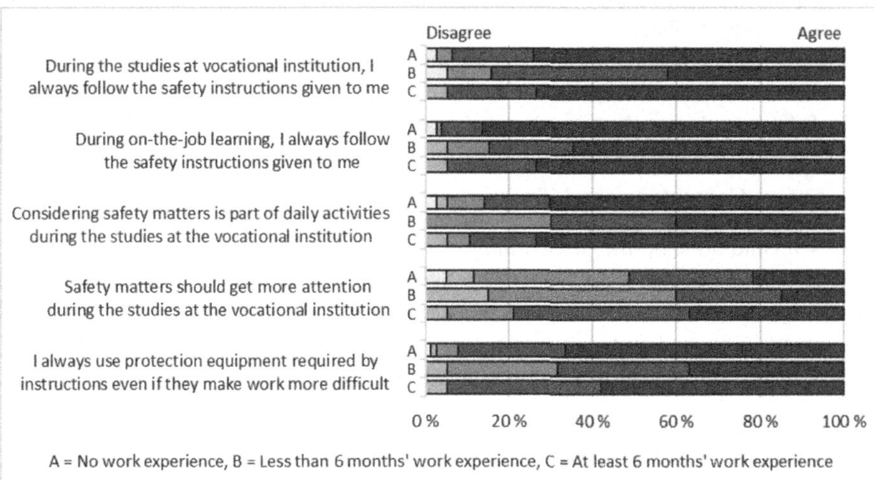

Fig. 2. Frequencies of responses grouped by length of work experience where the difference between groups was statistically significant.

3.3 Accidents

Almost all of the respondents stated that they try to learn from accidents. Nearly 80% of respondents thought that accidents can be at least partially prevented. More of those who had attended on-the-job learning strongly disagreed that nothing can be done to prevent accidents compared to those who had not yet learned on the job.

3.4 Safety Instructions

Over 80% of the respondents said that they always follow the safety instructions provided during on-the-job learning, and 69% of the respondents said they do so at their vocational institution. According to responses, fewer of those who had less than six months' work experience always followed safety instructions compared to those who had no work experience or at least six months' work experience. More of the respondents who had done on-the-job learning agreed that safety regulations make work more difficult compared to those who had not yet done on-the-job learning.

3.5 Communication

Over 70% of the respondents said they always inform their supervisor or workplace guidance personnel of hazards they notice during on-the-job learning, while 64% of respondents said they always inform teachers of hazards at their vocational institution.

3.6 Role of Safety

The majority of students thought that safety is an important part of their future work, where 80% of respondents disagreed and 9% partly disagreed that safety matters are not an essential part of the work of a process operator. Only 40% of those who had under six months' work experience agreed that consideration of safety matters is part of daily activities during studies at vocational institutions, while 71% of those who had no work experience agreed, and 74% of those with at least six months' work experience agreed. A large number (37%) of those who had at least six months' work experience agreed that safety matters should be get more attention during studies, compared to those who had less work experience, where 22% of those with no work experience and 15% of those with less than six months' experience agreed.

3.7 Safety Competence and Training

Those who had done on-the-job learning were a bit more confident that they know how to act in emergency situations, with 74% agreeing with the item I know how to act in emergency situations, while 50% of those with no on-the-job learning agreed. More respondents who had not done on-the-job learning would like to get more safety training before on-the-job learning, compared to those who had learned on the job.

3.8 Safety Performance

Most of the respondents said that they correct their behavior immediately if they are reminded about unsafe behavior. Those who had less than six months' work experience before studies were more likely to stop using protection equipment if the equipment made work more difficult, with 37% agreeing with the item I always use protection equipment required by instructions even if they make work more difficult, as compared to those who had no experience (67%) and those with at least six months' experience (58%).

4 Discussion

This study shows that process operator students had a mainly positive outlook on safety. The majority of respondents thought that safety matters are an essential part of their future work. They also try to learn from accidents and follow safety instructions.

More of those who had not learned on the job would like to get more safety training before on-the-job learning compared to those who had already learned on the job. This can be explained by the responses to the item On-the-job learning helped me to understand safety measures better than studies at vocational institutions, where 39% of those who had learned on the job agreed and 39% partially agreed.

Bigger part of those who had at least six months work experience in industry before the studies agreed that safety matters should get more attention during studies compared to those who had less or no work experience. It may be that they realized the importance of safety while working. It is interesting that those students who had less than six months of work experience had a less positive outlook on safety than those who had no work experience or those with at least six months of experience. This finding should be studied further. The same questionnaire could also be used in other industries. Interviews could reveal explanations for differences between groups.

This study contributes to vocational institutions and collaborating process industry companies as well as academics interested in enhancing vocational students' safety awareness. The study provides insight into both the development needs of safety training during vocational studies in general and the role of on-the-job learning in the formation of safety attitudes among vocational students. Even if the results do not show dramatic shortcomings in safety training, more attention could be paid to safety matters during the studies both at vocational institutions and in workplaces [7]. Students usually follow the safety instructions given to them, so it is important to provide proper safety instructions. Students also normally report hazards. In order to manage risks, it is crucial to handle these reports adequately and take corrective measures. Additionally, student work experience should be taken into account in designing safety training [15].

A limitation of this study is the small number of respondents. Also, half of them had just started their studies and had quite little experience. A benefit of the study is that respondents were studying at five different institutions in different cities.

Acknowledgments. The authors would like to thank the students for answering the questionnaire and the vocational institutions for delivering it. Furthermore, the authors appreciate the research funding provided by the Finnish Work Environment Fund, participating process industry companies, and Tampere University.

References

1. Rodríguez, M., Díaz, I.: A systematic and integral hazards analysis technique applied to the process industry. J. Loss Prev. Process Ind. **43**, 721–729 (2016)
2. Nazir, S., Sorensen, L.J., Øvergård, K.I., Manca, D.: How distributed situation awareness influences process safety. Chem. Eng. Trans. **36**, 409–414 (2014)
3. Tappura, S., Nenonen, S., Nenonen, N., Kivistö-Rahnasto, J.: Process safety competence of vocational students. In: Arezes, P. (ed.) Advances in Safety Management and Human Factors. AHFE 2019. Advances in Intelligent Systems and Computing, vol. 969, pp. 383–392. Springer, Cham (2019)
4. Salminen, S.: Have young workers more injuries than older ones? An international literature review. J. Saf. Res. **35**, 513–521 (2004)
5. Ministry of Education and Culture: Vocational Education and Training Will Be Reformed. Press Release, 10 August 2017
6. Tappura, S., Kivistö-Rahnasto, J.: Annual school safety activity calendar to promote safety in VET. In: Arezes, P.M. et al. (eds.) Occupational Safety and Hygiene VI: Proceedings of the 6th International Symposium on Occupation Safety and Hygiene (SHO 2018), pp. 131–135. CRC Press, London (2018)
7. Nenonen, N., Nenonen, S., Tappura, S.: Process operator students' abilities to assess OSH risks. In: Ahram, T., Karwowski, W., Pickl, S., Taiar, R. (eds.) Human Systems Engineering and Design II. IHSED 2019. Advances in Intelligent Systems and Computing, vol. 1026, pp. 566–572. Springer, Cham (2019)
8. Tappura, S., Nenonen, S., Nenonen, N.: Developing safety competence process for vocational students. In: Ahram, T., Karwowski, W., Taiar, R. (eds.) Human Systems Engineering and Design. IHSED 2018. Advances in Intelligent Systems and Computing, vol. 876, pp. 668–674. Springer, Cham (2018)
9. Andersson, M., Gunnarsson, K., Rosèn, G., Åberg, M.M.: Knowledge and experiences of risks among pupils in vocational education. Saf. Health Work **5**, 140–146 (2014)
10. Westaby, J.D., Lee, B.C.: Antecedents of injury among youth in agricultural settings: a longitudinal examination of safety consciousness, dangerous risk taking, and safety knowledge. J. Saf. Res. **34**, 227–240 (2003)
11. Cox, S.J., Cheyne, A.J.T.: Assessing safety culture in offshore environments. Saf. Sci. **34**, 111–129 (2000)
12. Kines, P., Lappalainen, J., Mikkelsen, K.L., Olsen, E., Pousette, A., Tharaldsen, J., Tómasson, K., Törner, M.: Nordic safety climate questionnaire (NOSACQ-50): a new tool for diagnosing occupational safety climate. Int. J. Ind. Ergon. **41**, 634–646 (2011)
13. Gunasekera, M., Khan, F., Ahmed, S.: Learning process safety principles through practice. Process Saf. Prog. **37**(3), 347–354 (2017)
14. Gong, Y.: Safety culture among Chinese undergraduates: a survey at a university. Saf. Sci. **111**, 17–21 (2019)
15. Stuart, A.: A blended learning approach to safety training: student experiences of safe work practices and safety culture. Saf. Sci. **62**, 409–417 (2014)

Safety Management in Accordance with Industry 4.0 Requirements: Analysis and Evaluation of the Level of Digitalization in the Slovak Companies

Hana Pacaiova, Renata Turisova, Anna Nagyova(✉), and Milan Oravec

Faculty of Mechanical Engineering, Safety and Quality Department, Technical University of Kosice, Letna 9, 04200 Košice, Slovakia
{hana.pacaiova, renata.turisova, anna.nagyova, milan.oravec}@tuke.sk

Abstract. Digitalization of industrial companies brings a lot of advantages for their management in the global entrepreneurial environment. Shortening of new product development, communication with customers already in their design phase, 'early' prevention of potential errors and discrepancies in processes, online status monitoring of important devices etc., enable to introduce effective management and meeting the stakeholders' needs. However, these advantages are conditioned by applying the suitable tools for processing and evaluating data in core processes of an organization. Industry 4.0 (I4.0) is a general term specifying the level of automation and digitalization implementation at the management of industrial as well as public activities within the advanced society.

Protection of human in work environment in the EU has been implemented as Frame Directives for 30 years. Protection of property – intellectual property, its importance in today's environment grows exponentially with the level of implemented digitalization. For this reason, it is important to apply such safety management tools, so that they enable a complex approach to the protection of physical but also intellectual property of an organization.

The objective of the study was to find the real level of Industry 4.0 factors implementation and asses the status of complex safety in the selected industrial organizations in the Slovak republic. The form of questioning stemmed from the exceptionality model – EFQM. Basic parameters of the requirements resulting from the I4.0 factor analysis were created. The area of results was structured in such a way so that it is possible to prove the level of digitalization achievement and complex safety management in companies operating in Slovakia and their readiness for current trends in I4.0 and their development.

Keywords: Integrated safety · Model EFQM · Industry 4.0 · Boxplot diagram

P. M. Arezes and R. L. Boring (Eds.): AHFE 2020, AISC 1204, pp. 33–39, 2020.
https://doi.org/10.1007/978-3-030-50946-0_5

1 Introduction

Most advanced industrial countries are nowadays intensively working on the onset of so called fourth industrial revolution, also called Industry 4.0. It is a concept relying on such elements as industrial internet, cybernetic-physical systems, artificial intelligence, etc. It seems that an early capture of the onset of the mentioned industrial revolution will, for individual industrially oriented countries, have a cardinal importance, not only from a view of their competitiveness [1, 2]. Also in Slovakia, which is industrially oriented, a lot of experts from the practice, academicians and unfortunately also politicians devote themselves to this problematic. They come up with various opinions, views, knowledge and determination to implement Industry 4.0 into various areas of the Slovak industry. Yet, there is one fundamental issue – whether the Slovak industrial companies are sufficiently prepared for Industry 4.0 implementation. According to the study by Grenčíková et al. [4] within 80 industries operating in mechanical sector, in 2017, only 66% of respondents stated that Industry 4.0 is very important for the future, but in 2018 only 59% of respondents maintained this opinion. On the other hand, compared to 2017, the number of companies stated to deal with I4.0 increased by 14% in 2018.

The industrial revolution is a connection of the virtual cybernetic world with the real world, where not only physical laws are valid, but also social, cultural, economic and other ones. This calls for the necessity to identify, get to know and understand the important interactions between individual systems and the entire society. For this reason, Industry 4.0 is spoken about as the cybernetical-physical and social revolution. Technological requirements and visions expect an industrial integration, deeply founded by knowledge, as a base of Industry 4.0 concept, which is based on information and cybernetical technologies. It must able to massively share loads of information (big data) and generally, in a real time, continuously communicate with autonomous robots, sensors, cloud and data storages. Above it all – as a thought and technological core, there must be the latest and suitably applied methods and procedures of cybernetics and artificial intelligence [5].

The problematics of the implementation of Industry 4.0 principles and the issue of its integration with regard to the existing and new technologies in the corporate practice is dealt with by a whole array of scientific and expert journals. Given the geopolitical position of Slovakia and its orientation towards the automotive industry, we prefer European, more particularly German ideas focused on the assessment of individual companies' readiness for Industry 4.0 [15].

Authors Bangemann et al. [6] tried to reach a harmony between the existing and new technologies and machinery equipment by means of the so-called mixed systems. Authors Müller and Voigt [7] in Nurnberg, on a sample of 177 small and medium-sized companies, looked into possibilities and readiness of respondents for the implementation of Industry 4.0 principles, from the point of view of activities, corporate relations, supplier-customer relations and their potential information connection. The problematics during Industry 4.0 implementation was dealt with by Pfeiffer [8]. The study by authors Veile and Kiel [9] is based on the empirical data gained from 13 semi-structured depth interviews with German experts who have experience with the implementation of Industry 4.0. It is one of the first documents stating specific cases of knowledge gained directly from the industrial application of Industry 4.0. Among others, it deals with the problematics of funding provision, integration of employees into the integration process and the creation of open flexible corporate culture. It also tackles the question of planning processes complexity, cooperation with external partners, correct usage of data interfaces, inter-disciplinary communication, organizational structure changes and data safety.

This study and the results of surveys from 2017 [9] inspired the researchers of department to assess the status of the readiness of Slovak companies with regard to selected aspects of I4.0. The questioning scheme stemmed from partly modified model focused on the company effectivity measurement EFQM [10], as the study objective was to verify the requirements of Slovak companies for Industry 4.0 implementation. From the mentioned EFQM model, only its Enablers part was used (without Results part). The survey was conducted on 36 organizations aimed at automotive industry, or automotive industry suppliers. A detailed description of questioning and the way of evaluation considerably exceeds the possible length of this contribution, which is why in the next one, we will mention only some results obtained from the survey in question.

2 Perception of Integrated Safety and Digitalization in the Automotive Industry

Digitalization is an inevitable part of a successful Industry 4.0 implementation, which is why, with a sort of exaggeration, it is possible to consider the digitalization of a particular company to be an absolute fundament of any assessment of the relevant company readiness for Industry 4.0. One of other fields, which was the subject of research of the mentioned questioning, is the integrated safety [11]. This term included not only the integration of safety into machinery/technology with regard to its life cycle, but also security aspect as an important part of integrated safety (Safety&Security). Should we consider digitalization a necessity, then the ability to integrate is an extension of the same importance of a successful implementation of Industry 4.0. Therefore, in this contribution, we will focus only on partial assessment of these two leading aspects and their comparison.

As it has already been mentioned in the previous part, in order to assess the readiness of companies, a modified EFQM model was used, more particularly its requirement part. Figures assigned to individual answers within the questioning have the same weight and structure as it is described in RADAR card within the evaluation of EFQM model [12]. The difference is only in the fact that, within EFQM model, the maximum point value (100%) accounts for 1000 points.

During our questioning, which stemmed only from the questions in requirement part, 100% accounts for only 100 points, while the proportionality of assigned points and weights is the same as during the evaluation of EFQM requirement part. The requirement part of EFQM model consists of five areas - Leadership, People, Strategy, Partnerships and Resources, Processes, Products and Services. Focusing the questions within the conducted survey has the same structure as the requirement part of EFQM model self-assessment. To be simple, the graphic description of individual results of questioning will not be shown in points but in percent. During each assessment, 100% means the maximum reachable point value. Figure 1 shows the 95% reliability interval for medium values of the total point evaluations of the questioning.

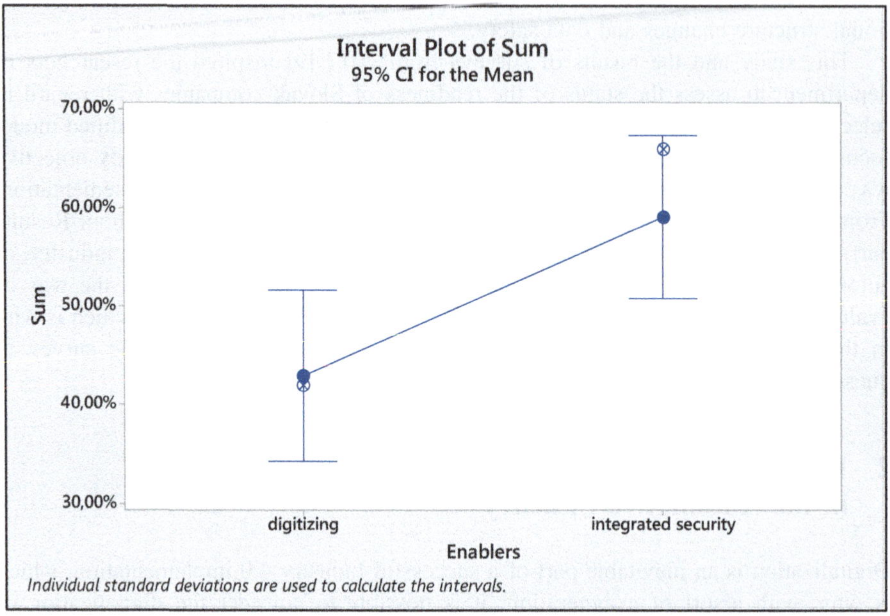

Fig. 1. 95% reliability interval of the total evaluation of digitalization and integrated safety by all respondents (axis x). The axis y shows the presented total % (points) reached in digitalization and integrated safety.

The interval sizes are approximately the same, so it is possible to consider the variance homoscedastic. At the digitalization aspect, almost all answers oscillated in the range between approximately 33% and 51%. The mentioned perception of digitalization of Slovak companies in the automotive industry can be considered acceptable given the fact that these are companies using entire-world state-of-the-art device and machinery equipment, hence generally with a high digitalization degree. The assessment of integrated safety reaches the values approximately from 51% to 68%. It is a relatively optimistic perception of the respondents' readiness in the field of integrated safety. The result of comparison of the total assessment is also interesting. It clearly shows that integrated safety is perceived by respondents on a significantly higher level than digitalization. To divide individual figures according to areas, we used Box-plot diagrams, see Fig. 2.

What results from them is that the biggest differences in respondents' assessment are in the area of human resources, planning and processes. There are smaller differences within the partnership assessment and the smallest as for the assessment of leadership. With the last two mentioned areas, there is a heteroscedastic variability, i.e.

variables of figures of individual respondents have a big range. The assessment of integrated safety with human resources, processes and planning is approximately on the same level and reaches high percentual values. On the contrary, digitalization reaches the lowest assessments in given areas, which causes the mentioned significant difference.

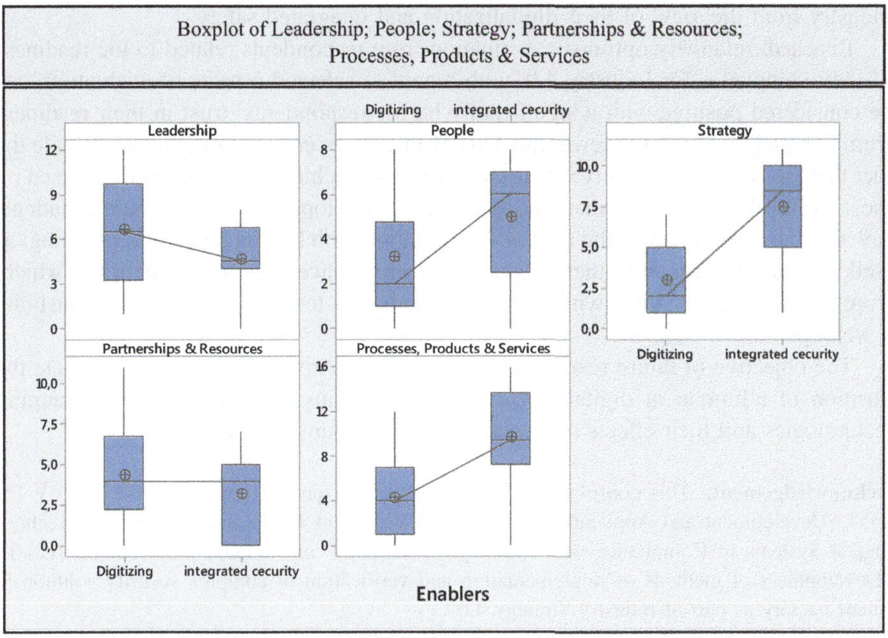

Fig. 2. Boxplot diagram of the integrated safety and digitalization assessment (axis *x*). The axis *y* shows % (points) reached in particular areas.

One of possible interpretations of the mentioned results is that the assessment of readiness perception for Industry 4.0 implementation in Slovakia can be considered optimistic. From the additional questions it is evident that the respondents consider the transition to Industry 4.0 to be a challenge and they feel ready for its implementation.

They trust themselves much more from the view of Integrated safety, namely in all areas which the questioning was focused on. They evaluate digitalization on significantly lower level, not only generally but also in all evaluated areas. They see the biggest digitalization insufficiencies from the view of human resources, processes and planning. There is a slightly higher trust in the digitalization mediated by partner organizations. From the leadership viewpoint, even though the answer variability is the biggest, the perception of leaders' importance during digitalization is on average the biggest.

3 Conclusion

Readiness for the onset of the fourth industrial revolution is an important factor for the support of top management decisions whether and to what content join this initiative. As we have shown, there are several studies [3, 13], which assess the readiness of individual countries or industry areas for Industry 4.0. The intention of this contribution was to assess the perception of readiness of the selected companies in the automotive industry from the view of their digitalization and integrated safety.

Reached, relatively optimistic evaluation from respondents related to the readiness of their companies for Industry 4.0, with regard to selected aspects of evaluation, can be considered positive with a significantly higher respondents' trust in their readiness from the integrated safety viewpoint. This evaluation is considerably higher despite the fact that the Slovak automotive industry possesses machinery equipment digitalized on the same level as e.g. in Germany or other highly developed countries. The respondents saw the biggest reserves in the area of digitalization with human resources, planning, as well as with the processes themselves. Basic areas have thus been identified, which, from the respondents' view within the digitalization in terms of Industry 4.0, constitute a weak point.

The objective of future research, based on current trends [14], is to investigate the position of a human in digital environment, risk analysis connected with changing technologies and their effects on social aspects and human health.

Acknowledgement. This contribution is the result of the project implementation APVV-15-0351 "Development and Application of a Risk Management Model in the Setting of Technological Systems in Compliance with Industry 4.0 Strategy" and VEGA project No.1/0121/18 "Development of methods of implementation and verification of complex security solution in Smart Factory as part of Industry Strategy 4.0".

References

1. Ślusarczyk, B.: Industry 4.0 – are we ready? Pol. J. Manag. Stud. **17**(1), 232–248 (2018). https://doi.org/10.17512/pjms.2018.17.1.19. BazTech
2. Ganzarain, J., Errasti, N.: Three stage maturity model in SME's toward Industry 4.0. J. Ind. Eng. Manag. (JIEM) **9**(5), 1119–1128 (2016). https://doi.org/10.3926/jiem.2073
3. Kagermann, H.: Change through digitization value creation in the age of Industry 4.0. In: Albach, H., Meffert, H., Pinkwart, A., Reichwald, R. (eds.) Management of Permanent Change, pp. 23–45. Springer (2015). https://doi.org/10.1007/978-3-658-05014-6_2
4. Grenčíková, A., Kordoš, M., Sokol, J.: The approach to Industry 4.0 within the Slovak business environment. Soc. Sci. **8**(104), 1–11 (2019). https://doi.org/10.3390/socsci8040104
5. Ramirez-Peňa, M., Sotano, A.J.S., Peréz-Fernandéz, V., Abad, F.J., Batista, M.: Achieving a sustainable shipbuilding supply chain under I4.0 perspective. J. Clean. Prod. **244**, 1–11 (2020). https://doi.org/10.1016/j.jclepro.2019.118789
6. Bangemann, T., et al.: Integration of classical components into industrial cyber–physical systems. Proc. IEEE **104**(5), 947–959 (2016). https://doi.org/10.1109/JPROC.2015.2510981

7. Voigt, K.I., Müller, J., Veile, W.J., Becker, W., Stradtmann, M: Industrie 4.0 – Risiken für kleine und mittlere Unternehmen. In: Geschäftsmodelle in der digitalen Welt, vol. 2, no. 3, pp. 517–538. Springer (2018). https://doi.org/10.1007/978-3-658-22129-4_26
8. Pfeiffer, S.: Berufliche Bildung 4.0? Ueberlegungen zur Arbeitsmarkt-und Innovationsfaehigkeit. Industrielle Beziehungen **23**(1), 25–44 (2016). https://doi.org/10.1688/IndB-2016-01-Pfeiffer
9. Veile, J.W., Muller, J.M., Voigt, K.I.: Lessons learned from Industry 4.0 implementation in the German manufacturing industry. J. Manuf. Technol. Manag. (2019). https://doi.org/10.1108/jmtm-08-2018-0270
10. Suárez, E., et al.: Quantitative research on the EFQM excellence model: a systematic literature review (1991–2015). Eur. Res. Manag. Bus. Econ. **23**(3), 147–156 (2017). https://doi.org/10.1016/j.iedeen.2017.05.002
11. Blecha, P., Durakbasa, N., Holub, M.: Digitized production – its potentials and hazards. In: Durakbasa, N., Gencyilmaz, M. (eds.) 2018 Proceedings of the International Symposium for Production Research, pp. 402–411. Springer, Cham (2019). https://doi.org/10.1007/978-3-319-92267-6_35
12. Andersen, H., Lawrie, G., Shulver, M.: The balanced scorecard vs. the EFQM business excellence model - which is the better strategic management tool? 2GC Active Management (2003). https://www.researchgate.net/publication/237117368_The_Balanced_Scorecard_vs_the_EFQM_Business_Excellence_Model_-_which_is_the_better_strategic_management_tool/link/5954d0eaa6fdcc16978cb538/download
13. Hanulakova, E., Dano, F.: Circular economy as a new managerial approach. AD ALTA-J. Interdisc. Res. **8**(1), 95–98 (2018). Magnanimitas
14. Urbančíková, N., Zgodavová, K.: Sustainability, resilience and population ageing along schengen's eastern border. Sustainability **11**(10), 1–18 (2019). https://doi.org/10.3390/su11102898
15. Tureková, I., Mračková, E., Marková, I.: Determination of waste industrial dust safety characteristics. Int. J. Environ. Res. Public Health **16**(12), 2103 (2019). https://doi.org/10.3390/ijerph16122103

Links Between Knowledge Transmission Programs and the Preservation of Occupational Health and Safety

Cláudia Pereira[1(✉)], Marta Santos[1], and Catherine Delgoulet[2]

[1] Faculdade de Psicologia e de Ciências de Educação da Universidade do Porto, Porto, Portugal
{cpereira,marta}@fpce.up.pt
[2] Conservatoire National des Arts et Métiers, Paris, France
catherine.delgoulet@lecnam.net

Abstract. Studies show that knowledge retention and transmission practices can have a positive impact on the occupational health and safety management. To understand the presence of occupational health and safety issues in knowledge retention/transmission practices in the last 15 years in professional contexts, a systematic review was conducted. The studies analyzed include initiatives such as mentoring programs and documentary repositories based on the explicit dimension of knowledge, and which do not focus on the participation and acknowledgement of workers or on the analysis of the role that their working conditions, health and safety behaviors have in the processes of knowledge transmission. The results allowed the identification of guiding principles to support companies in the implementation of practices that privilege the participation and acknowledgment of workers and their knowledge, for an action that guarantees production and quality for the company and that preserves the health and safety of workers.

Keywords: Knowledge transmission · Knowledge retention · Preservation of occupational health and safety · Systematic review

1 Introduction

The current scenario for professional contexts is characterized by the presence of several generations simultaneously and demographic changes that continues to be a challenge, due to the decrease in the proportion of young people, the increase in the number of people of working age and the number of elderly, and the progressive increase in retirement age in recent years [1, 2].

This scenario has posed challenges for companies to respond to: management of age diversity; scarcity of resources in the labor market; longevity of workers in companies; and loss of a significant number of workers due to their transition to retirement and, consequently, loss of critical knowledge acquired through experience and related to the preservation of health and safety at work [3]. In this sense, and taking into account the concerns about the loss of knowledge of older people in companies, with their retirement [4], and the difficulty in verbalizing the knowledge acquired during

P. M. Arezes and R. L. Boring (Eds.): AHFE 2020, AISC 1204, pp. 40–46, 2020.
https://doi.org/10.1007/978-3-030-50946-0_6

their professional experience, one of the concerns of companies is to identify ways to retain and transmit knowledge among workers of different ages and seniority, namely knowledge related to the preservation of safety and health at work, taking into account the intensification of time constraints at work and the acceleration of changes at work and in the companies themselves [5].

In this context, the Human Resources teams assume a support role in the implementation of knowledge transmission and retention initiatives [6]. However, these initiatives sometimes seem to be implemented without taking into account working conditions or the workers' perspective and, due to the demands of the Human Resources activity itself, they are not followed up or maintained over time [7, 8].

Moreover, the knowledge acquired through experience is important for the carrying out of the work, in particular for a safer performance, with the creation of ways of knowing and doing that preserves health and safety at work [9, 10]. Thus, managing knowledge in its contexts, through its retention and transmission, can have a positive impact on occupational health and safety management [11].

To understand the presence of occupational health and safety issues in knowledge retention/transmission practices, a systematic review was conducted in order to identify these initiatives implemented in the last 15 years in professional contexts.

The results obtained can support companies in the identification of guiding principles for the implementation of practices that privilege the participation and acknowledgment of workers and their knowledge, for an action that guarantees production and quality for the company and that preserves the health and safety of workers.

2 Method

2.1 Search Strategy

The research of the articles was carried out between October and November 2019 and was done in the SCOPUS database. The choice of this database is related to the type of articles published and to the fact that it integrates journals that explore themes related to the transmission and management of knowledge and to learning and training in the work context.

The keywords defined for the search were: "knowledge retention program" OR "knowledge transmission program" OR "knowledge transfer program" OR "knowledge management program". Although one of the focuses of this work was to understand the presence of occupational health and safety issues in knowledge retention/transmission practices, we chose not to integrate a search keyword explicitly related to "occupational health and safety" into the search, to make it broader and include articles that would be omitted in this way.

2.2 Inclusion and Exclusion Criteria

The inclusion criteria defined for the research of the articles were: (i) articles published between 2004 and 2019; (ii) empirical articles, with a qualitative analysis or case study component; (iii) articles whose title or abstract included the key research words; (iv) articles whose participants were workers in a professional context.

Articles that did not include in their title or abstract the key words defined for the research strategy, and whose methodology or results were not related to the intended theme or did not meet the objective of the systematic review (e.g. systematic review studies; presentation of theoretical models of knowledge management; studies conducted in an academic context) were excluded.

2.3 Number of Selected Articles and Categories Defined for Analysis

From the research conducted, 94 articles were found in SCOPUS. After a first reading of the abstracts, method and results of the articles, 84 were eliminated (for not meeting the inclusion criteria), making a total of 10 final articles for analysis.

The selected articles aim to understand: which actors are promoters/participants in these initiatives, what is the role of the Human Resources teams, what kind of knowledge is transmitted; if the preservation of occupational safety and health is taken into account in the transmission practices implemented; and whether the working conditions and the involvement of workers are elements taken into account in their construction, implementation and evaluation.

For each of the selected articles, data regarding the following categories were analyzed: professional context and country of the study; methods used and initiatives for retention and transmission of knowledge implemented and their duration; agents who implemented the initiatives; participants in the study; type of knowledge transmitted; main conclusions of the studies and impact of the initiatives.

3 Results and Discussion

The 10 studies analyzed (see Table 1) have a qualitative approach and were carried out in the following contexts and geographical areas: Health sector (Spain, United States, Malaysia), Financial and Insurance (Australia), Engineering (Hong Kong), Consulting (Malaysia), Technological Development (United Kingdom), Agriculture and Manufacturing (Australia) and Information (Geneva).

The analysis of the studies revealed that the practices implemented do not focus on the participation and recognition of workers nor on the analysis of the role that their working conditions have in the processes of knowledge transmission. Nor are references made to the involvement of Human Resources in the initiatives undertaken.

None of the studies mentioned knowledge sharing focusing on the preservation of occupational safety and health; nor did they mention this aspect as a positive impact on the work context, production and quality of work.

Seven types of knowledge retention and transmission practices were identified: mentoring; communities of practice; cognitive maps; training sessions; workshops to discuss problems and knowledge to be applied; digital communities of knowledge sharing; digital repositories of documents and information. In these initiatives, technical and core knowledge about the work activity was transmitted. Four of the studies lasted approximately one year, and one had a longer duration - two years; in the others it was not possible to recognize this information.

Regarding the main results of the implemented practices, the role of the leaderships as facilitators in the process of implementing the practices was mentioned; reference frameworks for knowledge sharing were identified; and materials were produced for the company (e.g. procedures manual, digital platforms for knowledge sharing).

Despite the difficulty that workers have in verbalizing the knowledge acquired, namely with experience [12], the methods used seem to help in structuring and raising awareness about knowledge and how to perform the work, which becomes useful for the transmission process. However, in only one of the articles it was referenced the use of observations of work and in none were working conditions mentioned (e.g. overload, pace of work, hyper solicitation), which corroborates results of previous studies that indicate that initiatives for transmission and retention of knowledge are sometimes implemented without understanding the working conditions or the workers' perspective [7, 8].

The design and implementation of the practices identified was mainly undertaken by researchers outside the work context and, in some cases, by some key actors in companies with leadership roles (e.g. managers). Although Human Resources assume a fundamental role in professional contexts, in order to promote the development and training of workers [13], these actors were not integrated in the studies found.

In five of the 10 articles mention is made of the impact that the implemented practices have had in their contexts: change in the strategies used to carry out the work; greater understanding by workers of the knowledge needed to carry out the work; greater confidence during the performance of the work; a feeling of recognition by workers for having been involved in the initiatives; reduction of errors in carrying out the work; creation of a culture of knowledge sharing; expansion of the business of the company and recruitment of new workers.

These results show us that in most of these studies there are actors who are not privileged/not taking an active role in the implemented practices, such as Human Resources teams and workers. Moreover, the type of knowledge privileged in the initiatives seems to be limited to technical/procedural knowledge, and is presented, in most articles, in a subtle way. In other words, the final product of the initiatives is visible (e.g. construction of digital repositories; workshops), but in most cases no information is given on the process of construction or preservation of knowledge in the contexts.

The results allow us to systematize some principles that companies should take into account when designing and implementing knowledge retention and transmission practices: (i) identify key interlocutors, such as managers/departmental leaders, actors whose critical knowledge is recognized to them, and support departments - clarifying the role they should play in the process; (ii) integrate leadership to support, recognize workers' experience, and guide the implementation of the practices themselves; (iii) actively involve workers, from the design to the evaluation of initiatives, in order to value their contributions and recognize their knowledge and the importance they have for the company; (iv) implement initiatives, in a combined manner, and in a way that integrates the analysis of the characteristics and conditions of work; (v) integrate the analysis and intentionalize the sharing of knowledge focused on the preservation of occupational safety and health (vi) assume that the implementation of initiatives of this type implies the allocation of time, costs, human resources, and must be clear, from the outset, its purpose and objectives.

4 Conclusion

In this study we have presented knowledge retention and transmission initiatives carried out over the last 15 years in professional contexts.

We have shown that the implementation of knowledge retention and transmission initiatives plays an important role in the development of contexts and workers. We also highlight the lack of understanding of working conditions in the knowledge retention and transmission processes implemented and the lack of studies mentioned knowledge sharing focusing on the preservation of occupational safety and health or on the impact that this may have on the work context, production and quality of work. With regard to the latter aspect, we consider that it may be an interesting fact to deepen in future studies. For example, to understand why safety and health issues are not considered critical knowledge to transmit and why they are not taken into account in the analysis of the impact of the implementation of the practices.

The results obtained allowed us to identify some guiding principles to support companies in the implementation of programs that privilege the participation and acknowledgment of workers and their knowledge, for an action that guarantees production and quality for the company and that preserves the health and safety of workers.

Acknowledgments. The authors would like to thank FCT for the Grant of: Cláudia Pereira (PD/BD/143112/2018).

Appendix

Table 1. Identification of the articles selected for the analysis.

	Title	Year	Authors	Context and country
1	Knowledge management and international organizations: perspectives on information professionals' role	2008	Makani, J. [14]	International Federation of the Red Cross, Geneva
2	The organizational knowledge iceberg: an empirical investigation	2009	Haider, S. [15]	SoftNetCo (software company), United Kingdom

(*continued*)

Table 1. (*continued*)

	Title	Year	Authors	Context and country
3	Knowledge elicitation in reliability management in the airline industry	2009	Kwong, E. & Lee, W.B. [16]	Engineering Department of an airline, Hong Kong
4	How to implement a knowledge management program in hospital-in-the-home units	2010	Cegarra-Navarro, J.-G. & Cepeda-Carrión, G. [17]	Home care unit - Regional hospital, Spain
5	Transferring knowledge about knowledge management: Implementation of a complex organisational change programme	2012	Pollack, J. [18]	Financial and Insurance Company, Australia
6	The role of knowledge intermediaries in developing firm learning capabilities	2014	Parker, R. & Hine, D. [19]	Small and Medium Enterprises in Agriculture and Manufacturing, Australia
7	Implications of technology transfer in the design and construction of load-bearing masonry buildings	2015	Abdullah, C.S., Bahaudin, A.Y., Nawi, M.N.M., Baluch, N.H., Kamaruddeen, A.M., Mohtar, S., Mohamed Udin, Z., Zulhumadi, F., & Abu Bakar, Z. [20]	Consulting Firm, Malaysia
8	Using Kotter's eight stage process to manage an organisational change program: presentation and practice	2015	Pollack, J. & Pollack, R. [21]	Financial and Insurance Company, Australia
9	Knowledge management in the era of digital medicine: a programmatic approach to optimize patient care in an academic medical center	2016	Shellum, J.L., Nishimura, R.A., Milliner, D.S., Harper Jr., C.M., & Noseworthy, J.H. [22]	Medical Clinic, United States of America
10	Pediatric palliative care in kelantan: a community engagement model	2016	Taib, F., Rostenberghe, H.V., & Muhammad, N. A. [23]	Foundation for people with disabilities and terminal illnesses, Malaysia

References

1. United Nations: World population aging 2013. United Nations, New York (2013)
2. United Nations: World population ageing 2017. United Nations, New York (2017)
3. OSHA: Guia eletrónico sobre "Envelhecimento e Trabalho" (2016). https://eguides.osha.europa.eu/all-ages/PT_pt/1-envelhecimento-e-trabalho-0
4. European Union: Demography report 2010: older, more numerous and diverse Europeans. Publications Office of the European Union, Luxembourg (2011)
5. Volkoff, S., Delgoulet, C.: L'intensification du travail, et l'intensification des changements dans le travail: quell enjeux pour les travaillers expérimentés? Psychol. du Travail et des Organ. **25**(1), 28–39 (2019)
6. Soliman, F., Spooner, K.: Strategies for implementing knowledge management: role of human resources management. J. Knowl. Manag. **4**(4), 337–345 (2000)
7. Joulain, M., Martin, N.: La qualité de vie au travail dans les services de resources humaines. Psychol. du Travail et des Organ. **19**(1), 71–87 (2013)
8. Oltra, V.: Knowledge management effectiveness factors: the role of HRM. J. Knowl. Manag. **9**(4), 70–86 (2005)
9. Floyde, A., Lawson, G., Shallow, S., Eastgate, R., D'Cruz, M.: The design and implementation of knowledge management systems and e-learning for improved occupational health and safety in small to medium sized enterprises. Saf. Sci. **60**, 69–76 (2013)
10. Cru, D., Dejours, C.: Les savoir-faire de prudence dans les metiers du batiment. Les Cahiers Medico-Sociaux **3**, 239–247 (1983)
11. Champoux, D., Brun, J.P.: Occupational health and safety management in small size enterprises: an overview of the situation and avenues for intervention and research. Saf. Sci. **41**, 301–318 (2003)
12. Oddone, I.: Experiência. Laboreal **3**(1), 52–53 (2007)
13. Staniewski, M.: The element of human resources management supporting knowledge management. Amfiteatru Econ. **10**, 283–291 (2008)
14. Makani, J.: Knowledge management and international organizations: perspectives on information professionals' role. Libri: Int. J. Libr. Inf. Stud. **58**(3), 144–154 (2008)
15. Haider, S.: The organizational knowledge iceberg: an empirical investigation. Knowl. Process Manag. **16**(2), 74–84 (2009)
16. Kwong, E., Lee, W.B.: Knowledge elicitation in reliability management in the airline industry. J. Knowl. Manag. **13**(2), 35–48 (2009)
17. Cegarra-Navarro, J.-G., Cepeda-Carrión, G.: How to implement a knowledge management program in hospital-in-the-home units. Leadersh. Health Serv. **23**(1), 46–56 (2010)
18. Pollack, J.: Transferring knowledge about knowledge management: Implementation of a complex organisational change programme. Int. J. Proj. Manag. **30**, 877–886 (2012)
19. Parker, R., Hine, D.: The role of knowledge intermediaries in developing firm learning capabilities. Eur. Plann. Stud. **22**(5), 1048–1061 (2014)
20. Abdullah, C.S., Bahaudin, A.Y., Nawi, M.N.M., Baluch, N.H., Kamaruddeen, A.M., Mohtar, S., Udin, Z.M., Zulhumadi, F., Bakar, Z.A.: Implications of technology transfer in the design and construction of load-bearing masonry buildings. Jurnal Teknologi (Sci. Eng.) **77**(5), 127–134 (2015)
21. Pollack, J., Pollack, R.: Using Kotter's eight stage process to manage an organisational change program: presentation and practice. Syst. Pract. Action Res. **28**, 51–66 (2015)
22. Shellum, J.L., Nishimura, R.A., Milliner, D.S., Harper Jr., C.M., Noseworthy, J.H.: Knowledge management in the era of digital medicine: a programmatic approach to optimize patient care in an academic medical center. Learn. Health Syst. **1**, e10022 (2016)
23. Taib, F., Rostenberghe, H.V., Muhammad, N.A.: Pediatric palliative care in Kelantan: a community engagement model. Bangladesh J. Med. Sci. **15**(1), 51–56 (2016)

The Characteristics of Industrial Safety Risk Management

Gyula Szabó(✉)

Bánki Donát Faculty of Mechanical and Safety Engineering, Óbuda University, Budapest, Hungary
szabo.gyula@bgk.uni-obuda.hu

Abstract. In recent decades, risk management practices and methodologies in many other areas have evolved, and a merge has begun. In this article, we first introduce a risk concept and then interpret the specifics of occupational safety risk management. Risk assessment is not only a primary tool or modern management systems but, like in occupational health and safety, it is an obligation. The distinctive feature of the occupation health and safety risk assessment is that the protected value is the health of the worker, and the employer's activities cause the hazards making it the problem owner. At OSH risk assessment, the various scenarios appear in the wording of the hazards and exclude intentional harm. Meeting minimum health and safety requirements and compliance with limit values means often characterize safe conditions. The occupational safety risk set is usually fragmented, typically consisting of several assessment areas that include procedures for monitoring changes.

Keywords: Workplace safety and health · Integrated risk management · Financial OSH risk assessment · Risk profil · OSHAS

1 Introduction

The risk concept plays a central role in safety science. In the field of occupational safety, a thorough risk assessment is the first step in defining preventive measures, from eliminating hazards, through technical measures, to the training of workers and ordering the use of personal protective equipment by workers.

Occupational safety risk management methodologies barely cover all the elements specified in the risk management standard [1] and taken into account in other fields of applications.

2 Occupational Safety Risk Management

In the last 25 years of workplace safety, it has become natural that occupational health and safety decisions based on risk assessment. Knowledge of the risks enables both employers and workers to take appropriate measures together to prevent accidents and occupational diseases and to reduce exposure at workplaces [2].

P. M. Arezes and R. L. Boring (Eds.): AHFE 2020, AISC 1204, pp. 47–52, 2020.
https://doi.org/10.1007/978-3-030-50946-0_7

Most importantly in occupational safety, on the one hand, the protected value is the health and well-being of workers and those who appear in the workplace; on the other hand, hazards arise from work, working arrangements, workplace. The workplace is in the employer's interest; the company controls the resources, has the right to set rules; consequently, the CEO is the owner of workplace safety and responsible for risk management [3].

Accepting the risk approach, that is, the likelihood of an event occurring and the level of risk, in occupational safety signifies the acceptance that somebody, somewhere, will somehow have an accident, with just a little or fatal outcome. In occupational safety this raises an enormous ethical issue since we do not accept morally any death as the consequence of the job, i.e. in workplace safety communication there is a "zero accident goal", "everyone must return home" and "never with anyone can this happen".

Being the problem owner, it is the employer's responsibility and expense to meet the health and safety requirements. At the social level, the resolution of safety requirements and the tolerable burden on businesses develops through tripartite cooperation, i.e. employees, employers and the state. In Europe, some minimum conditions explicitly required in the field of occupational safety, e.g. for tools, workplaces or workstations with screen terminals. Concerning the risk, there is no defined mandatory maximum level set, and the company should continuously strive to improve the occupational safety situation.

3 Risk Profile

Determining the detailedness of risk assessment also appears in OSH practice, as it may range from a full-fledged investigation to the presentation of an OSHAS certificate.

For example, one personal demand is that a truck driver has a valid driver's license, so its absence is usually identified as a hazard and then checked for sampling to determine the level of risk. One technological requirement is that the machinery should be in working order and maintained, including proper guards in place so inadequate or missing guards are regarded as a hazard, and all the machinery checked accordingly.

However, the organization may have a process, an organization, a person in charge of the above issues, that is, a human process and HR department responsible for ensuring that employees have the necessary qualifications and rights or a maintenance department responsible for the state of the machinery. At this level, there is a hazard that these departments may not be doing their job well (license expires unnoticeably, overlooked maintenance) or there are security breaches elsewhere, for example, unauthorized driving a forklift or using machines for the job. The risk assessment should then focus on these hazards, i.e. firstly, which OSH requirements are covered, e.g. forklift trucks drivers must have a valid license, and there is maintenance. After that, the entire function - i.e. maintenance, personnel - can be treated as one unit to evaluate its operational risks. In the examples, we did not cover production management, which could include the process, ensuring that only authorized persons perform specific tasks.

The operational hazards of production, personnel or other departments like maintenance can be identified and mitigated by developing and reviewing control processes.

Once validated functions within each department have been implemented, risk assessment can move to a higher level in the area and address the risks of reviewing control processes, whether they identify malfunctions in the unit and take action to address them. For example, the unit notices when the license expires, when maintenance is missed, and they take structural action to resolve these malfunctions.

Going up in the organisation the functions integrate, and the various threats become more and more organisational and ultimately lead to a senior management risk management function. Verification of company-wide compliance can be performed on request by an independent third party, e.g. management system review, customer audit or even occupational safety and health inspection.

As shown in the examples above, even without a management system like ISO 45000, risk assessment can operate across several levels and areas. Higher-level risk management can rely on sectors risk assessments, using residual risks identified preferably in the form of a well-selected performance indicator. This mechanism enables the identification of problem owners at all levels and the management of risks locally [4].

However, just having an OSHAS certification is not enough; it must be backed up by real, verified risk management. Risk assessment at different levels and departments should consist of an occupational safety risk profile that covers all the risks to the health and safety of individuals arising from the operation of the business and provides opportunities for management. Returning to the example above, somewhere, somebody consciously must consider the possibility of driving a truck without a valid driver's license or that the controls may be defective. It is not necessarily the job of health and safety professionals to personally inspect them or to take care of them at a higher level. However, it remains the responsibility of the health and safety professionals to ensure that at all levels, the relevant OSH knowledge and the relevant risk profile is available.

For example, lack of guards and unauthorized truck driving on workshop level is due to lack of cover replacement activity in the maintenance workshop relates to the neglect of occupational safety review or the lack of maintenance in one area at management level. Dividing the risk profile into levels and areas also means that risk communication and employee consultation can be customized.

In the matter of occupational safety risk assessment, the risk profile is usually not very detailed, and the risks are typically independent of each other. To include the chance of interference of various risks, OSH people regularly add a circumstantial 'multiple risks', or evaluate these hazards independently regarding the interacting hazard as a contributing risk factor for each. Confirmed causality permits the application of quantitative risk methods, e.g. risk assessment of dangerous substances, or fire prevention.

4 Integrated Risk Profile

Integrated management systems require risk-based processes, decision-making and implementation across all fields.

Responsible decision-making requires an accurate understanding of the situation with comparable decision options. For top managers adapting the risk approach, the integrated risk profile must present the situation and the expected consequences of the decision alternatives in a uniform, senior management-relevant scale. However, the difficulty is to obtain an integrated risk management system is too high, and, in practice, many risk management systems combined.

In practice in a given situation, for several objectives, various methods can be used to assess the risk, and in order to achieve high-level objectives mutually and completely the solution is to create several risk assessment, or complementary, e.g. machines, hazardous substances and psycho-social risks in the risk assessment of work safety, or independently of one another, e.g. corporate financial, environmental and occupational safety risk assessments.

Although the reality is the same, that is, the subject matter of the assessment is the operation of the same company, the results of the risk assessments differ significantly depending on the purpose of the risk assessment, the methods used, the knowledge, experience of assessors.

4.1 Fragmentation of the Integrated Risk Profile

The integrated risk profile fragmented because on the one hand independent risk assessments can contain similar hazards with the same or different wording many times, and on the other hand despite multiplied efforts, the completeness of the hazards is not justified. It is like putting together puzzle only recognising that the pieces fit the same picture but produced by different manufacturers and suspecting that some items absent.

For example, an oil spill from a machine failure may appear, e.g. in fire risk assessment, work tools risk assessment, environmental risk assessment, and facility risk assessment for slip injury. Inconsistent phrasing may occur from a diverse perception of the phenomenon or the probability of the event, circumstances and consequences, for example in the above example they may indicate a hazard of material fatigue, container crack, floor unevenness, deficiency of maintenance or cleaning, fuel emissions. However, the coverage of the phenomenon was not comprehensive, no mention, e.g. the expense of compensating oil or the IT vulnerability of the drain valve remote control.

4.2 Homogeneity of the Integrated Risk Profile

Homogeneous is the risk profile that presents the results of different risk assessments using the same vocabulary, metrics, and scales. It does not allow for this and does not take full advantage of the risk approach by using inhomogeneous risk assessment, which uses multiple risk metrics, such as frequently occurring severe work accidents, infrequent environmental catastrophes, and occasional financial losses.

The uniform definition of risk levels would ensure the homogeneity of the integrated risk picture and the comparability of the individual risks. The main obstacle to the creation of an integrated risk picture is that each method produces the result in a different form. One of the critical elements of integrated risk assessment is the standardized description of incident probabilities, which is impossible or difficult to perform subsequently.

Aiming an integrated risk profile, the complementary qualitative risk assessments methods need to use at least the same scale of frequencies from the beginning. As typical incidence rates vary across risk areas, in practice, by merging them, it is advisable to define one single scale with the magnitude across multiple scales and to apply the relevant sub-range to each risk assessment.

It is much more challenging to generate a standard scale of outcomes, i.e. the severity of events or the significance of consequences because they expressed according to the purpose of the risk assessment and to the value protected. A minor consequence is usually the identification of a signal that indicates the inception of a sequence of events leading to injury, the most severe consequence being the unique or multiple destructions of a protected value. For example, in workplace safety, the consequences are expressed with the severity of an accident-related illness or injury, from an event without personal injury to multiple fatalities, in environmental risk assessment from harmless environmental impact to extinction of animal species, in IT security from an operator error without a consequence to total system crash and in finance from a marginal cost to bankruptcy.

5 Financial Occupational Safety Risk Assessment

Many times, businesses do not regard employee health and safety as part of their core business, but as a form of social responsibility, charity, ethical commitment, moral expectation, and responsiveness. In addition to the personal involvement of managers, their attitudes depend on the cultural environment, ownership structure, size of the business, industry sector, economic trends, and other business policy considerations, and thus the degree of maturity at work safety and health of companies can vary widely.

Bottom-up cost evaluation of accidents and occupational illnesses has become a decisive element of the OSH argumentation, including, for example, medical care, sick pay, indemnity, labour replacement, substitution, reduction in production, fine [5].

In occupational safety risk assessment, hazards origins to work and the value to be protected is the health and safety of workers. Accordingly, occupational health and safety address the work-related events and express the unwanted outcome with the effect on the worker on the scale of the severity of employee's injury. However, this is not exclusively necessary and should be other scales relevant to the business and comprehendible by management [6].

Evaluating the economic consequences of work-related illnesses and injuries is nothing else than an assessment of the company's occupational safety risk, which uses the financial consequences rather than the severity of the injury or illness. Because the OSH risk profile relates to work, the apparent metric for integrated corporate risk assessment today is primarily cost. The corporate level integrated risk profile should include occupational safety and health risks severities similarly to other risk fields, expressed in terms of the expected costs.

5.1 Practice of Occupational Safety Risk Assessment

The occupational health and safety risk profile cannot support the enterprise-level decision-making not only due to its employee-injury-based-metrics but because fragmentation and inhomogeneity already occur within the field. There are diverse risk assessment methods just in the field of occupational safety using different information sources and different scales of severity and frequency. Separate risk assessments of work equipment, hazardous substances, biological agents or psycho-social factors solely provide a basis for the art of workplace intervention planning. The OSH related decisions must have the same foundations as all the rest in the organisation, including financial analysis of alternatives to the individual OSH intervention sub-packages, i.e. based on the financial OSH risk assessment.

In light of the moral principles that represent respect for human life and employees' self-determination, financial risk assessment of occupational safety and health can be questionable. However, it can be effectively used also by employees to improve health and safety at work.

The overlapping of occupational safety and health risk profile using employee-injury-based-metrics overlaps with the one using cost as a measure means than employees and employers have a mutual understanding of the safety situation of the workplace and have the same interest.

6 Summary

A thoughtful enterprise-level decision requires an accurate understanding of the situation, which is the risk assessment according to the risk approach. The difficulty in composing such an integrated risk profile is due to the diversity of objectives, methods and metrics, which, in consequence, leads to fragmented and inhomogenous integrated risk profile.

The limitation of the comparison occupational safety and health risks to other risk fields is due to its fragmentation, inhomogeneity and employee-injury-based-metrics, which is difficult to interpret by the business and management.

References

1. ISO 31000:2018 Risk management—guidelines
2. Council directive of 12 June 1989 on the introduction of measures to encourage improvements in the safety and health of workers at work (89/391/EEC)
3. évi XCIII. törvény a munkavédelemről és a végrehajtásáról szóló 5/1993. (XII. 26.) MüM rendelet
4. ISO 45001:2018 Occupational health and safety management systems—requirements with guidance for use
5. Tompa, E. et al.: The value of occupational safety and health and the societal costs of work-related injuries and diseases. European Agency for Safety and Health at Work, Bilbao (2019). ISBN 978-92-9479-138-2. https://doi.org/10.2802/251128
6. Szabó, G.: Distinctiveness of occupational safety and health risk management. Bánki reports, Budapest, vol. 3, no. 1 (2020)

REPAIRER Reporting System User Analysis for SMS Compliance in Aviation Maintenance

Mark Miller(✉) and Bettina Mrusek(✉)

Worldwide College of Aeronautics, Embry-Riddle Aeronautical University, Daytona Beach, FL, USA
{millmark, mrusekb}@erau.edu

Abstract. To resolve the issue of human error in maintenance the REPAIRER reporting system is revisited as it has great potential by combining a human factors analysis with a risk management safety reporting mechanism. It is also timely as a human factors centered safety reporting method like the REPAIRER could now be feasibly implemented through the new mandatory FAA (Federal Aviation Administration) FAR 121 requirement to use SMS (Safety Management System) pillars and through the new FAA MxHF human factors training. With the current FAA support in place and the ever growing need to add human factors to combat human error in aviation maintenance, the REPAIRER model would seem attractive to many aviation maintenance organizations. To illustrate this, the researchers' intention is to take the REPAIRER model to a point of hypothetical use in an aviation maintenance organization to gain an understanding of its potential benefits. To accomplish this, a thorough look at the economic gains were first identified in the form of cost savings through safety and less accidents, but then also in the form of possible efficiency gains. The REPAIRER was then looked at as a tool to achieve employee motivation and gain a just culture. The last area of the REPAIRER added value was the ease of implementing it into various types and sizes of organizations.

Keywords: REPAIRER reporting system · SMS · MxHF · Safety · Efficiency · Motivation · Just culture

1 Introduction

In 2018 and 2019, the authors presented two papers respectively at the AHFE conference. The first entitled, "The REPAIRER Reporting System for Integrating Human Factors into SMS for Aviation Maintenance" [1] was centered around developing a better way to manage safety in aviation maintenance through a human factors analysis integrated into a risk management style reporting system that would comply with FAA (Federal Aviation Administration) SMS (Safety Management System) mandates recently placed on the US (United States) commercial aviation industry operating under FAR121. The acronym of REPAIRER [2] first needs to be revisited as a process that by design has great potential to make a difference if used properly in the aviation maintenance arena. First and foremost it starts off with the first 'R' for reporting and rating the hazard. Then it is joined by a human factors analysis portion to analyze the Hazard

P. M. Arezes and R. L. Boring (Eds.): AHFE 2020, AISC 1204, pp. 53–60, 2020.
https://doi.org/10.1007/978-3-030-50946-0_8

with 'EPAIR'. The 'E' stands for the environment the maintainers are working in and it is followed by the 'P' which stands for the people involved. Where the REPAIRER becomes unique is in 'AIR' portion of the human factors analysis. It is here that the 'A' requires the actions of the people involved to be studied to determine what they did and did not do. Also important to maintenance is the step represented by the 'I' which calls for an investigation of the proper procedures to the maintenance action, as most aviation maintenance errors stem from not following the written procedures properly. To end the human factors analysis portion, the REPAIRER requires the use of a second 'R' to look closely into the resources used by the maintainers involved and to see if there were any issues with those resources. Once the human factors analysis is completed a second 'E' is tactically placed in the REPAIRER model to execute mitigation strategies. The execute portion is critical as the REPAIRER is not solely focused on analyzing serious problems in maintenance and identifying solutions, but also allows for the implementation of solutions. The final 'R' is used to reevaluate those solutions, allowing for future modification if needed. These eight steps not only integrate the steps that should be found in any modern aviation safety system but are also greatly enhanced by edition of a human factors analysis. The second paper, presented in 2019, entitled, "Implementing the REPAIRER Human Factors Safety Reporting System through MRM(MxHF) to meet SMS Compliance in Aviation Maintenance" [3] reiterated the necessity for implementing the REPAIRER reporting system into aviation maintenance organizations, noting the recent shift from the FAA's traditional Maintenance Resource Management (MRM) program to an online Human Factors training program for maintainers, now referred to as MxHF.

2 The REPAIRER Reporting System and Economic Gains

In an industry where profits are heavily linked to the number of occupied seats on the plane, it comes as no surprise that every dollar spent on safety must have clear financial benefits. While safety is paramount in aviation, there are other critical costs that must also be considered such as daily labor, fuel, and maintenance as they are often used to offset filled seat margins. Maintaining the aircraft is considered one of the three major costs in commercial airline operations, making safety within this area incredibly important. In aircraft maintenance, human error can drastically and quickly challenge established safety standards, harming personnel and/or equipment resulting with revenue losses. Although personnel injury can be very costly, an incident that causes an aircraft to be delayed can contribute to passenger frustration and overnight hotel costs, in some instances. Additionally, accidents related to maintenance malpractice can take these costs to a much higher level, especially if lives are lost or the aircraft is damaged. While monetary losses from human error in aviation maintenance are costly to businesses, what often goes unnoticed are the costs incurred to maintenance technicians that must cover the work for an injured coworker or disappointment of the customer that was unable to arrive on time. In extreme circumstances, an aircraft accident can have lifelong consequences. The lasting grief of family, friends, and crew members can be challenging to overcome. These organizations must work to earn that trust back from its various stakeholders.

Years of research in aviation maintenance clearly points out that the culprit of these unfortunate events is often human error. In many instances that human error is caused by a failure to stay within the guidelines of established maintenance procedures. If human factors is the best way to combat human error in aviation maintenance, then it is highly logical that human factors could lower that cost. However, it is imperative that the method of integrating human factors in maintenance must have clear benefits, from both a safety and efficiency perspective. A human factor reporting system such as the REPAIRER could achieve this economic balance by reinforcing the importance of reporting safety hazards and identifying efforts to mitigate those circumstances.

3 Using a Human Factors Safety Reporting System in Maintenance

A safety reporting system that incorporates human factors allows employees to actively participate in the identification and reporting of hazards. In doing so, they learn the many elements that comprise human error including physical, psychological, and ergonomic factors; all of which directly contribute to aviation accidents. By completing the free online training at the FAA MxHF training site, the aviation maintainer will not only be knowledgeable in reporting hazards, but also help make larger strides to conquer them. For instance, a company working to complete a major maintenance phase during night shift encounters personnel issues. One of the experienced maintainers is home with an injury. The oncoming shift supervisor inspects the job done by the night crew and notices several things wrong. The night supervisor explains that one experienced maintainer is at home with a bad back and the others with less experience had to continue without him. As a result, the night supervisor could not oversee all maintenance work completed during the shift. A scenario like this could be handled in many ways. However, with a REPAIRER reporting system both shift supervisors would recognize the potential for human error, setting the reporting system into motion. A risk management assessment code is then applied to address the poor maintenance completed during the night crew shift.

4 Using a Human Factors Analysis to Identify Root Causes: 'EPAIR'

In the process both supervisors discuss the problem and break it down in terms of human factors. The 'E' environment was broken down, noting the time the work was completed as well as the pressure to finish the inspection phase on time. During the shift, the night crew supervisor noticed that many of the personnel were getting fatigued early in the morning and stress was playing a role. When investigating the 'P' for people, having the experienced maintainer out with a bad back did not help and clearly contributed to the stress. It forced work to be done without proper experience. Further investigation revealed that the experienced maintainer was injured by using improper platform equipment and fell. This is a separate incident and needs to be reported as well. In both cases, the supervisors determine that night crew personnel did not follow proper procedures; the first from confusion in the maintenance manual and the second from the company platform regulations. The 'I' for investigating the proper procedures revealed that the personnel using the procedures manual that night misinterpreted the directions due to their inexperience and lack of supervision. It was also found that the experienced maintainer used the wrong equipment platform because the correct one was being used elsewhere and the other was broken. The 'R' for resources showed that the proper experienced personnel and the proper equipment were not available.

5 Gaining Safety Data to Study Trends and Manage or Eliminate Risk

From the 'REPAIR' portion of the REPAIRER, significant steps have been taken to improve future safety efforts, while also reducing costs, representing economic gain. By simply reporting the hazard, the information is going into a data base where it can be collected and studied with similar data over time. Rating the hazard through risk assessment means that the hazard can now be scientifically managed, including the probability of future occurrences. While human error in aviation maintenance cannot be completely eliminated, minimizing these threats greatly improves the overall safety of the organization. Reviewing safety trend data allows aviation maintenance managers to take a proactive approach to managing safety. Instead of waiting for repetitive incidents to turn into accidents, these hazardous trends can be faced early on with preventative measures and/or safety techniques, thus reducing the likelihood of a catastrophic event. This represents significant cost savings for the organization. The data from the human factors 'EPAIR' analysis is intended to get to the root cause of the accident in terms of human factors issues. These elements are collected over time and emerge as human factors data categories, allowing hazards to be proactively identified. In the example, fatigue could have been identified as a contributing factor to both incidents. Additionally, other incidents reported from the night shift previously highlighted fatigue as a contributing factor. Therefore, fatigue has become an underlying human factors causation that will be worthwhile to rectify. A solution that addresses the fatigue hazard at night represents cost savings by reducing incidents related to this specific human factor error. These trends can help guide management through intentional decision making. In the example provided, excessive fatigue exacerbated by stress during night shift could be viewed as a trend. Additionally, through the identification of human factor errors, the REPAIRER method can also help to identify and resolve costly injuries resulting from improper adherence to EPA (environmental protection standards) OSHA (occupation safety and health administration) standards. These can add up over time and be just as costly as traditional aircraft incidents. Although the main thrust of the REPAIRER is to reduce costly maintenance mistakes, the value placed on human factors represents cost savings that could benefit all aviation maintenance endeavors. This is due to the synergistic value in the 'REPAIR' portion of the reporting system by combining the popular risk management technique (identify, track and manage hazards) with the details of why they are occurring in terms of human factors.

6 Adding the Process of Continuous Improvement

As powerful as the steps in the 'REPAIR' are for the economic health of the maintenance department, the 'REPAIR' gains its total value by adding the 'ER' steps at the end, bringing together the full REPAIRER system. With the reporting completed, the hazard rated, and the human factors data collected, the REPAIRER has the opportunity to make a greater impact by adding the last 'ER' for process of improvement. The second 'E' in REPAIRER stands for 'execute the mitigation strategy'. The 'R' stands for 'reevaluate' how the mitigation strategy is working. By collecting human factors data, it is now possible to identify a better strategy to correct the hazard. It is in this 'execution mitigation strategy' step that human factors find its way back to its historic beginnings in the form of ergonomics by the studying and improving the man-machine interface. While the REPAIRER reporting system idea is founded on the premise of identifying a safety solution based on accepted levels of risk, there is also a way to turn the safety risk into an efficiency gain. Human factors and the ergonomics of improving the man-machine interface started in the 1800's during the industrial revolution. It utilized scientific management to improve the efficiency of assembly line work. The scientific manager was essentially a human factors expert studying the work of each employee (man-machine interface) on the assembly line via frame by frame pictures until the most efficient way of doing the work was created. The original intent was to make the factory assembly line as efficient as possible, producing maximum economic gains. Procedures to make assembly line work safer did not come until sometime the early 1900's. In the case of REPAIRER and modern aviation maintenance safety, the execution of a safety mitigation strategy comes first. However, if efficiency gains exist then the principles of modern TQM (total quality management) in the form of Lean Six Sigma continuous improvement are welcomed. Under this premise, there are cost savings and economic gains through the execution and reevaluation of the mitigation strategy, which could be greater if the hazard problem has an efficiency solution as well. Given that aviation maintenance is found on standardized procedures, which also creates the opportunity for human error, the use of such a reporting tool is incredibly valuable. In some cases, it is a matter of simply identifying a mitigation strategy to ensure procedures are properly followed. There are cases where new technology could be infused to make the maintenance procedure safer and more efficient, such as altering the way the procedures are completed. In the example provided, given the lack of experience on night shift, the final two 'ER' steps can be completed. Bringing on additional experience from the day crew shift would address the hazard and ensure that proper procedures are followed. Complicated procedures can be scrutinized and altered for efficiency gains. The current working platform must be remedied, and reevaluation goals identified for later adjustments, if needed.

7 REPAIRER Becomes a Lean Six Sigma Improvement Method

With the potential for efficiency gains and safety gains, the REPAIRER reporting system now mirrors the 5 phases of Lean Six Sigma continuous improvement process: Define, Measure, Analyze, Improve and Control [4]. The DMAIC acronym starts by 'Defining' the problem which matches the 'Report a hazard step' in REPAIRER. 'Measuring' the problem is accomplished by 'Rating the hazard using risk management' in the REPAIRER. Analyzing the problem is accomplished through the 'Human Factors analysis' in the (EPAIR)' portion of the REPAIRER. 'Improve' the problem is initiated with the 'Executing mitigation strategy' in the REPAIRER. 'Controlling' the problem is completed through 'Revaluating the strategy' at the end of the REPAIRER. DMAIC is used in Lean Six Sigma as an established business practice to improve speed, quality and cost. The phases of DMAIC which define a problem through implementing solutions linked to underlying causes, and then establish best practices to make sure the solutions stay in place are powerful [4]. Using the DMAIC process encourages creative thinking within the scope of the current system, but keeps the basic process, product or service [4]. Although the primary mission of the REPAIRER is safety in aviation maintenance, the same guiding principles used in DMAIC for continuous improvement are also fundamentals in the REPAIRER system.

8 Can Safety and Efficiency Gains Work Together? The Boeing One Plan

One of the best examples of creating a safe, but efficient workplace in an aviation maintenance environment can be found at Boeing's EHS (Environment, Health, Safety) One Plan. It goes beyond a safety plan. It emphasizes continuous performance improvement and sustainability in the areas of zero injuries, environmental leadership and operational excellence. Lean Six Sigma Workshops are continuously conducted throughout Boeing to teach continuous improvement techniques such as accelerated improvement shops, value stream mapping, Kaizen events and structured problem solving [5]. The centralized EHS team moves throughout the company to inspire local solutions as teams identify solutions to safety, environment and operational problems. The solutions are then tested for possible implementation as a new company wide standard. Utilizing safety and health as a pillar of quality supports the use of environmental leadership and operational excellence through the identification of efficiency gains which can be built into industry-wide safety programs [5]. Perhaps the most intriguing point about Boeing's initiative is the resulting impact on workforce motivation. Employees feel they can make a difference to any of Boeing's companies through safety, environmental and operations improvements. The Lean Six Sigma training combined with the encouragement in the formation of work teams to address potential issues is positively transforming the Boeing culture [5]. The REPAIRER system could accomplish similar goals in a maintenance by addressing both safety and efficiency improvements and in the process greatly motivate the workforce.

9 Establish Intrinsic Value, and a Continuous Improvement Just Culture

By the virtue of the REPAIRER having the opportunity to establish both safety and efficiency solutions to cut costs for economic gains, it could also be of intrinsic value. From this the REPAIRER is similar to the enthusiasm recently found at Boeing via the EHS One Plan. The REPAIRER system has the potential to improve different aspects of the workplace, creating an environment where maintenance personnel support and find value in a culture of continuous improvement. The cultural shift experienced at Boeing was rooted in the intrinsic value for its employees. Their behavior was driven by internal organizational rewards, creating a sense of satisfaction.

In Maslow's Hierarchy of Motivational Needs [6], Maslow described this as the top of the motivational pyramid by gaining esteem and ultimately self-actualization as the individual realizes that their full personal potential has been achieved. This is the highest form of motivation and the internal reward therefore comes from a person knowing that they are making a difference. The REPAIRER could also be viewed as an instrument for maintainers to attain esteem, self-fulfillment and be motivated intrinsically. As with Boeing, this motivation could form a culture that leverages the REPAIRER in order to make a difference in aviation maintenance organizations. This type of environment establishes a culture that is not based on fear, blame, or punishment from coercive leadership. Instead it stems from the creation of a just culture; discovering what went wrong and motivating people to come up with innovative ways to fix and learn from their mistakes. With the strength of human factors analysis integrated into the REPAIRER system, the establishment of a just culture is integrated into the culture through an unbiased and no-fault inquiry of identifying what went wrong, but even more importantly properly identifying what the human errors were that caused the problem. Through this lens, the REPAIRER system is poised to improve organizational success in aviation maintenance through economic and efficiency gains, adding intrinsic value to employees' work, and through the formation of a powerful just culture. However, perhaps its greatest value is in its simplicity. The REPAIRER steps could be customized and integrated into any type or size aviation maintenance organization, then tailored as necessary for optimal output. Whether the maintenance organization is a small General Aviation organization, an outsourced venue, a military unit, or a large commercial entity, the REPAIRER and its important elements is an opportunity to make a positive difference in that maintenance organization.

References

1. Miller, M., Mrusek, B.: The REPAIRER reporting system for integrating human factors into SMS for aviation maintenance. In: Proceedings of the AHFE 2018 in Advances in Safety and Management Systems, vol. 791. Springer (2018)
2. Miller, M.: The repairer reporting system strategy for aviation maintenance: integrating human factors and risk management into aviation maintenance for SMS compliance. Presentation to FAA Aviation Maintenance Mega Conference Honolulu (2016)

3. Miller, M., Mrusek, B.: Implementing the REPAIRER human factors safety reporting system through MRM(MxHF) to meet SMS compliance in aviation maintenance. In: Proceedings of the AHFE 2019 in Advances in Safety and Management Systems, vol. 969. Springer (2019)
4. George, M., Maxey, J., Price, M., Rowlands, D.: The Lean Six Sigma Pocket Tool Book. McGraw-Hill (2005)
5. Boeing Company: EHS Excellence in Boeing's Second Century. Application for the National Safety Council's Robert W. Campbell Award (2018). https://www.campbellaward.org/wp-content/uploads/2019/01/RWC-Boeing-Application.pdf
6. McLeod, S.: Maslow's Hierarchy of Needs, Simply Psychology (2018). https://www.simplypsychology.org/maslow.html

Occupational Exposure to Biological Agents

Joana Santos[1(✉)], Carla Ramos[1], Manuela Vaz-Velho[1], and Marta Vasconcelos Pinto[1,2]

[1] CISAS, Escola Superior de Tecnologia e Gestão, Instituto Politécnico de Viana do Castelo, Rua Escola Industrial e Comercial de Nun'Álvares, 4900-347 Viana do Castelo, Portugal
{joana, cramos, mvazvelho}@estg.ipvc.pt, martavasconcelos@estescoimbra.pt
[2] Instituto Politécnico de Coimbra, ESTESC–Coimbra Health School, Saúde Ambiental, Coimbra, Portugal

Abstract. Biological risks were mentioned by the "European Agency for Safety and Health, at Work's Expert Forecast on Emerging Biological Risks related to Occupational Safety and Health", the second most important emerging risk. In an era when the "one health approach" is considered the right way to address public health issues occupational safety continues to look at biological agents as a minor issue compared to other agents. The food industry is known to be very committed to food safety assurance but less concerned with the safety of food operators unless they may compromise the quality of the food product. However, the focus on biological agents is essential not only due to air spread or contact with organic dust but we must not overlook another crucial aspect which is the antimicrobial resistance that may be associated with biological agents, nowadays one of the biggest threats to global health, food security and development.

Keywords: Biological agents · One health approach · Food industry · Antibiotic resistance

1 Introduction

In an era when the "one health approach" is considered the right way to address public health issues, occupational health continues to look at biological agents as a minor issue compared to other agents, although the former include microorganisms capable of causing infections, allergies and/or toxicity to humans, and their presence in the workplace implying an increased risk to workers. Biological risk assessment is usually neglected by the occupational health professionals due to the lack of systematized information about the biological agents involved in many work activities. The publication *bias* and a large amount of confounding factors don′t help the risk assessment activity although, since microorganisms are omnipresent in the environment, exposure to biological agents in various contexts is inevitable. The occupational safety professional is always facing gaps in hazard assessment of the risks, engendered by biological agents, such as: (i) lack of information on biohazards in the workplace; (ii) lack of training of workers; (iii) lack of maintenance of ventilation systems/air conditioning/water systems,

P. M. Arezes and R. L. Boring (Eds.): AHFE 2020, AISC 1204, pp. 61–67, 2020.
https://doi.org/10.1007/978-3-030-50946-0_9

and; (iv) lack of knowledge of combined effect of biological and chemical agents. When biological agents are introduced into workplaces, as part of the process, for example in a biotechnology laboratory, workers usually show high levels of awareness of the biological hazards to which they are exposed, knowing and using correct protective measures. However, this level of consciousness does not exist in other workplaces where biological agents are present, as a core of the activities performed, but as contaminants.

Several studies focus on the theme of environmental contamination by bacteria, fungi and their metabolites, seeking to establish a relationship between occupational exposure and the appearance of different symptoms among employees, but the role of biological agents in disease development and worsening of symptoms of workers is poorly understood since it is not entirely clear which components are really responsible for the health effect [1–4].

2 Exposition to Biological Agents

It is consensual that biological agents are microorganisms, including genetically modified, human cell cultures and endoparasites that may cause infection, allergies or toxicity. Although the term biological agent is very comprehensive universally, when addressing the theme "biological risk" attention focuses on the following elements: bacteria, fungi, viruses and parasites. These pathogens can be transmitted from a variety of sources: human, animals and environmental sources and can be acquired through inhalation, ingestion, skin and mucous membranes and the placenta. Humans have a variety of bacterial, fungal and viral infections that are transmitted from person to person through body fluids (hepatitis through a needle stick, as an example), secretions (tuberculosis through sneezing or coughing) and cross-contamination on surfaces (influenza virus, through touch on the nose and subsequent contamination of surfaces). From animal sources, occur severe infections of fungal, bacterial and viral origin, which are transmitted through contact with animals (animal-to-person) (e.g. rabies and leptospirosis) named zoonosis. From environmental sources (e.g. water, soil, air) and related to exposure to biological agents' diseases such as Legionnaire's, dust mites' allergies, etc. can be transmitted. The transmission and infection pathways are identified and the basic mechanism of infection is through the direct passage from the reservoir to the host, the prevention involves cutting the chain of transmission.

Biological agents can be classified into four risk groups, according to their level of pathogenicity, virulence, transmission, existence of effective prophylactic measures, the availability of effective treatment and others, depending on the country and/or organization [5] (e.g. World Health Organization, Canada, European Union (EU), European Federation of Biotechnology). In EU the Biohazard Prevention is mandatory [6] but there are several challenges regarding biological agents that should be considered: (i) the diversity of agents involved; (ii) they are not visible to the naked eye; (iii) the occupational exposure limits are not defined; (iv) they are a hazardous at very low concentrations; (v) the toxic effect exist due to metabolites they produce; (vi) reproduce themselves, (vii) rapid mutation, (viii) incubation period and infectiousness and rapid multiplication.

Bioaerosols are considered the main responsible for the mobility of microorganisms by air. According to Cox and Wathes (1995), bioaerosols are fully defined as a collection of aerosol-shaped biological particles. However, exposure to bioaerosols and dust is not the only risk factor to consider, and the transmission of biological agents can be effective through contact with surfaces [2, 7, 8]. Microorganisms, including pathogens that cause disease in humans, can be transmitted from one location to another in various forms. The transmission of disease agents may be carried by droplets, direct contact or indirect contact [9]. However, inhalation represents the main route of exposure related to adverse effects on human health, being the most relevant in the occupational context [10]. The effects related to exposure to biological agents are diverse as: irritant effect (eyes, nose, throat and skin); toxic reactions (mycotoxins, endo and exotoxins); allergic Reactions (Rhinitis, Sinusitis, Asthma) and Infections (Legionnaires Disease, Tuberculosis, Pneumonia, Cryptococcosis) [1].

3 Air Sampling Methods

Bioaerosol sampling involves the removal and collection of particles of biological origin from the atmosphere [11]. There is a considerable variety of bioaerosol sampling and analysis methods. But there are several limitations, including the lack of standardized methods and protocols that are effective in sampling and analyzing all types of bioaerosols [12, 13]. Therefore, comparing results between research studies becomes a difficult task due to the use of substantially different sampling and analysis methods.

The methods more commonly used for bioaerosol sampling are: impact, that promotes particle separation from a forced air stream and uses particle inertia to force deposition on a solid, semi-solid or liquid collection surface; filtration, that captures airborne particles that are trapped in a porous material as air passes through it and the sampling is done using different filters with different constitution and porosity, with the main objective of sampling different particle sizes; and sedimentation, that uses gravity to quantify sediment particles on adhesive surfaces. This last method is a non-quantitative method recurrently used due to its simplicity and low cost, a plate with nutrient medium that will identify the colonies arising from the growth of spores retained during the time of exposure to air.

All methods used for surface sampling have limitations, but those commonly used are: contact plates; swabs, tape, dust collection and material collection.

4 Biological Risks Related to Occupational Safety and Health

In 2007, biological risks were mentioned by the "European Agency for Safety and Health, at Work's Expert Forecast on Emerging Biological Risks related to Occupational Safety and Health" [14], the second most important emerging risk. Once the state of knowledge about biological hazards is still relatively immature, the development of methodologies for research and counting of biological agents and the harmonization of methodologies so that laboratory comparison can exist, urges.

The first emerging risk identified in the above mentioned report was Occupational Risks Related to Global Epidemics: SARS - Acute Respiratory Syndromes: Bird Flu, Cholera, and Yellow Fever. Several risk factors are associated with increased outbreaks, such as globalization together with social changes, intercontinental traveling of humans and animals influencing both the geographical distribution and abundance of ticks and pathogens [15], and the ability of microorganisms to adapt to behavioural factors and environmental changes (climate change).

Healthcare workers, workers in contact with animals, laboratory workers and refuse workers seem to have the highest risk of infection by a variety of pathogens [16]. As more than three quarters of these diseases are zoonoses, workers at risk are those who are in contact with infected live or dead animals with aerosols, dust or surfaces contaminated by animal secretions, in animal trade, breeding and slaughtering units, operators cleaning and disinfecting contaminated areas, veterinary services, research laboratories, customs, zoos and pet stores.

4.1 Transmission of Biological Agents in Occupational Environment-Agriculture and the Food Processing Industry

Regarding agriculture and farmers, there are several studies and, more recently, some studies have focused on this aspect namely the use of microalgae with the ability to biosynthesize a number of metabolites that can be considered biological agents for the control of organisms harmful to soil and plants [17] but that can indirectly affect man. A recent study on perception, knowledge and preventive measures by agricultural workers, despite the high probability of exposure to biological agents, revealed the lack of perception and knowledge of the risk, as well as the poor awareness of available preventive measures among agricultural worker [18]. An epidemiological study on occupational diseases among farmers conducted in Poland in 2016 demonstrated the need to introduce periodic health examinations programs focusing on agricultural workers to monitor health and well-being and improve working conditions and the working environment [19].

The food industry is known to be very committed to food safety assurance but less concerned with the safety of food operators unless they may compromise the quality of the food product. It is common in studies on occupational risk assessment of the food industry that the perspective of zoonosis as a source of risk is not referenced in occupational health and safety plans. In the latest report from the European Food Safety Authority of 2018 zoonoses are among the most common foodborne diseases (commonly referred to as food poisoning) and can also be spread to humans either by direct contact with a sick animal or by contact with a vector (or carrier) [20]. However, the focus on of biological agents is essential in food processing industry namely in activities such as slaughterhouses and poultry and meat processing industries not only due to air spread or contact with organic dust. However, we must not overlook another crucial aspect which is the antimicrobial resistance that may be associated with biological agents, nowadays one of the biggest threats to global health, food security and development [21, 22]. Risk assessment in professional activities, where exposure to biological agents can be frequent and intense, considers the potential levels of exposure, and the level of infectious risk of the potentially existing agents. However, current risk assessment models are inadequate to evaluate the effect of antibiotics and antibiotic resistance genes (ARGs) on resistance emergence and selection, especially in non-clinical environments [23]. Biological risk assessment has become even more complex because, in addition to detecting biological agents with potential for risk for workers, it is necessary to assess the antimicrobial resistance that can be seen to bring to the workplaces being an aggravated danger to public health issues. The emergence of antimicrobial-resistant bacteria in mass animal husbandry has raised further concern regarding occupational exposure and its potential public health consequences. Particular attention has been paid to methicillin-resistant *Staphylococcus aureus* (MRSA) and multidrug-resistant Gram-negative enteric pathogens (GNEP) [24–27].

5 Conclusion

The state of knowledge about biological risks in work environments is still relatively immature. Publication *bias*, a large number of confounding factors and the lack of harmonization of methodologies to research and count biological agents is notorious. This brings another problem to the occupational health professionals in addition to the nonexistence of exposure limit values for biological agents. Since the presence of gram-negative bacteria is the most health-relevant concern due to the presence of endotoxins, the detected bacteria should also be categorized according to gram stain and their antimicrobial resistance evaluated when appropriate. In conclusion there is a need to develop uniform and concise guidance on risk assessment of biological risks (e.g. sampling, harmonization of methodologies for analysis of biological agents). The list of classifications of biological agents should be reevaluated and the information on the dose-effect would also help to establish occupational exposure limits at least for the most common microorganisms.

References

1. Douwes, J., Thorne, P., Pearce, N., Heederik, D.: Bioaerosol health effects and exposure assessment: progress and prospects. Ann. Occup. Hyg. **47**(3), 187–200 (2003)
2. Corrao, C.R., Mazzotta, A., La Torre, G., De Giusti, M.: Biological risk and occupational health. Ind. Health **50**(4), 326–337 (2012)
3. Dutkiewicz, J., Cisak, E., Sroka, J., Wojcik-Fatla, A., Zajac, V.: Biological agents as occupational hazards - selected issues. Ann. Agric. Environ. Med. **18**(2), 286–293 (2011)
4. Dutkiewicz, J., Jablonski, L., Olenchock, S.A.: Occupational biohazards - a review. Am. J. Ind. Med. **14**, 605–623 (1988)
5. International classifications schemes for micro-organisms based on their biological risks. https://www.biosafety.be/content/contained-use-international-classifications-schemes-micro-organisms-based-their-biological
6. Directiva 2000/54/CE do Parlamento Europeu e do Conselho de 18 de Setembro de 2000 relativa à protecção dos trabalhadores contra riscos ligados à exposição a agentes biológicos durante o trabalho. J. Oficial da Comunidades Europeias L 262, 0021–0045 (2000)
7. Montano, D.: Chemical and biological work-related risks across occupations in Europe: a review. J. Occup. Med. Toxicol. **9**(28), 1–13 (2014)

8. Wan, M.P.: Modeling the pathogen exposure and infection risk associated with fomite transmission in an aircraft cabin mock-u. In: 2nd International ISCM Symposium and the 12th International EPMESC Conference, pp. 1576–1582. American Institute of Physics (2010)
9. Tortora, G.J., Funke, B.R., Case, C.L., (eds.): Microbiologia. 6th edn. Artes Médicas Sul, Porto Alegre (2000)
10. Stetzenbach, L.D.: Introduction to aerobiology. In: Hurst, C.J., Crawford, R.L., Knudsen, G. R., McInerney, M.J., Stetzenbach, L.D. (eds.) Manual of Environmental Microbiology, 2nd edn, pp. 801–813. ASM Press, Washington DC (2002)
11. Pillai, S.D., Ricke, S.C.: Bioaerosols from municipal and animal wastes: background and contemporary issues. Can. J. Microbiol. **48**(8), 681–696 (2002)
12. Goyer, N., Lavoie, J., Lazure, L., Marchand, G.: Bioaerosols in the workplace: evaluations. In: Control and Prevention Guide, Québec (2001)
13. Buttner, M., Willeke, K., Grinshpun, S.: Sampling and analysis of airborne microorganisms. In: Hurst, C.J., Crawford, R.L., Knudsen, G.R., McInerney, M.J., Stetzenbach, L.D. (eds.) Manual of Environmental Microbiology, pp. 814–826. ASM Press, Washington DC (2002)
14. Emmanuelle, B.: Expert forecast on emerging biological risks related to occupational safety and health. European Agency for Safety and Health at Work (2007). https://osha.europa.eu/en/publications/reports/7606488
15. Medlock, J.M., Hansford, K.M., Bormane, A., Derdakova, M., Estrada-Pena, A., George, J. C., Golovljova, I., Jaenson, T.G.T., Jensen, J.K., Jensen, P.M., et al.: Driving forces for changes in geographical distribution of Ixodes ricinus ticks in Europe. Parasites Vectors **6** (2013). https://doi.org/10.1186/1756-3305-6-1
16. Haagsma, J.A., Tariq, L., Heederik, D.J., Havelaar, A.H.: Infectious disease risks associated with occupational exposure: a systematic review of the literature. Occup. Environ. Med. **69**, 140–146 (2012)
17. Costa, J.A.V., Freitas, B.C.B., Cruz, C.G., Silveira, J., Morais, M.G.: Potential of microalgae as biopesticides to contribute to sustainable agriculture and environmental development. J. Environ. Sci. Health B **54**, 366–375 (2019)
18. Tamburro, M., Anzelmo, V., Bianco, P., Sammarco, M.L., Ripabelli, G.: Biological risk in agriculture and construction workplaces: a survey on perception, knowledge and prevention measures. G. Ital. Med. Lav. Ergon. **40**, 195–202 (2018)
19. Szeszenia-Dabrowska, N., Swiatkowska, B., Wilczynska, U.: Occupational diseases among farmers in Poland. Med. Pr. **67**, 163–171 (2016)
20. EFSA and ECDC: The European Union One Health 2018 Zoonoses Report. EFSA J. (2019)
21. Antibiotic Resistance: World Health Organization. https://www.who.int/news-room/fact-sheets/detail/antibiotic-resistance
22. Anderson, M., Clift, C., Schulze, K., Sagan, A., Nahrgang, K., Ait Ouakrim, D., Mossialos, E.: European observatory on health systems and policies. In: Policy Brief 32. World Health Organization (2019)
23. Berendonk, T.U., Manaia, C.M., Merlin, C., Fatta-Kassinos, D., Cytryn, E., Walsh, F., Burgmann, H., Sorum, H., Norstrom, M., Pons, M.N., et al.: Tackling antibiotic resistance: the environmental framework. Nat. Rev. Microbiol. **13**, 310–317 (2015)
24. Alba, P., Feltrin, F., Cordaro, G., Porrero, M.C., Kraushaar, B., Argudin, M.A., Nykasenoja, S., Monaco, M., Stegger, M., Aarestrup, F.M., et al.: Livestock-associated methicillin resistant and methicillin susceptible Staphylococcus aureus sequence type (CC)1 in European farmed animals: high genetic relatedness of isolates from Italian cattle herds and humans. PLoS ONE **10**, 1–10 (2015)

25. Peeters, L.E.J., Argudin, M.A., Azadikhah, S., Butaye, P.: Antimicrobial resistance and population structure of Staphylococcus aureus recovered from pigs farms. Vet. Microbiol. **180**, 151–156 (2015)
26. Butaye, P., Argudin, M.A., Smith, T.C.: Livestock-associated MRSA and its current evolution. Curr. Clin. Microbiol. Rep. **3**, 19–31 (2016)
27. Mehndiratta, P.L., Bhalla, P.: Use of antibiotics in animal agriculture & emergence of methicillin-resistant Staphylococcus aureus (MRSA) clones: need to assess the impact on public health. Indian J. Med. Res. **140**, 339–344 (2014)

Health Protection Criteria for Airborne Infrasound Exposure: An International Comparison

Fabio Lo Castro[1(✉)], Sergio Iarossi[1], Massimiliano De Luca[1], Maria Patrizia Orlando[1,2], Claudia Giliberti[3], and Raffaele Mariconte[3]

[1] CNR-INM – Sect. of Acoustics and Sensors O.M. Corbino, Via del Fosso del Cavaliere 100, 00133 Rome, Italy
{fabio.locastro,sergio.iarossi, massimiliano.deluca}@cnr.it, mariapatrizia.orlando@uniroma1.it

[2] Dipartimento Organi di Senso, Università Sapienza Rome, Rome, Italy

[3] INAIL, Dipartimento Innovazioni Tecnologiche e Sicurezza Degli Impianti, Prodotti ed Insediamenti Antropici, Rome, Italy
{c.giliberti,r.mariconte}@inail.it

Abstract. Noise risk in the audible frequencies in the range 20 Hz ÷ 20 kHz is well defined at international level, but in the infrasound region just below 20 Hz, there is not an unique criteria and international agreement to define the infrasound risk. This study is aimed to analyze and compare different country legislations, standards and guidelines issued by international organizations in order to provide a review, pointing out positive features and limitations, as well as perspectives for further development.

Keywords: Infrasound · Noise risk · Airborne infrasound · Legislation

1 Introduction

Airborne infrasounds are elastic waves having frequencies below 20 Hz, namely frequencies just below 20 Hz known as near infrasound and frequencies below 1 Hz often called far infrasound. They are produced by a variety of natural (wind, thunder, heartquakes) and man-made sources and when their amplitude is sufficiently high, humans can feel the near infrasound both as hearing and as tactile sensation or vibration.

Focusing our work on man-made infrasound the principal sources are road traffic industrial plants, wind turbine, aircraft, watercraft, train, weapon and etc.

The first studies on the effects of the human exposition to high level infrasound go back to the World War II [1, 2], while the effects due to low level infrasound, below the hearing threshold, are still under investigation. So far, there is not an agreement on the exposure limit or threshold to avoid hearing impairment and extra-auditory system long term effects.

In the next paragraphs, after highlight the infrasound health effect, will be analyzed and compared different country legislations, standards and guidelines issued by

P. M. Arezes and R. L. Boring (Eds.): AHFE 2020, AISC 1204, pp. 68–75, 2020.
https://doi.org/10.1007/978-3-030-50946-0_10

international organizations in order to provide a review, pointing out positive features and limitations, as well as perspectives for further development.

2 Infrasound Health Effect

2.1 Infrasound Hearing Threshold

The threshold of human hearing and the equal-loudness curves, in the audible frequency range are well documented, but outside this range they are not defined internationally, even if there are different studies in literature [3–8].

In particular for the infrasound in Fig. 1 is reported the hearing threshold level according to the study of Van den Berg et al. [9] where the frequencies below 20 Hz are not audible at typical acoustic levels and become audible at levels above about 80 dB. Also, the hearing threshold have 12 dB/octave roll-off for frequency below 16 Hz.

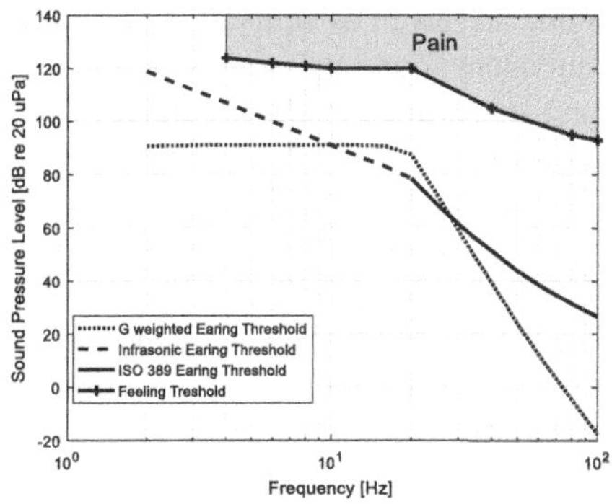

Fig. 1. Infrasonic hearing threshold level: according to ISO 389-7 (solid line), according to Van der Berg (dashed line), G weighted earing threshold (dotted line) and feeling threshold (solid-dotted line)

2.2 Auditory and Extra-Auditory Infrasound Noise Effects

The effects of infrasound on the human body are not fully understood and are still under investigation, but it is assumed not only a discomfort sensation in different organs as a result of exposure to high levels of infrasound, but also a noticeable effect on a person's health [10].

Focusing our attention on high pressure level infrasound, one of the first effects is the distortion of the acoustic wave when arrives in the middle ear, causing harmonics in the audible band. For example, a pure tone of 8 Hz is perceived as a wave composed of harmonics of 16 Hz, 24 Hz, 32 Hz, 40 Hz and etc. [11].

In 1973 Nixon noted that sounds with a frequency between 4 ÷ 20 Hz and 140 dB levels caused the hearing threshold to increase by 10 dB to around 1000 Hz [12].

Infrasounds between 2 ÷ 20 Hz at high levels 120 ÷ 140 dB, levels higher than the audibility threshold, are able to vibrate parts of the human body that resonate at their characteristic frequencies. The presence of gas or air inside the exposed part such as in the lungs, intestines, stomach and middle ear intensify the phenomena [13]. Above 125 ÷ 130 dB, pain in the ear and discomfort have been reported, due to perceptions of pressure in the ear [13, 14].

In particular higher than 130 dB, communication disturbances may occur due to voice modulation [11].

Johnson in 1980 suggested not to subject the ear to infrasound greater than 150 dB in order to avoid rupture of the eardrum and damage to the middle ear [11].

But also, low pressure level infrasound, under the audible threshold could cause annoyance [15–17] and rattling [18].

3 Materials and Methods

Different assessment criteria for protecting worker health and safety have been considered, in force in eight countries Germany [19], Poland [20, 21], Slovak [22], Denmark [23], Sweden [24], Japan [18], Russian Federation [25] and USA [26, 27].

Usually infrasound limits are given as equivalent G-weighted sound pressure level or on the un-weighted equivalent sound pressure level for some infrasound frequency band. Some regulations are not limited to the workplace but include the living environment too, others differentiate the limits depending on human effect type.

Also, the criteria act whether in the time domain or in the frequency domain. Sound pressure limits can be expressed as weighted values or un-weighted one.

Signals, if filtered, are G-weighted according to the ISO 7196. This weighting curve has the property of leveling the auditory threshold curve as shown on Fig. 1.

In order to compare and find the most restrictive criteria, two ways have been followed. The first one transforms the limits defined in frequency domain, as shown in Fig. 2, in an interval of levels G weighted. The extreme of the interval are defined by the minimum and maximum value of the G-weighted limits given for each frequency band, as shown in Fig. 3. In this way the lower limit corresponds to the more severe criteria. The second method is based on the evaluation of recorded infrasound sources. Some soundtracks, with a specific spectrum and SPL, are related to different real working situation and other to synthetized signal.

The first two synthetic signals are a pink noise corresponding respectively to 85 dBA and 137 dBC with a spectrum ranging from 2 Hz to 20 kHz, level values coming from acoustic limits whose spectrum is extended in the infrasonic interval. The second two signals have an SPL of 85 dBA and 137 dBC with a spectrum ranging from 2 Hz to 20 Hz. The last signal has a SPL of 80 dBA with a spectrum ranging from 2 Hz to 20 Hz. Again the values of 80 dBA is taken from the acoustic risk assessment regulation. The recorded infrasound signals are produced by electric power generator (105.2 dBG), office (63.5 dBG), industrial area outdoor (79.6 dBG), cabin offshore boat (94.0 dBG), inside bus (104.2 dBG), inside vehicle (94.2 dBG). Each of these signals has different characteristic both in spectrum and in level in order to figure out differences in the compliance with exposure limits. Limits given as stated above in dB(G) or in dB (Lin). The compliance has been evaluated also considering the most recent studies available in literature on infrasound health effects.

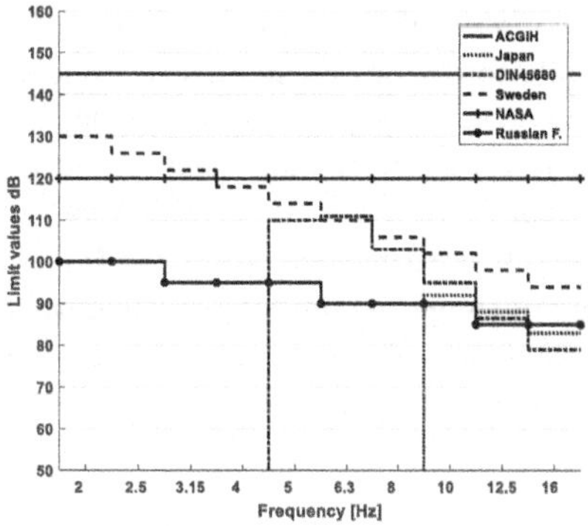

Fig. 2. Limits given for some frequency band for different countries

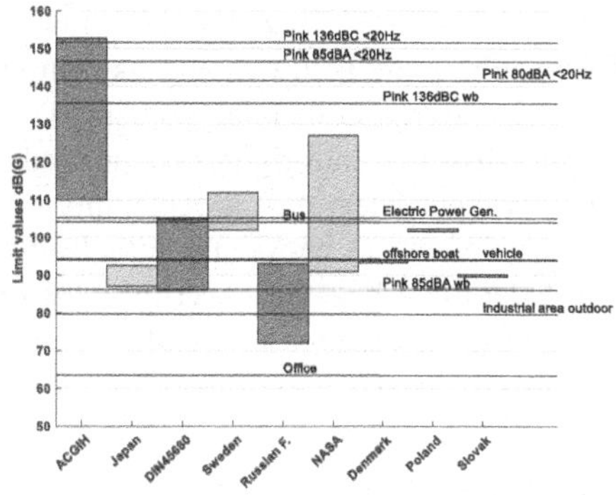

Fig. 3. Limits of Fig. 2 transposed in the G unit, plus limits stated directly in dB(G). The horizontal lines represent the LGex, 8 h of the test sources.

4 Results and Discussion

The comparison of some aspects of each national regulation, relative to how, where and who is exposed to the infrasound noise, highlight that there not a uniformity on the features treated, some of which are shown on Table 1.

Table 1. Features selected from different regulations.

Features	ACGIH	NASA	Denmark	Germany	Japan	Slovak	Poland	Sweden	Russian F.
Linear unit	x	x		x	x			x	x
G weighted unit			x		x	x	x		
Frequency band limit	x			x	x			x	x
Broadband limit		x	x		x	x	x		x
Continuous signal	x	x	x	x	x	x	x	x	x
Impulse signal	x	x	x	x		x			x
Risk classes for workers		x	x		x		x		x
Lower safety limit for special categories							x		x
Noise zones			x	x					x

In addition to the risk for the auditory system for worker the other selected parameters are for example the presence of the lower safety limit for special categories or the subdivision of the territory in noise zones like in environmental acoustics. In dividing the territory in zones, it's important to observe that the attenuation of the airborne infrasound is low during its path, especially for the lower frequencies, due to their long wave (also hundred meters). This means that zones cannot be too small.

Risk classes are not rigorously defined, evaluations are given following the in/out criteria, without interval of acceptability or uncertainty.

These differences between criteria lead to different assessments as shown in Fig. 3.

In order to evaluate the severity of the imposed or suggested limits, they have been leveled to the same unit and applied both a synthetic signal and a suitable recorded signal as shown in Fig. 3. Table 2, a pass/no pass table summarizes the results.

Table 2. Rejected and passed Signal according to different country regulations.

	ACGIH	Japan	Din 45680	Sweden	Russian F.	NASA	Denmark	Poland'	Slovak
Electric Power Gen.	0	0	0	0	0	1	1	1	1
Office	0	0	0	0	0	0	0	0	0
Industrial area outdoor	0	0	0	0	0	0	0	0	0
Inside offshore boat	0	1	1	0	0	1	1	0	1
Inside Bus	0	1	1	0	0	1	1	1	1
Inside vehicle	0	0	1	0	0	1	1	0	1
Pink 85dBA wideband	0	0	0	0	0	0	0	0	0
Pink 85dBA < 20 Hz	0	1	1	1	1	1	1	1	1
Pink 80dBA < 20 Hz	0	1	1	1	1	1	1	1	1
Pink 136dBC wideband	0	1	1	1	1	1	1	1	1
Pink 136dBC < 20 Hz	0	1	1	1	1	1	1	1	1

Remark: 1 rejected; 0 passed

In some case such as Sweden, Germany and Japan regulations, the choice of the infrasound limit level has been suggested from the short gap between the earing threshold and the pain threshold as shown in Fig. 1. In Dutch regulations is also assumed that infrasound only slightly above the hearing threshold may be annoying.

Those imply that the limit is a curve parallel to the auditory threshold. Other countries as Poland, Denmark and Slovak have simplified the evaluation of the limit crossing adopting the G-weighted curve as filter. So that it easier for the instrument and the evaluator checking in real time the exceeding of the limit.

Considering that average hearing threshold for infrasound corresponds to tones each one having in the infrasound region a G-weighted level of approximately LpG = 96 dB or a wideband level equal to 106 dBG, the limit level is set below that values. For example, in Poland it is set to 102 dBG just around 6 dB over the earing threshold.

Other criteria, as the NASA line guide [27], assume as limit the pain threshold that is approximately constant, in frequency, above about 125 dB till just below earing damage equal to150 dB as stated by ACGIH [26] that set the limit level at 145 dB for each frequency band and 150 dB for broadband sound. Not all country considers intermittent sound generally considered more annoyed and dangerous and so have lower limit around 5 dB less.

Denmark has established noise level inside dwelling, school and office up to 85dBG but in enterprises room level is increased of 5 dB considering 8 h work instead of 24 h.

The limit of 85 dBG is set approximately 10 dB lower the auditory threshold, value that take in account the 10% more sensitive of the population.

One question is due, are levels of infrasound below hearing threshold potentially harmful? If this is true for secondary effect as the annoyance, are there safe levels?

It has to be kept in mind that setting limits to high or lower values implies different costs to employer and the society. Lower limits mean high cost for the employer in term of best machinery, manufacturing process and reduced exposure, while higher limits mean increased cost for the society in term of health care services.

5 Conclusions

Since the end of the War Word II, around 75 years ago, many studies have been focused on the effect of the infrasound on human being, but there not still arrived to an international agreement on how treat them in order to reduce the risk for workers as in the audible properly noise. Their assessment criteria have been applied both on infrasound recorded in some real case studies and on specific synthetic signal in order to figure out differences in the compliance with exposure limits, generally defined for protection of worker health. The tested methods are not completely comparable, because the context of their application is not properly the same, so our evaluation is useful to test their severity in case of application all for the worker.

The outcome of this study could give an indication on which infrasound exposure limit have a high level of severity more than others, and that other countries don't even have any law or line guide about them, as in Italy.

References

1. Eldredge, D.H., Parrack, H.O.: Biological effects of intense sound. J. Acoust. Soc. Am. **21**, 55 (1949)
2. Romani, J.D., Bugard, P.: Nouveaux essais sur l'action des bruits sur le système endocrinien. Acustica **7**, 91–93 (1957)
3. Yeowart, N.S., Bryan, M.E., Tempest, W.: Low-frequency noise thresholds. J. Sound Vibr. **9** (3), 447–453 (1969)

4. Yeowart, N.S., Bryan, M.E., Tempest, W.: The monaural M.A.P. threshold of hearing at frequencies from 1.5 to 100 c/s. J. Sound Vibr. **6**(3), 335–342 (1967)
5. Yeowart, N.S., Evans, M.J.: Thresholds of audibility for very low-frequency pure tones. J. Acoust. Soc. Am. **55**(4), 814–818 (1974)
6. Broner, N.: The effects of low frequency noise on people - a review. J. Sound Vibr. **58**(4), 483–500 (1978)
7. Yamada, S., Kosaka, T., Bunya, K., Amemiya, T.: Hearing of low frequency sound and influence on human body. In: Proceedings of the Conference on Low Frequency Noise and Hearing, pp. 95–102 (1980)
8. Moller, H., Pedersen, C.S.: Hearing at low and infrasonic frequencies. Noise Health **6**, 37–57 (2004)
9. van den Berg, F., Passchier, G.P., Vermeer, W.: Assessment of low frequency noise complaints. In: 1999 Proceedings of the Internoise (1999)
10. Gregorio, C.A., Maull, E.A., Carson, B.L., Haneke, K.E.: Infrasound: Brief Review of Toxicological Literature. Edited by National Institute of Environmental Health Sciences (2001)
11. Johnson, D.L.: The effects of high level infrasound. In: Conference on Low Frequency Noise and Hearing (1980)
12. Nixon, C.W., Johnson, D.L.: Infrasound and hearing. In: International Conference on Noise as Public Health Hazard (1973)
13. Landstrom, U.: Human effects of infrasound. In: Proceedings of the Conference Internoise (2000)
14. Johnson, D.L.: Auditory and physiological effects of infrasound. In: Proceedings of the Conference Inter-Noise (1975)
15. Andresen, J., Møller, H.: Equal annoyance contours for low frequency noise. J. Low Freq. Noise Vibr. Act. Control **3**, 1–9 (1984)
16. Møller, H.: Annoyance of audible infrasound. J. Low Freq. Noise Vibr. **6**, 1–17 (1987)
17. Kjellberg, A., Tesarz, M., Holmberg, K., LandstrOm, U.: Evaluation of frequency-weighted sound level measurements for prediction of low-frequency frequency noise annoyance. Environ. Int. **23**(4), 519–527 (1997)
18. Japan - Office of Odor, Noise and Vibration, Air Quality Management Division, Ministry of The Environment. Handbook to Deal with Low Frequency Noise (2004)
19. DIN 45680:1997-3: Messung und Bewertung tieffrequenter Geräuschimmissionen in der Nachbarschaft (Measurement and evaluation of low-frequency noise letting in the neighborhood) (1997)
20. PN-Z-01338 acoustics - measurement and assessment of infrasonic noise at the workplaces (2010)
21. Rozporządzenie Ministra Pracy i Polityki Socjalnej z dnia 29 listopada 2002 r.w sprawie najwyższych dopuszczalnych stężeń i natężeń czynników szkodliwych dla zdrowia w środowisku pracy (Regulation of the Minister of labor and social policy, 2002-11-29. On the maximum allowable concentrations and intensity of harmful factors in the work environment.) Dz. U. No. 217, Item. 1833 (2002)
22. Decree No. 549/2007 Coll. Decree of the Ministry of Health of the Slovak Republic laying down details on permissible values of noise, infrasound and vibration and requirements for objectification of noise, infrasound and vibration in the environment (2007)
23. Miljøministeriet Miljøstyrelsen, Orienteringfra Miljøstyrelsen Nr. 9 1997 Lavfrekvent Stoj, infralydog vibrationer i eksternt Miljø (The Danish Environmental Protection Agency: Environmental review no. 9 low frequency noise, infrasound and vibration in the environment) (1997)

24. AFS 2005:16 BULLER Arbetsmiljöverkets föreskrifter om buller samt allmänna råd om tillämpningen av föreskrifterna (Work environment authority regulations on noise and general advice on application of the regulations) (2005)
25. СанПиН 2.2.4./2.1.8.583-96. Физические факторы окружающей природной среды инфразвук на рабочих местах,в жилых и общественных помещениях и на территории жилой застройки (Physical factors environment: Infrasound on jobs, in residential and public premises and territory residential area). Russian Ministry of Health, Moscow (1997)
26. ACGIH® Threshold Limit Values (TLVs®) and Biological Exposure Indices (BEIs®) Appendix B (2012)
27. NASA: Man-Systems Integration Standards, Natural And Induced Environments NASA-STD-3000 Volume I, Section 5, Revision B (1995)

Psychosocial Risk and Turnover Intention: The Moderating Effect of Psychological Wellbeing

Michael Akomeah Ofori Ntow(✉), David Kwaku Abraham, Noble Osei Bonsu, Ophelia Delali Dogbe Zungbey, and Evans Sokro

Department of Human Resource Management Central Business School, Central University, P.O.Box 2305, Tema, Ghana
{ntow.michael, dkabraham72}@gmail.com, {Nosei.bonsu, odzungbey, esokro}@central.edu.gh

Abstract. Nature and conditions of work has necessitated research in the area of psychosocial risk, as employees spend more time on their work and workplace than ever. As an emerging concept, research indicates that psychosocial risk resulting from work demand and stress has effect on organisational outcomes. However, most research on psychosocial risk is centered on the mining sectors ignoring other sectors like the manufacturing sector where psychosocial risk factors seem to be threatening the safety and wellbeing of workers. This present study seeks to examine the moderating role of psychological wellbeing in the relationship between psychosocial risks and turnover intention. A cross-sectional survey was used to collect data from participants within selected manufacturing companies in Ghana. Multi-staged sampling technique was used to select 351 respondents for the study. Hierarchical correlation coefficient was used to test hypotheses and mediation. The study found that psychosocial risk is statistically significant in predicting turnover intention. Psychological wellbeing is also found to moderate the relationship between psychosocial risk and turnover intention, an indication that high levels of psychological wellbeing experienced from the work as a result of lower psychosocial risk will reduce turnover intent at the workplace. Therefore, within the Ghanaian manufacturing sector, workers perceive psychosocial risk as major contributor to turnover intention.

Keywords: Psychosocial risk · Psychological wellbeing · Turnover intention · Work

1 Introduction

Over the years, organisations and firms have witnessed new development at the workplace relating to employee's safety and work-related risk captured as psychosocial risk [1]. An emerging yet dangerous form of risk with the possibility of affecting workers' contribution to work and productivity. A myriad of factors such as increased workload or responsibility, information dissemination, increasing work-related demands and also job insecurity among others have contributed to the issue of psychosocial risks and its dangers at the workplace [2]. The extent to which work settings and conditions affect employees' psychological wellbeing deserve special attention taking cognizance of the importance of an individual's wellbeing to the

P. M. Arezes and R. L. Boring (Eds.): AHFE 2020, AISC 1204, pp. 76–83, 2020.
https://doi.org/10.1007/978-3-030-50946-0_11

accomplishment of organisational goals. [3] maintain that psychosocial risks, work design, content, context of work setting are likely to cause harm to individuals at the workplace. These are conditions that may result in psychological, physical and social outcomes such as stress, depression among others.

Extant literature in the field of psychosocial risk has recognised the economic impact of psychosocial risk on organisations [4]. However, few studies have found psychosocial risk to be associated with certain behavioural outcomes such as burnout [5, 6], anxiety [7, 8], poor job performance, inefficiency [9, 10] and job dissatisfaction [11, 12].

In addition, some studies have shed light on how psychological wellbeing affect organisational outcomes. For example, [13] argued that employee wellbeing as a result of work and job stress has an effect on organisational overall performance. These researchers further posit that poor work conditions could lead to low productivity. [14] also suggested that employee wellbeing alongside job alternatives, moderate the relationship between stressors and turnover intention. Similarly, [15–17] found that psychosocial work such as strain, fatigue and lack of direction have consequences for employee behavior outcomes. Presently, very little empirical evidence is known about the role of a possible moderator such as psychological wellbeing. Thus, research on psychological wellbeing as moderator in the relationship between psychosocial risk and employee turnover intention is clearly undefined or eluded management and organisational behaviour scholars. Most of these studies are unable to explain the extent to which psychological wellbeing is affected by psychosocial risk factors and consequently trigger or reduce turnover intent. This gap in the literature justifies the need for the current study.

Contextually, most of the studies on psychosocial risk factors and turnover intention for the past decades have focused more on emerging and developed economies such as China and Europe with little emphasis on developing economies like Ghana. For this reason, the researchers situate the context in Ghana as a developing economy with high demands in the manufacturing sector.

The present study is based on the Job-Demand-Control Model by [18] in explaining how work and conditions of work influence organisational and employee behaviour outcome. According to the model, people with excess job demand, with little or no control over the job, experience fatigue and job stress. The current study argues that an experience of psychological wellbeing will statistically moderate the relationship between psychosocial risk and turnover intention. This study therefore proposes that:

H_1: There will be a positive relationship between psychosocial risk and turnover intention.
H_2: There will be a negative relationship between psychosocial risk and psychological wellbeing.
H3: Psychological well-being moderates psychosocial risk and turnover intention relationship

2 Method

Researchers adopted cross-sectional survey design for this study. Six (6) manufacturing companies drawn from the Ghana club 100 list of companies were selected using convenient sampling technique to select participants for the study. Three hundred and forty-four respondents were sampled using purposive technique. Scale used for data collection were piloted to ensure validity. Ethically, researchers were given clearance or consent from sampled institutions and participants involved in the study.

3 Psychosocial Risk Scale

The 22-items scale was adopted from Copenhagen Psychosocial Questionnaire (2004–2005) which is a 22 item scale self-designed and scored on a Likert scale with a range of 1–5 where 1 = Strongly Disagree and Strongly Agree. The scale recorded Cronbach Alpha value of .797.

4 Psychological Wellbeing Scale

The researchers adopted the 42-item Psychological Wellbeing Scale developed by [19]. The scale measured all facets of psychological wellbeing (autonomy, environmental mastery, personal growth, positive, purpose in life, and self-acceptance). It is scored on a Likert scale ranging from 1–5 where 1 = strongly agree; 2 and 5 = strongly disagree. The scale was piloted, tested and scored a cronbach alpha value of 0.073.

5 Turnover Intention Scale

The Turnover Intention Scale used for this study was developed by [20]. This 8-item scale was scored on a Likert scale ranging from 1–5 where 1 = Strongly Disagree and 5 = Strongly Agree. This scale also scored an alpha value of 0.632.

6 Correlational Analysis

This section aimed at analysing the relationship between psychosocial risks and turnover intention and to further moderate the relationship with psychological well-being. For this purpose, correlation among main variables were analysed and discussed (Table 1).

Table 1. Correlation among main variables

	Mean	Standard deviation	1	2	3
1. Psychosocial risk	2.6538	.62557	1	.	
2. Psychological wellbeing	3.2753	.73888	−.045	1	
3. Turnover Intention	2.8827	.76005	.243**	−.021	1

**. Correlation is significant at the 0.01 level (2-tailed).

Table 1 depicts and confirms that statistically, all variables (psychosocial risk, psychological wellbeing and employee turnover) are significant as correlated among each other (sig. level $p < 0.01$). Additional, turnover intention was found to have a weak positive correlation with Psychosocial risk ($r = .243$), meaning that a proportionate increase in psychosocial risk leads to proportionate increase in turnover intention and vice versa. On the other hand, psychological wellbeing is negatively correlated with psychosocial risk ($r = -.045$). The correlation coefficient indicates an insignificant negative relationship between the variables. Lastly, employee turnover shows week negative correlation with psychological wellbeing ($r = -.021$) indicating that the relationship between these two variables is an insignificantly negative one. Thus, an increase in psychological wellbeing leads to an insignificant decrease in turnover intention and vice versa.

7 Research Hypotheses

The research hypotheses tested in this study are in relation to psychosocial risks and employee turnover and the moderating role of psychological wellbeing.

8 Hypotheses Testing

Hypothesis one confirms a significant positive relationship between psychosocial risk and turnover intention ($r = 0.243$ $p < 0.01$). Also, psychosocial risk was found to predict turnover intention ($\beta = 0.243$, $p < 0.02$). Second, the data shows that a negative relationship exists between psychosocial risk and psychological wellbeing ($r = -.045$, $p < 0.01$), while predictability yielded $\beta = -.045$, $p > 0.566$.

Table 2. Process results for moderation analysis among Psychosocial Risk, Psychological Wellbeing and Turnover Intention

Model 1. Outcome variable: Turnover Intention

Summary:	R	R-sq	MSE	F	df1	df2	p
	.2819	.0794	.5418	2.8533	3.0000	159.0000	.0391

Variables	ß	se	t	p	LLCI	ULCI
Constant	2.8864	.0589	48.9886	.0000	2.7700	3.0028
Psychological wellbeing	.1237	.1072	1.1538	.2503	-.0880	.3354
Psychosocial risk	.2680	.1025	2.6154	.0098	.0656	.4704
Interaction term (Psychoso × Psycholo)	.1787	.1538	1.1619	**.2470**	-.1250	.4823

R-square increase due to interaction(s):	R^2-chng	F	df1	df2	p
Psychoso × Psycholo	.0192	1.3499	1.0000	159.0000	.2470

Conditional indirect effects of X on Y:	Psycholo	Effect	se	LLCI	ULCI	t	p
Psychoso -> Psycholo -> TI	-.7389	.1360	.1489	-.1581	.4301	.9135	.3624
	.0000	.2680	.1025	.0656	.4704	2.6154	.0098
	.7389	.4000	.1570	.0900	.7101	2.5480	.0118

Note: p is significant at .05; ß denotes beta coefficient, LLCI, lover limit confidence interval, ULCI, upper limit confidence interval, Psycholo, Psychological Wellbeing, TI, Turnover Intention, Psychoso, Psychosocial risk.

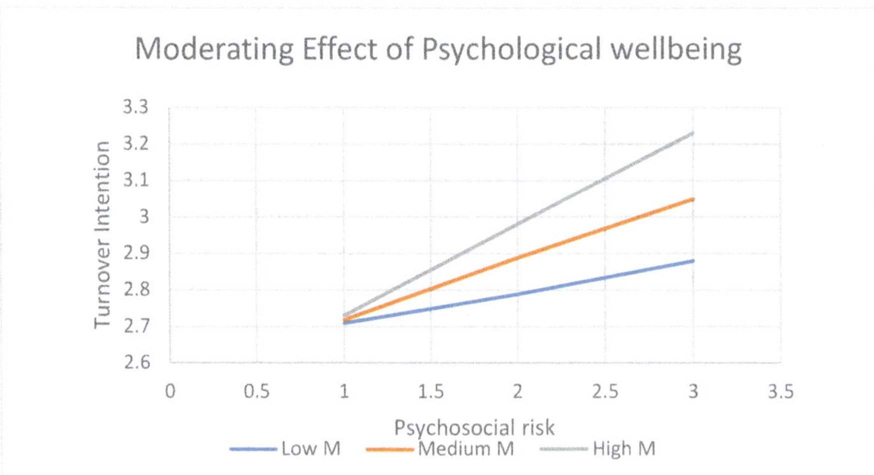

Fig. 1. Moderation analysis

Figure 1 depicts graphical moderation analysis. According to [21], the introduction of a third moderator variable is important if it increases the strength of the independent variable in predicting dependent variable. Also [22] explained that a moderator variable would result either in increasing or decreasing the strength of independent variable on the dependent variable. In testing for moderation, the researchers conducted the three-step process recommended by [23] and [24]. First, both psychosocial risk and turnover intention were included. These variables were to have increased amount of variance in turnover intention, $R^2 = .059$, $F = 10.125$, $p < .002$. In model 1, it was confirmed that there is a positive relationship between psychosocial risk and intention ($\beta = .0296$, $p < 0.02$). In order to test for moderation, an interaction term was introduced (psychosocial risk and psychological wellbeing), this however contributed to increasing the variance, $R^2 = .054$, $F = .331$, $p < .566$. The results show that the introduction of the interaction term maintained the significance of the previous significant relationship (shown in Table 2) and increased the variance by 24.7%. An indication that psychological wellbeing increased the strength of psychosocial risk in predicting turnover intention by 1.92% (refer to Table 2). In support, the data shows beta coefficient (ß) of psychosocial risk (0.2680) and psychological wellbeing (0.1237).

9 Discussion

Findings from the study indeed confirm the strength of psychological wellbeing as moderator in the analysis. Thus, it confirms and explains the extent to which psychological wellbeing strengthens or weakens psychosocial risk in predicting turnover intention. This supports the argument of [9, 25] that risk associated with workplace setting and the job itself has an effect on the wellbeing of the individual. Also, [3] observed that psychosocial risk and work environment affect individual behaviour outcomes such as absenteeism and commitment. This is consistent with the notion that psychological wellbeing results from satisfaction derived from the work and conditions attached [7]. As argued earlier, the Job Demand Control model suggests that work setting with no role clarity, lack of autonomy, poor delegation and demanding jobs contribute to job strain. Researchers argue that organisations will experience high turnover intent if the work performed by individuals does not guarantee psychological wellbeing and inner satisfaction.

This paper suggests that psychological wellbeing is an important factor in explaining the rate at which psychosocial risk is able to predict turnover intention. Further, an experience of low or poor wellbeing from work could increase the impact of psychosocial risk on turnover intention. Increased workload, job demands, manipulative structures and lack of role clarity are typical psychosocial risk factors that place much strain on the wellbeing of individuals [15]. Various studies have linked psychosocial risk to both physical and mental diseases. For instance, [6] found that psychosocial risk factors are major causes of back pain and mental illness. Psychologically, individuals are stressed to meet deadlines with limited resources and often sit for longer hours to deliver on set targets. The researchers argue that an important prerequisite factor in controlling turnover intent at the workplace is by ensuring that employees have certain degree of control and resources to effectively deliver beyond expectation. This could eventually lead to high levels of job satisfaction and psychological wellbeing. This study explains how managing employees' wellbeing is critical in reducing the impact of work-related risk on turnover intent. With a rapidly changing world, a healthy workforce gives an organization competitive advantage. The costs associated with reduced performance, absenteeism and lack of commitment due to reduced psychological wellbeing, is enough grounds for organizations to keep their employees healthy and well by providing the needed support and create an atmosphere that engenders psychological wellbeing. Our findings suggest that, discussing the impact of psychosocial risk on turnover intent is inconclusive without considering the effect of psychological wellbeing.

References

1. EU-OSHA: European Survey of Enterprises on New and Emerging Risks: Managing safety and health at work. Publications Office of the European Union, Luxembourg (2010a)
2. Bartram, D., Yadegarfar, G., Baldwin, D.: Psychosocial working conditions and work-related stressors among UK veterinary surgeons. Occup. Med. **59**, 334–341 (2009)
3. Cox, T., Griffiths, A.: The nature and measurement of work stress. In: Wilson, J.J.R., Corlett, E.N. (eds.) Evaluation of Human Work. 2nd edn., pp. 783–803. Taylor & Francis, London (1995)
4. Arnoux-Nicolas, C., Sauvet, L., Lhotellier, L., Bernaud, J.-L.: Development and validation of the meaning of work inventory among French workers. Int. J. Educ. Vocat. Guidance **17**, 165–185 (2016)
5. Baillien, E., De Witte, H.: Why is organizational change related to workplace bullying? Role conflict and job insecurity as mediators. Econ. Ind. Democr. **30**(3), 348–371 (2009)
6. Lloyd, C., King, R., Chenoweth, L., et al.: Social work, stress and burnout: a review. J. Ment. Health **11**, 255–265 (2002)
7. Bakker, A.B., Van Emmerik, H., Euwema, M.: Crossover of burnout and engagement in work teams. Work Occup. **4**, 464–489 (2006)
8. Hu, Q., Schaufeli, W.B., Taris, T.W.: The job demands-resources model: an analysis of additive and joint effects of demands and resources. J. Vocat. Behav. **79**, 181–190 (2011)
9. Driskell, J.E., Salas, E.: Stress and human performance. Psychology Press, Mahwah (2013)
10. Schat, A.C., Kelloway, E.K., Desmarais, S.: The Physical Health Questionnaire (PHQ): construct validation of a self-report scale of somatic symptoms. J. Occup. Health Psychol. **10**, 363–381 (2005)
11. Dierdorff, E.C., Morgeson, F.P.: Getting what the occupation gives: Exploring multilevel links between work design and occupational values. Pers. Psychol. **66**, 687–721 (2013)
12. Rich, B.L., Lepine, J.A., Crawford, E.R.: Job engagement: antecedents and effects on job performance. Acad. Manag. J. **53**, 617–635 (2010)
13. Jain, A.K., Giga, S.I., Cooper, C.L.: Organizational support as a moderator of the relationship between stressors and organizational citizenship behavior. Int. J. Organ. Anal. **21**, 313–334 (2009)
14. Huppert, F.A.: Psychological well-being: evidence regarding its causes and consequences. Appl. Psychol.: Health Well-Being **1**, 137–164 (2009)
15. Larsman, P., Sandsjö, L., Klipstein, A., Vollenbroek-Hutten, M.M., Christensen, H.: Perceived work demands, felt stress, and musculoskeletal neck/shoulder symptoms among elderly female computer users. The NEW study. Eur. J. Appl. Physiol. **96**(2), 127–135 (2006)
16. Idris, M.A., Dollard, M.F., Winefield, A.H.: Integrating psychosocial safety climate in the JD-R model: a study amongst Malaysian workers. SA J. Ind. Psychol. **37**(2), 29–39 (2011)

17. Tett, R.P., Meyer, J.P.: Job satisfaction, organizational commitment, turnover intention, and turnover: path analyses based on meta-analytic findings. Pers. Psychol. **46**, 259–293 (1993)
18. Karasek Jr., R.A.: Job demands, job decision latitude, and mental strain: implications for job redesign. Adm. Sci. Q. **24**(2), 285–308 (1979)
19. Ryff, C.D., Keyes, C.L.M.: The structure of psychological well-being revisited. J. Pers. Soc. Psychol. **69**, 719–727 (1995)
20. Jackofsky, E.F., Slocum, J.J.S.: A causal analysis of the impact of job performance on the voluntary turnover process. J. Occup. Behav. **8**(3), 263–270 (1987)
21. Hayes, A.F.: An index and test of linear moderated mediation. Multivar. Behav. Res. **50**, 1–22 (2015)
22. Dardas, L.A., Ahmad, M.M.: Coping strategies as mediators and moderators between stress and quality of life among parents of children with autistic disorder. Stress Health: J. Int. Soc. Inv. Stress **31**(1), 5–12 (2015)
23. Baron, R.M., Kenny, D.A.: The moderator-mediator variable distinction in social psychological research: conceptual, strategic, and statistical considerations. J. Pers. Soc. Psychol. **51**, 1173–1182 (1986)
24. Hayes, A.F.: Introduction to Mediation, Moderation, and Conditional Process Analysis: A Regression-Based Approach. The Guilford Press, New York (2013)
25. Wu, W., Rafiq, M., Chin, T.: Employee well-being and turnover intention: evidence from a developing country with Muslim culture. Career Dev. Int. **22**(7), 797–815 (2017)

Psychosocial Risk Assessment by Fine Kinney and ANFIS Method: A Case Study in a Metal Processing Plant

Nalan Baç[(✉)] and Ismail Ekmekci

Graduate School of Science and Engineering, Occupational Health and Safety Program, Istanbul Commerce University, Istanbul, Turkey
nalanbac@gmail.com, iekmekci@ticaret.edu.tr

Abstract. Occupational health and safety practices continue to improve, but accidents and occupational health issues continue to cause problems for management and employees. Although psychosocial risks and hazards have substantial adverse effects, they are often overlooked in most risk assessment practices. The objective of this work is to assess the employee psychosocial risks by a case study using the COPSOQII questionnaire at a metal processing plant. Inputs from COPSOQII questionnaire were used to assess psychosocial hazards in Fine Kinney analysis. The Fine-Kinney method contains fuzzy decisions. This uncertainty is resolved by using the ANFIS (Adaptive Neuro-Fuzzy Inference System) module. Results of the ANFIS model indicate that when actual risks and those predicted by the model are compared, the model could predict the risk scores with high accuracy.

Keywords: Psychosocial risks · COPSOQII questionnaire · Fine kinney analysis · ANFIS

1 Introduction

The factors that impair occupational health and safety (OHS) at the workplace are usually grouped as, physical, biological, ergonomic and psychosocial. Psychosocial risks are among the new emerging risks that most institutions have yet to assess their effects in the risk analyses at their workplace. Human factors play an important role in the safe operation of a facility. A significant part of human factors and employee errors occur during the maintenance phase. The negative impact of work accidents and occupational diseases on employees causes work stress. Psychological and physical health of employees can be adversely affected by poorly designed or managed work environment, a traumatic event at work, violence at work, bullying or harassment, and excessive or prolonged work pressures.

Thus, Copenhagen Psychosocial Questionnaire (COPSOQ) has become a widely used tool for assessment of psychosocial risks in the work environment [1, 2].

A study of the safe working culture and risk perceptions of employees at a large steel production facility in Sweden examined that the safety of the working performance is obvious, but the pressure of the production has a negative effect on the safety.

P. M. Arezes and R. L. Boring (Eds.): AHFE 2020, AISC 1204, pp. 84–90, 2020.
https://doi.org/10.1007/978-3-030-50946-0_12

They emphasized the importance of communication between employees and managers on security performance issues [3].

A study related to applications of health and safety strategies in the European steel industry indicated that the main success factors addressing psychosocial constraints are raising awareness, supporting workers who deal with psychosocial difficulties, and establishing a common approach that includes all stakeholders (management, workers and worker representatives) [4]. Due to the different functions and professions in steel companies, it is considered that transferring the necessary measures from one operation to another is difficult.

A study analyzing human factors in maintenance suggested a design/methodology approach. They made a general review of the human factors in maintenance and stated that the human role at different stages of life cycle of a product, including design, installation, production and maintenance, is remarkable and cannot be replaced by any new technology systems. As a result, they stated that the important aspect of human factors in maintenance should be taken into account from the design phase of equipment and, management ensures continuity, such that maintenance workload and downtime, fatigue and work injuries, and the possibility of human error are reduced, while employee satisfaction is improved [5].

In principle, all employees are responsible for best practices of OHS, but in most cases blue collar workers may be more prone to work accidents. In this study, the second version of the COPSOQ questionnaire, COPSOQII, was administered to all blue-collar maintenance workers to assess psychosocial risk factors as a case study in a metal processing plant in Istanbul, Turkey

2 Methods

2.1 Copenhagen Psychosocial Questionnaire

COPSOQ II has been developed to include the items such as reward, trust, justice and discrimination that were lacking in COPSOQI [2]. COPSOQ II Copenhagen Psychosocial Questionnaire II involves 7 domains composed of: demands at work; work organization and job contents; interpersonal relations and leadership; work and individual interface; values at the workplace; health and well- being and offensive behavior. There are 41 scales in the 7 domains. The COPSOQ II has been approved by the Danish National Institute of Occupational Health and has become a mandatory cornerstone in workplace risk assessments performed by Danish companies every three years. COPSOQ seeks to address the uncertainty and breadth of psychosocial factors by implementing a multidimensional approach with a wide range of aspects [1].

A study to identify psychosocial risk factors in workplaces due to high levels of tension and attrition notices from school principals used the COPSOQ II survey. As a result of their research, they noted that COPSOQ II closed the large gap in measuring psychosocial risk factors [6].

In a study of labor market participants using COPSOQ questionnaire in Canada employees are classified with gender, gender roles and age factors. They noted that using a standard survey, such as COPSOQ, allows comparisons between different industries and worker populations over time [7].

2.2 Fine Kinney Risk Analysis

Fine-Kinney method uses risk scores to establish the level of acceptance of hazards, and priority of risks. The three parameters used to assess the risks are exposure, probability, and consequence. The product of the three parameters gives the risk score [8, 9]. Fine-Kinney method contains uncertainty due to the criteria it contains. The risk scores may depend on the person who makes the decision. The change of risks according to each individual assessor rather than definite numerical results causes an assessor or a team to be uncertain about the magnitude of risk in question.

2.3 ANFIS Method

The uncertainty in the Fine Kinney Method can be resolved by using the ANFIS (Adaptive Neuro-Fuzzy Inference System) module included in the MATLAB program. The ability of artificial neural networks to learn using available data, to be able to store the knowledge it learns, to establish self-relations according to data, and to work with a large number of variables, and lack dependence on human experience is the biggest advantage of ANFIS [10].

Neural networks and fuzzy logic methods are two methods that complement each other, and ANFIS is a hybrid method in which these methods are used together.

3 Results

The COPSOQ II questionnaire was adapted such that the answers are in terms of five-point Likert scales. Cronbach Alpha (α) coefficient which is a measure of internal consistency is calculated as 0.995 indicating the high reliability of the questionnaire. In addition, the Cronbach Alpha values of the seven domains are as follows: Demands at work: 0.969; Work organization and job contents: 0.961; Interpersonal relations and leadership: 0.993; Work-individual interface: 0.970; Values at the workplace: 0.920; Health and well-being: 0.985 and Offensive behavior: 0.937.

The domains and items in COPSOQII questionnaire were used as inputs of psychosocial hazards in Fine Kinney analysis. Psychosocial risk analysis was performed for the first time at the metal processing plant studied using the Fine Kinney method. This was done by consulting with the workplace physician, psychologist and occupational safety expert. The results of risk scores and risk classes are given in Table 1. Risk classes are defined as 1: acceptable; 2: possible; 3: severe, 4: high; 5: extremely high risk.

Since psychosocial risk assessment was done for the first time, probability values in the Fine Kinney methods was selected as the worst probability. The frequency of exposure was decided by taking into account the frequency of exposure of all maintenance workers, and responses to the COPSOQ II questionnaire. In addition, the

Table 1. Risk scores and risk classes by Fine Kinney Method

	Domain Codes	Scale	Fine Kinney Risk class	Fine Kinney Ave score
1	QD	Quantitative Demands	5	468,00
2	WP	Work Pace	5	570,00
3	CD	Cognitive demands	3	102,00
4	ED	Emotional demands	2	33,00
5	HE	Demands for hiding emotions	2	62,00
6	IN	Influence	2	59,00
7	PD	Possibilities for development	3	84,00
8	VA	Variation	3	78,00
9	MW	Meaning of work	2	62,00
10	CW	Commitment to the workplace	2	63,00
11	SS	Social support from supervisors	3	92,00
12	SC	Social support from colleagues	2	28,00
13	SW	Social community at work	3	84,00
14	PR	Predictability	3	75,00
15	RE	Recognition	2	22,00
16	CL	Role clarity	2	35,00
17	CO	Role conflicts	1	17,00
18	QL	Quality of leadership	3	81,00
19	JI	Job insecurity	3	136,50
20	JS	Job satisfaction	2	68,00
21	WF	Work–family conflict	4	264,00
22	FW	Family–work conflict	2	31,50
23	TE	Mutual trust between employees	2	28,00
24	TM	Trust regarding management	2	64,50
25	JU	Justice	3	112,50
26	SI	Social inclusiveness	2	47,25
27	GH	General health perception	2	54,00
28	SL	Sleeping troubles	2	36,00
29	BO	Burnout	3	72,00
30	ST	Stress	2	63,00
31	DS	Depressive symptoms	2	55,00
32	SO	Somatic stress	2	69,00
33	CS	Cognitive stress	2	66,00
34	SE	Self-efficacy	2	51,00
35	SH	Sexual harassment	3	120,00
36	TV	Threats of violence	3	84,00
37	PV	Physical violence	5	480,00
38	BU	Bullying	3	180,00
39	UT	Unpleasant teasing	3	180,00
40	CQ	Conflicts and quarrels	5	480,00
41	GS	Gossip and slander	3	84,00

severity factor was evaluated by consulting the workplace physician and psychologist and the possible serious results were evaluated with a little pessimism.

The uncertainties based on the experience and opinions of the experts who performed the evaluation during the evaluation can only be examined by fuzzy tools, so the ANFIS module is used. The results from the COPSOQ II survey were normalized and constitutes the 7-input data of the 7 main sections that make up the ANFIS Model. The output of the ANFIS model is the psychosocial risk score.

ANFIS rule base is given in Fig. 1, and the comparison of ANFIS risk scores and actual values is given in Table 2. Table 2 indicates that ANFIS model predicts the risk scores accurately.

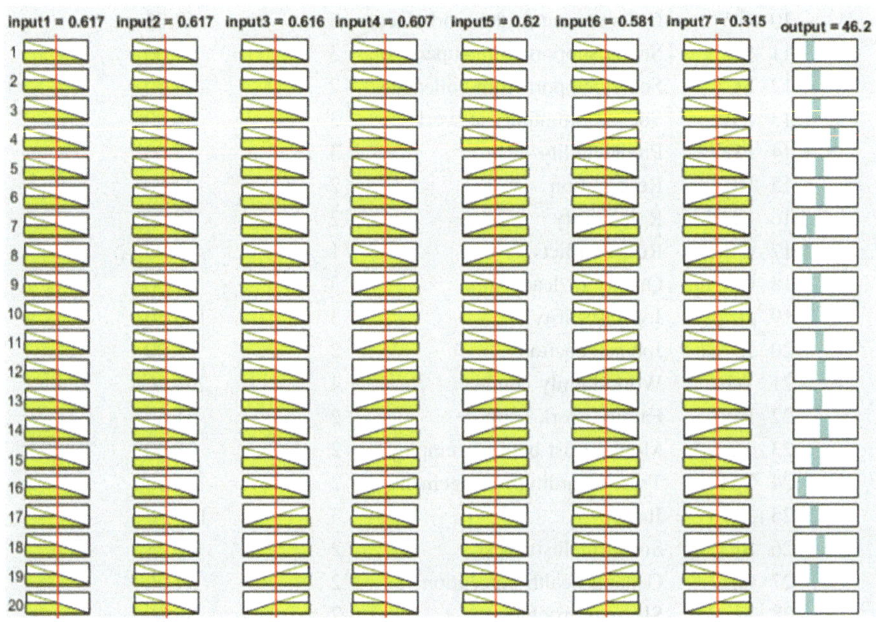

Fig. 1. ANFIS rule base

COPSOQ II 7 Domains: Demands at work (Items 1–5); Work organization and job contents (6–10); Interpersonal relations and leadership (11–18); Work-individual interface (19–22); Values at work (23–26); Health and well-being (27–34) Offensive Behavior (35–41)

Table 2. Actual values and ANFIS model predictions

Actual value	Model prediction
89.644	89.639
81.291	81.288
74.732	74.727
71.602	71.602
68.914	68.912
66.639	66.603
64.732	64.719
61.326	61.313
58.508	58.511
55.742	55.738
52.642	52.624
51.377	51.301
48.842	48.886
50.346	50.225

4 Conclusion

Psychosocial risk analysis that is usually overlooked was performed for the workplace studied for the first time. The hazards are identified by the COPSOQ II questionnaire and risks are assessed by the Fine Kinney method. An ANFIS model was used to remedy the uncertainty of Fine Kinney Method, and the model predicted the psychosocial risks accurately.

References

1. Pejtersen, J.H., Kristensen, T.S., Borg, V., Bjorner, J.B.: The second version of the copenhagen psychosocial questionnaire. Scand. J. Public Health **38**, 8–24 (2010)
2. Kristensen, T.S., Hannerz, H., Hogh, A., Borg, V.: The Copenhagen Psychosocial Questionnaire (COPSOQ)- a tool for the assessment and improvement of the psychosocial work environment. Scand. J. Work Environ. Health **31**, 438–449 (2005)
3. Nordlöf, H., Wiitavaraa, B., Winblad, U., Wijk, K., Westerling, R.: Safety culture and reasons for risk-taking at a large steel-manufacturing company: investigating the worker perspective. Saf. Sci. **73**, 126–135 (2015)
4. Agostini, M., Bauchiere, D., Beaufort, P., Montreuil, E., Van Criekingen, L., Weingarten, J.: Industrial relations practices related to psychosocial constraints at work in the steel sector. Cornell University ILR School DigitalCommons@ILR (2014)
5. Sheikhalishahi, M., Pintelon, L., Azadeh, A.: Human factors in maintenance: a review. J. Qual. Maint. Eng. **22**, 218–237 (2016)
6. Dicke, T., Marsh, H.W., Parker, P.D., Guo, J., Horwood, M.: Validating the Copenhagen Psychosocial Questionnaire (COPSOQ-II) using Set-ESEM: identifying psychosocial risk factors in a sample of school principals. Quant. Psychol. Measur. Sect. J. Front. Psychol. **9**, 584 (2018)

7. Ramkissoon, A., Smith, P., Oudyk, J.: Dissecting the effect of workplace exposures on workers' rating of psychological health and safety. Am. J. Ind. Med. **62**, 412–421 (2019)
8. Fine, W.T.: Mathematical evaluation for controlling hazards. J. Saf. Res. Defin. **3**, 157–166 (1971)
9. Kinney, G.F., Wiruth, A.D.: Practical Risk Analysis for Safety Management, vol. 5865. Naval Weapons Center, China Lake (1976)
10. Jang, J.S.R.: ANFIS: adaptive-network based fuzzy inference systems. IEEE Trans. Syst. Man Cybern. **23**, 665–685 (1993)

Ergonomic Risk and Safety Assessment of Typical Household Products

Xu Qian[1], Huimin Hu[2(✉)], and Nan Li[1]

[1] Capital University of Economics and Business, Beijing, China
18201020246@163.com, 929814405@qq.com
[2] SAMR China National Institute of Standardization, Beijing, China
huhm@cnis.ac.cn

Abstract. According to the date of Chinese children, such as the physical and psychological characteristics, preferences of them, and user needs of them. Based on the functional characteristics of typical children's furniture products currently on the market in China, as well as the availability and use safety problems during using it, researching typical and the ergonomic risks of furniture products. In order to provide basic method guidance for improving and enhancing the safety of the use of furniture products in China. The research products including the children's bunk beds, children's tables and chairs, and children's wardrobes. This research uses theoretical analysis, literature survey, human-computer interaction behavior analysis, and market research to study the structural and functional characteristics and target users of typical furniture products. During the investigation, it was found that the reasons for the existence of high-risk children's furniture in the market are mostly related to their safety dimensions, such as the relevant dimensions of children's bunk beds, including the height of the safety fence, the width of the ladder, the width of the pedal, etc., using the comfort analysis in the Jack simulation software and force analysis, we found that when a minor climbs into bed, the pedal width is too narrow and the ladder distance is too large, it will cause significant discomfort to the minor, and it is very easy to cause a fall accident, and the height of the safety fence is too low which will even cause a fall accident. In the investigation, I found that the frequency and frequency of falling accidents on and off the bed has become one of the important reasons for minor injuries across the country. Based on this phenomenon, we use the means of usability testing, simulation analysis, physical testing, theoretical analysis, data statistics, and user experience evaluation to study the main safety influencing factors of minors in the use of furniture products and guidelines for improving the use of safety. Preliminary establish the target user characteristic parameters of furniture products and functional unit parameters of furniture products. Based on this, the methods of human-computer interaction analysis and simulation are used to study the main safety influencing factors of the target user in the use of furniture products and the improvement directions to improve the safety level of use, and determine the safe size range suitable for minors at various stages. And finally form a guide for prevention and evaluation of various ergonomic risk indicators. Provides evaluation basis and reference for the safety and experience of furniture products during use it.

Keywords: Children furniture · Ergonomic risk · Index · System

P. M. Arezes and R. L. Boring (Eds.): AHFE 2020, AISC 1204, pp. 91–96, 2020.
https://doi.org/10.1007/978-3-030-50946-0_13

1 Introduction

With the improvement of people's living standard, parents' requirements for children's furniture comfort and safety are increasing day by day, but the safety of commonly used children's furniture needs to be improved. Commonly used children's furniture includes children's bunk bed, children's table and chair and children's wardrobe. According to the sixth national census results, the number of people aged 0–14 accounted for 16.60 percent of the total population. However, accidental injury poses a great threat to minors in China, second only to disease. Therefore, it is very important to analyze and evaluate the ergonomic risk of typical children's furniture to ensure that children have a safe, healthy and comfortable environment [1].

The size of children furniture is directly related to the safety of the children's daily life safety. In recent years with the improvement of children furniture injury accident, the safety of children's furniture is becoming more and more important. The occurrence of accidents is due to the unsafe behavior of people and the unsafe state of things, and the increase of ergonomic risk will inevitably lead to the unsafe behavior of people. Therefore, this study is mainly based on theoretical analysis and user research to study the factors related to children's furniture casualty accidents in recent years. Starting from the safety of typical children's furniture, this paper studies and establishes the ergonomic risk assessment indicators that affect the safety of children's furniture, so as to evaluate the safety risk of children's furniture and reduce the probability of accidental injury when children use children's furniture.

2 Construction of Evaluation Index System

In order to evaluate the ergonomic risk of children's typical furniture objectively and accurately, it is necessary to select appropriate indicators from the current typical children's furniture and establish an objective and reasonable indicator system.

2.1 Principles for Establishment of Indicator System

Scientific Principle. The structure of index system should be scientific and reasonable, which can reflect the essence of ergonomic risk of children's typical furniture normatively and accurately. The meaning of index should be clear, simple and easy to measure. Only when the evaluation method is scientific and standard, can the evaluation result be true and objective.

The Principle of Representative. There are many factors of affecting the risk of children furniture ergonomics. Choosing indicators should not only fully consider its systemic, avoid the phenomenon of incomplete information coverage, and to avoid for index system of the pursuit of the completeness of the index system is too complex, should highlight the key index. The index of the source data to define clear, on the premise of reflecting the basic connotation, as far as possible, reduce the number of indicators, indicators between are independent of each other, avoid inclusion or similar phenomenon.

Hierarchy Principle. The evaluation system should have clear levels and logical relations.

Principle of Operability and Comparability. The choice of each index should be easy to realize in real life, universal and easy to compare with children's furniture in different countries and similar areas in different cities.

2.2 Preliminary Summary of Indicators

In the literature survey, we have consulted the research on the evaluation index system of children's furniture and referred to many related articles

In the literature, according to relevant domestic standards and norms, 20 typical children's furniture ergonomic risk evaluation indicators are preliminarily summarized, and the specific indicators are shown in Table 1 [2].

Table 1. Preliminary evaluation index list

Evaluation index system of typical children furniture		
Children's bunk bed	Safety barriers	Safety clearance
		Safety Barrier Gap
		Safety barriers high
	Ladder	Ladder tilt angle
		Width of ladder
		Ladder spacing
	Footstep	Tread width
		Pedal depth
		Bed height
		Seam bed
Desks and chairs	Top of a table	Desktop height
		Angle of inclination
	The back of a chair	Height of seat back
		Leaning back
	Seat surface	Seat depth
		Seat dip
Wardrobe		Handle height
	Cabinet	Height
		Forced anti-dumping device
		Degree of depth

2.3 Filter Method Selection

There are a variety of methods for screening indicators, including expert evaluation method, investigation method and so on. The expert evaluation method is mainly based on the expert experience to extract the target indicators. The survey method collects a large number of indicators available through questionnaires and summarizes the feedback information. Because children and their parents have rich experience in using children furniture, their experience knowledge plays a very important role in the determination of indicators. Moreover, the ergonomic risk evaluation index of typical children's furniture is closely related to human factors, so this paper selects the method of investigation on target users to screen and investigate the indicators preliminarily summarized.

Methods: typical event survey, market survey, literature survey network questionnaire survey. So, we can summarize and analyze to screen the ergonomic risk evaluation indexes of typical children's furniture.

2.4 Survey Overview

Children's Bunk Bed. Children's bunk bed is a kind of children's furniture that is common in children's daily life. In recent years, children have frequent accidents when using the bunk bed. The accident occurs in different units. In the process of investigation, it is found that the accident of children falling off the bed has become one of the main accidents of children accidental injury. The causes of the falling bed accident include the following:

(1) Safety guardrail is too low. In the market survey, the height of the safety guardrail for some children's upper and lower beds is not up to the standard.
(2) Safety guardrail gap is too large. During the visit of the users, some children fell off the bed due to the gap in the safety fence.
(3) The pedals are too far apart. In recent news reports, one of the reasons for falling off the bed is that the ladder is too far apart during the process of falling off the bed, so that children cannot immediately step on the ladder during the process of falling off the bed, resulting in falling off the bed accident.
(4) The pedal width is too narrow. In the market survey, most parents and children responded that the pedal width of the upper and lower bed was too narrow, the foot of the child pedal have obvious discomfort, need to buy extra foot pad to stick on the pedal.

In addition to the falling bed injury, children in the use of the upper and lower bed, injury also occurs from time to time, in the market survey, there are more than 30% of children because of the bed seam is too small, the occurrence of varying degrees of injury.

Tables and Chairs for Children. Children's table and chair in children's daily life is an indispensable part, in addition to meeting the needs of children's daily activities, children's table and chair also subtly affect the physical health of children, children's body is in the stage of growth and development, so the safety of children's table and chair cannot be underestimated [3].

(1) Table height too high. According to the investigation of relevant literature, the height of the desktop should make the visual range of people within the best operating range of the perspective, too high will make arms uncomfortable.
(2) Desktop Angle is too low. If desktop inclination Angle is too low, when reading and writing, bow, can destroy the natural curvature of the neck, the effect of a long time will cause neck pain and discomfort.
(3) The back of the chair is tilted at an improper Angle. Chair backrest inclination also has very big concern with child's spine development, which can lead to curvature of the spine in a developing child.
(4) Improper seat depth. If the seat is too deep, causing the back to be unable to lean back, the knee and popliteal fossa will also be in front of the seat compression.
(5) Improper inclination of seat surface. If the seat is too deep, the front of the seat can also produce pressure on the child's thigh. At the same time, if the seat surface inclination is too large, it will lead to the upper body forward bending increase, so that the curvature of the spine increase, long-term operation will cause muscle strain, and lead to spinal disease. In addition, too large inclination will also be in the child's abdomen for a long time in the compressed situation, the child's internal organs have adverse effects [4].

Children's Wardrobe. Since 2014, at least eight children have died from climbing over cabinets that collapsed, and more than 90 children have been injured in the United States alone.

(1) The cabinet is too high and has no fixing device. This is the main reason of most accidents, which caused the child to climb over the wardrobe unattended, causing the wardrobe to collapse and eventually death. In market research, most wardrobes are only told to have dumping risk, not mandatory requirement fixed, many families may choose not fixed accordingly. So, ark height is one of this furniture ergonomic risk evaluation index.
(2) The depth of the cabinet. On the other hand, in extreme case, if ark is deep enough, although ark height exceeds 60 centimeters not to be fixed, also won't fall, accordingly, cabinet depth also serves as one of evaluation index.
(3) Cabinet handle height is not appropriate. In the market research, the height of the handle of the wardrobe is also one of the harmful factors, younger children often hit the head, resulting in injuries.

2.5 Determine the Ergonomic Risk Evaluation Index System

Based on the analysis of the above survey results, the following 9 categories and a total of 16 typical children's furniture ergonomic risk indicators have a great impact on children's use safety, as shown in Table 2.

Table 2. List of final evaluation indicators

Evaluation index system of typical children furniture		
Children's bunk bed	Safety barriers	Safety Barrier Gap
		Safety barriers high
	Ladder	Width of ladder
		Ladder spacing
	Footstep	Tread width
		Pedal depth
		Seam bed
Desks and chairs	Top of a table	Desktop height
		Angle of inclination
	The back of a chair	Leaning back
	Seat surface	Seat depth
		Seat dip
Wardrobe		Handle height
	Cabinet	Height
		Forced anti-dumping device
		Degree of depth

3 Conclusion

According to the four principles of the index system, a preliminary summary is made through literature research and data review 20 indicators of ergonomic risk dimensions of typical children's furniture were compared. After comparing various methods of screening indicators, the research method was selected to screen the indicators, and 16 indicators of ergonomic risk evaluation of typical children's furniture were selected to complete the establishment of ergonomic risk evaluation index system of typical children's furniture. For the analysis of index weight, the design and development of ergonomic risk experiment of typical children's furniture should be prepared.

Acknowledgements. This research is supported by 2017 National Quality Infrastructure (2017NQI) project (2017YFF0206506 and 2017YFF0206603) and China National Institute of Standardization through the "special funds for the basic R&D undertakings by welfare research institutions" (522018Y-5984).

References

1. Wei, J., Xu, J.: Research on human-computer ergonomics in children's furniture design. Art Des. **2**, 148–150 (2009)
2. Hong, P.: Study on Ergonomic Design and Evaluation Index System of Office Chair. Zhengzhou University (2019)
3. Xingshuang, S.: Research on Humanized Design of the Poly-Functional Furniture for Student's Flats in University. Central South University of Forestry and Technology (2007)
4. Zhang, X.: Ergonomics in Students Apartment Furniture Design and Research. Taiyuan University of Technology (2010)

Human Failures on Production Line as a Source of Risk of Non-conformity Occurrence

Anna Nagyova(✉), Zuzana Kotianova, Juraj Glatz, and Juraj Sinay

Faculty of Mechanical Engineering, Safety and Quality Department, Technical University of Kosice, Letna 9, 04200 Kosice, Slovakia
{anna.nagyova, zuzana.kotianova, juraj.glatz, juraj.sinay}@tuke.sk

Abstract. Recently, the production organizations and their suppliers have been facing big pressure concerning an increase in their performance, minimization of costs and, finally yet importantly, meeting of their customers' needs. In order to keep up the pace with the requirements, the production organizations are using various tools and methods of their improvement. Recently, the application of Industry 4.0 elements has become a key tool for them, and they are based on the need of automation application within production processes. Despite that, there are several activities and processes, which cannot be fully replaced with automation elements, and for their functioning, it is necessary to use a human-machine interface. Because of that, the activities and processes present the risks that do not have to be foreseeable in advance and their late identification and management are combined from discordant product generation, which automatically leads to a loss of competitiveness, or drop of the company profit.

This article is oriented on the risk analysis as an influence value of non-conformity, which originated by the manual placing of components in the process of production organization operating within the automotive area. Using the quality tools, the causes of the non-conformity origination were identified, and system solutions for its elimination were proposed. In line with the strategy Industry 5.0, training, as well as investment into the operators′ training program, created a part of the latter solutions within the context of a significant influence of human factor in the human-machine system.

Keywords: Non-conformity · Measure · Improvement tools · Industry 5.0

1 Introduction

An important element in the management of an organization is to find ways to produce high quality and low-cost products. It is common practice for organizations (especially manufacturing ones) to introduce sophisticated tools and methods into their processes to meet these requirements. However, it is not possible to prevent organizations from producing defective products, which are often the result of a human factor failure. A well-known approach to preventing human errors is the Toyota Production System (TPS), which began to be used in the 1980s in Japan. Today, as technology and human

P. M. Arezes and R. L. Boring (Eds.): AHFE 2020, AISC 1204, pp. 97–103, 2020.
https://doi.org/10.1007/978-3-030-50946-0_14

collaboration are increasingly used, this approach is also a way of ensuring the compatibility of this cooperation [10].

According to [8], the effective collaboration between technology and man depends on the skills and capabilities of employees who radically affect productivity, performance, and the added value of production flows. An important part of higher productivity and efficiency in today's industrial enterprises facing increasing automation and digitization of production processes is strongly linked to employees' ability to prepare for communication with the latest digitized production technologies [7]. With the introduction of industrial concepts such as Industry 4.0 and 5.0, great emphasis is placed on the development and efficient workplaces associated with human digital literacy. People in conjunction with collaborative robots manage the production more efficiently of value-added tasks [9]. The introduction of new technologies also depends on the nature of production in individual industrial enterprises: piece production, series production, mass production, continuous production (Fig. 1).

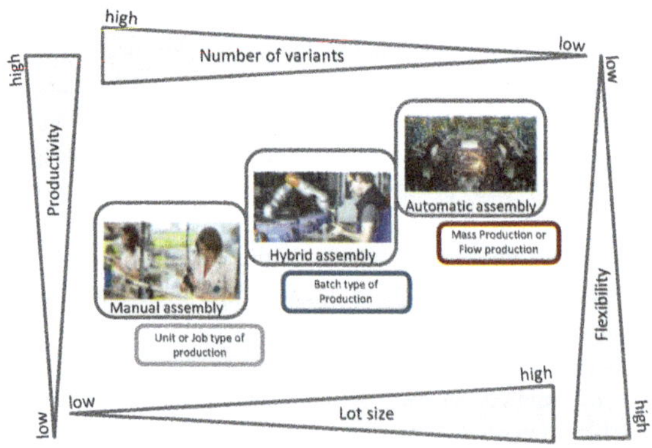

Fig. 1. The level of Industry 4.0 elements introduced, depending on the nature of production

In this case too, the level of technology deployed depends on the nature of production, as well as the maturity of the organization and its resources (financial, human and material) [6, 11].

2 Poka Yoke as a Tool for Ensuring Human Reliability

Although manufacturing organizations minimize the human factor through the introduction of various automation elements, it cannot be totally excluded in the production process. For this reason, it is necessary to ensure its reliability to the extent that the functionality and quality of the product itself are not compromised. In the event of a human factor failure, it is important to immediately identify the possible causes of this and then implement effective measures. Reducing the probability of error is one of the responsibilities of the management of the organization, which can reduce the probability by means of appropriate tools. A human error can be defined from the point of view of the production process as a phenomenon resulting from the failure to fulfil a specified task (or performance of a prohibited activity), which can cause disagreement or disruption of the production process [1]. The causes of human error may therefore be different. These may be inappropriate management practices, lack of qualifications, employee stress or negative intentions. Considering the maximum functionality and at the same time the effort to ensure the flawless functioning of the process, it is necessary to focus on man as an element directly entering the process. The level of employee reliability in the production process depends on three basic factors:

- physical properties,
- psychical properties,
- knowledge level [2].

Psychic and mental ability, as well as the physical capacity of an individual, cannot be a guarantee of a job well done if there is a lack of adequate education to perform the task. The aim of implementing various methods and tools in a human-machine environment is to provide the employee with the comfort of the environment so that the activities he performs are relieved of unnecessary consideration of whether he performs the activity correctly or not. One such tool is Poka Yoke, which in practice is often used in combination with other methods. The history of the first mention of the use of this instrument dates to 1961 [3]. Japanese industrial engineer Shigeo Shingo visited the Japanese Yamada power plant, which had the problem of incorrectly inserting the springs into the switches. The spring insertion process was performed by operators who incorrectly and in incorrect number inserted these springs, resulting in errors and malfunction of the switch. The problem was remedied by creating a template in the crate, in which two springs were placed behind each other, so the operator knew how and how many springs to insert into the switch [4]. Thus, he introduced the use of this tool in multiple operations, minimizing the human factor failures in performing simple manual operations. According to Shingo, the more a worker controls his/her activity, the sooner he/she makes a mistake. Poka Yoke thus relieves workers from constant repetitive checks, allowing them to devote themselves fully to their work. According to literary sources, Poka Yoke can be used as a tool against disagreements and unintentional errors, performing 3 basic functions:

- the stoppage of operation - in this case Poka Yoke performs so-called control function. This means that if an abnormality occurs in the process, the tool stops the operation.
- the check - in this case, Poka Yoke is used to check the exact number of movements or storage of the product.
- the caution - this type uses light or sound signalling when an abnormality occurs but does not stop the operation [5].

In all the above cases, POKA YOKE is not aimed at controlling the operation or detection of cases of non-conformity.

3 Case Study of the Use of Poka Yoke

The case study was carried out in a manufacturing organization dealing with the production of components for the automotive industry. Among other things, the organization has 15 separate production lines, known as Surface Mounting Technology (SMT), where a Printed Circuit Board (PCB) is produced.

The organization encounters various nonconformities in the manufacture of these PCB components, each of which may have several different causes. To identify these discrepancies, a combination of several statistical tools is used, but also the internal software DATABASE, which can clearly identify the frequency of these discrepancies. Using this software, it is easy to identify on which of the 15 production lines an error occurred and to identify the cause of the error by means of the Ishikawa diagram. Within this case study, analyses for the years 2018 and 2019 were carried out upon which in total 12958 errors were identified.

The lines labelled SMT06, SMT04 and SMT03 accounted for 43% of the total number of errors and for this reason the organization decided to carry out a more detailed analysis on these lines. The line with the largest number of nonconformities was chosen first by management. In particular, the SMT 04 line with 1933 nonconformities.

By applying the Pareto analysis (also performed through QUALITYDATABASE software) the frequency was determined, and the 10 most common causes of non-conformities were defined.:

1. Incorrect component
2. Damaged PCB
3. Incorrect PCB size
4. Wrong code position
5. Wrong references
6. Wrong component position
7. Incorrect PCB numbering
8. Part of the PCB is missing
9. Incorrect code assignment to PCB
10. Incorrect foregoing PCB repair

Significant errors identified by Pareto analysis results were: Wrong component composition, wrong component, Wrong code position, incorrect PCB numbering, damaged board. Based on the Pareto analysis, the most common cause was identified - Wrong component position, which represented 22% of the total number of errors (Fig. 2). Wrong component position means that there were errors in manual placement of the components on the PCB while the components were not embedded in correct place.

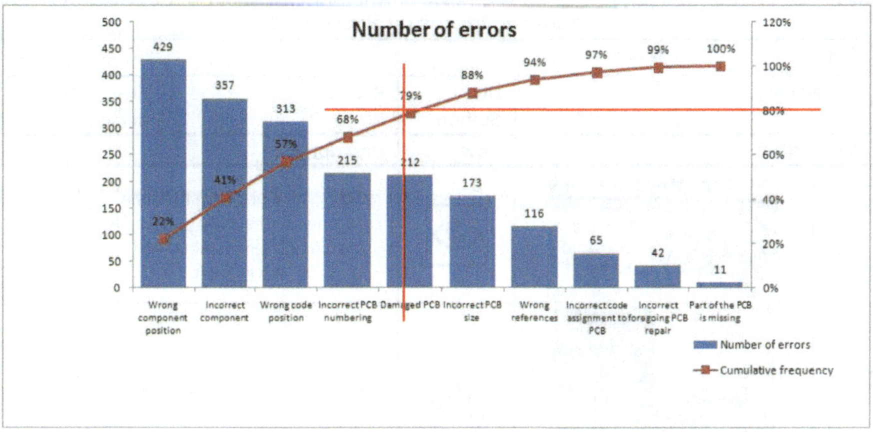

Fig. 2. Pareto analysis of number of errors

Following this analysis, management took corrective actions to minimize the occurrence of non-conformities due to the wrong component position.

4 Corrective Actions to Reduce Non-conformity

The first step in introducing corrective actions to prevent the recurrence of non-conformity was modelling of the fixture made of polymethyl methacrylate (plexiglass) using a visual program. This fixture is intended to serve as a Poka Yoke tool and has been introduced into production as a standard aid in manual PCB fitting. After the fixture was placed on the production line, it was noticeable that the operator could not make the mistake, as the fixture does not "allow" the component to be fitted, and the operator had to position the component exactly at the marked position. In order to ensure the correct use of the fixture, operators of the line were retrained in a second step and a standard operating procedure (SOP) was created, i.e. a set of detailed instructions for performing the selected operation (Fig. 3).

Following the implementation of the new fixture, error records continued to be made, and the data uploaded after two weeks of monitoring showed that the cause of the "bad component" no longer occurred in the QUALITYDATABASE on the SMT04 line. Of course, it is necessary to continue to monitor whether this cause will occur in the future or to verify whether the proposed corrective action is sufficiently effective. In this case, it was also necessary to address other causes which were classified as significant errors.

<table>
<tr><th colspan="6">Standard Operating Procedure</th></tr>
<tr><td colspan="2">Date:</td><td colspan="4">Name:</td></tr>
<tr><td colspan="2">Department:</td><td>Shop:</td><td>Line:
Station:</td><td>Project:</td><td>Responsible person:</td></tr>
<tr><td>Nr.</td><td colspan="2">Visual photo:</td><td>How:</td><td>Description:</td><td>Time:</td></tr>
<tr><td>1.</td><td colspan="2"></td><td></td><td>Check that you have fixed the components. Check the part number (the component number) and its designation.</td><td>5</td></tr>
<tr><td>2.</td><td colspan="2"></td><td></td><td>Take the component and place it one by one in a designed place.</td><td>7</td></tr>
<tr><td>3.</td><td colspan="2"></td><td></td><td>Check the placement of all components.</td><td>6</td></tr>
<tr><td>4.</td><td colspan="2"></td><td></td><td>Push the component until it fits.</td><td>1</td></tr>
</table>

Fig. 3. Standard operating procedure

5 Conclusion

Production workers with repetitive routine actions, in an ergonomically difficult environment or with high accuracy requirements make many mistakes, and this work is exhausting and unattractive. In such an environment, their motivation often decreases and at the same time, human potential is wasted. In this case study, a measure aimed at minimizing human errors leading to significant losses was proposed. The introduction of Industry 4.0 elements and the promotion of robot-human collaboration as part of the Industry 5.0 approach aims not only at making production more effective but also at minimizing losses due to human factor failure.

Acknowledgement. This contribution is the result of the implementation of VEGA project No. 1/0121/18 "Development of methods of implementation and verification of complex security solution in Smart Factory as part of Industry Strategy 4.0".

References

1. Sinay, J., Pačaiová, H., Glatz, J.: Bezpečnosť a riziká technických systémov. Košice: TU, 246 s (2009). ISBN 978-80-553-0180-8
2. Sinay, J., Kamenický L.: Vzťah medzi kvalitou produkcie a posudzovaním chýb zamestnanca vo výrobnom procese/Juraj Sinay, Lukáš Kamenický - 2013. In: Novus Scientia 2013: 12. ročník medzinárodnej vedeckej konferencie doktorandov strojníckych fakúlt technických univerzít a vysokých škôl, Košice: TU, S. 1-4. (2013). ISBN 978-80-553-1381
3. Tsou, J.-C., Chen, W.-J.: The impact of preventive activities on the economics of production systems: modeling and application. Appl. Math. Model. **32**(6), 1056–1065 (2008). https://doi.org/10.1016/J.APM.2007.03.005
4. Lazarevic, M., Mandic, J., Sremcev, N., Vukelic, V.D., Debevec, M.A.: Systematic literature review of poka-yoke and novel approach to theoretical aspects. Strojniški vestnik J. Mech. Eng. 454–467 (2019). https://doi.org/10.5545/sv-jme.2019.6056
5. Mateides, A.: Manažérstvo kvality – história, koncepty, metódy (2006). ISBN 80-8057-656-4
6. Ready for Industry 5.0?: Furniturk Industry (2020). https://furniturkonline.com/2018/08/06/ready-for-industry-5-0/
7. Juhászová, D.: Application of SPC in short run and small mixed batch production: case of bakery equipment producer. Qual. Innov. Prosper. **22**(3), 55–67 (2018). https://doi.org/10.12776/QIP.V22I3.1174
8. Bravi, L., Murmura, F., Santos, G.: The ISO 9001:2015 quality management system standard: companies' drivers, benefits and barriers to its implementation. Qual. Innov. Prosper. **23**(2), 64–82 (2019). https://doi.org/10.12776/QIP.V23I2.1277
9. Zgodavova, K., Bober, P., Sutoova, A., Lengyelova, K.: Supporting sustainable entrepreneurship in injection molding of plastic parts by optimizing material consumption. Przem. Chem. **98**(3), 399–408 (2019). https://doi.org/10.15199/62.2019.3.9
10. Girmanova, L., et al.: Application of six sigma using DMAIC methodology in the process of product quality control in metallurgical operation. In: Acta Technologica Agriculturae. Roč. 20, č. 4 (2017), pp. 104–109 (2017). ISSN 1338-5267 https://doi.org/10.1515/ata-2017-0020
11. Zgodavova, K., Sutoova, A., Cicka, M.: Launching new projects in industry 4.0: best practices of automotive suppliers. In: Proceedings of the 4th International Conference on the Industry 4.0 Model for Advanced Manufacturing, (Švajčiarsko), pp. 183–191. Springer, Cham (2019). https://doi.org/10.1007/978-3-030-18180-2_14

The Relationship Between Company Survival, Site Risk and Accidents in Construction Industry

José M. Carretero-Gómez(✉), Francisco J. Forteza, and Bàrbara Estudillo Gil

University of the Balearic Islands, Palma, Spain
{josem.carretero, francisco.forteza}@uib.es, barbara.estudillo@uib.cat

Abstract. This paper examines the relationships between the accident rate of a company and the likelihood that a construction firm survives in the industry. We have done 409 assessments of risk levels on site from 344 Spanish firms from 2004 to 2010. With these assessments, accidents rates and survival/mortality data from SABI database we constructed a panel of data. Our general hypothesis is that accident rate makes less likely that a company survives in the sector. We estimate the relationships among these variables using probit regression methodology. Our results are relevant to highlight the imperative of define effective policies to control accident rates, and improve the construction sector health and its sustainable competitiveness.

Keywords: Occupational accidents · Firm survival · Construction industry

1 Introduction

Construction sector is characterized by a poor performance in health and safety (H&S), which is confirmed being one of the sectors with the highest accident rate [1–4]. This industry is also of great economic relevance as many studies have confirmed that construction accounts traditionally for approximately 3-8% of the value added in the gross domestic product (GDP) [5].

There is a broad consensus on the need for investing in H&S in organizations, and the evidence has shown that higher investment in prevention reduces accident rates [6]. However, what we do not know yet is to what extent the relationship between preventive investment and accident rate can affect the company performance and/or its survival. Our paper is aimed to shed more light in this last relevant issue.

Consequently, it seems relevant to ascertain to what extent H&S cots affect economic performance of companies, become urgent to know their business results and combine H&S and "productivity" arguments [7]. Going on with this line of research, a relevant research question is to know whether the survival likelihood of a construction firm can be predicted as a function of the accident rate that results from its managerial decisions in H&S. If the answer to this question is affirmative, with a confirmed significant relation between accident rate and likelihood of firm survival, we can

P. M. Arezes and R. L. Boring (Eds.): AHFE 2020, AISC 1204, pp. 104–110, 2020.
https://doi.org/10.1007/978-3-030-50946-0_15

propose that economic collapse can be partly prevented via reducing the quantity of occupational accidents. The benefits for economy and society seem clear.

The impact of occupational H&S in economic terms has been addressed in the literature, with a number of studies aimed to get its total cost estimation. [8–18]. Evidence of a negative association between accident rates in the one hand, and firm incomes and GDP, in the other hand, has been also found [19, 20]. Unfortunately, there are not many empirical tests of the relationship between accidents rate and firm profitability. One is Argilés-Bosch et al. [21] where was found a negative relationship of accident rate with economic performance.

2 Methodology

There is a general consensus in the research community that the consequences of poor development in health and safety systems, reflected in high accident rates, can cause high costs that can offset firm benefits and even expose the company at risk to the point of compromising its survival.

In the line of previous researches both theoretical [22, 23] and empirical [21, 24], our study proposes an empirical test of the relationship between accidents rates and the probability of bankruptcy of a company. Therefore, our hypothesis can be stated as follows:

H1: The higher accident rates of a construction firm, the more probable is the firm bankruptcy.

a. Empirical design

   Our objective is to model the likelihood of bankruptcy versus the survival of a company. Table 1 summarizes the set of variables we have considered in our analyses.

Table 1. Codes and definitions of variables.

Variable	Code	Description/values
Event of firm survival	estate	0 bankrupt 1 insolvent ("concurso de acreedores") 2 firm alive
	estate_0_2	0 bankrupt 1 firm alive
Firm accident rate	acc-rate	Total number of accidents divided by total number of workers
Return on assets	roa	Measure of firm profitability = firm net income divided by firm average total assets
Sells volume	sales	Indicator of firm activity = firm revenue obtained from selling goods (products/services)
Asset turnover	asset-turnover	Proxy of organizational efficiency = firm sales volume divided by firm total assets

Following we present the model specification to analyze hypothesis 1:

$$estate_{i,t} = \alpha + \beta_1 . acc - rate_{i,t} + \beta_2 . roa_{i,t} + \beta_3 . sales_{i,t} + \beta_4 . asset - turnover_{i,t} + \varepsilon_{i,t} \quad \text{(model1)}$$

$$estate\_0\_2_{i,t} = \alpha + \beta_1 . acc - rate_{i,t} + \beta_2 . roa_{i,t} + \beta_3 . sales_{i,t} + \beta_4 . asset - turnover_{i,t} + \varepsilon_i \quad \text{(model2)}$$

The dependent variable in model 2 is dichotomous, taking different values like is shown in Table 1.

b. Independent variables

These variables are included also in Table 1.

It is obvious that a number of factors different to accidents rates or general safety performance may impact the firm survival or bankruptcy. Because of that, we have included several control variables commonly considered in empirical economics literature, which can capture other managerial and financial issues [25, 26].

Sample and data collection

On one hand, we use official records from Labor Authority to obtain firm's accidents rates, our dependent variables. On the other hand, we use SABI database (Bureau van Dijk), to obtain firm financial data, the independent variables in our model.

3 Results

The descriptive statistics of the variables considered in this research are shown in Table 2.

Table 2. Summary statistics.

Variable	N	Mean	Std. Dev.	Min	Max
acc-rate	2242	0.1081802	0.2025597	0	2.67
roa	1734	−0.0128951	23.95159	−429.19	68.19
sales	1662	59756.3	344163.2	0.22	3564441
asset-turnover	1662	4.273408	116.2348	0.0004089	4739.75

As we can see in Table 3, the analysis of the Pearson correlation matrix shows that the estate binary dependent variables in model 1 is significantly correlated with most of economic variables as roa and sales ($p < .1$), and asset-turnover ($p < .01$). Regarding estate dependent variable, that measures the event of firm survival, the analysis of significant correlations suggests that firms with better economic performance and lower ratio of assets turnover is related with less bankruptcy events. The correlation between firm estate and the accident rate turned out to be not significant of low value and positive, all together against our hypothesized relationships.

Table 3. Model 1 and model 2 – hypothesis: the higher accident rates (acc-rate) of a construction firm, the more probable is the firm bankruptcy (estate and estate_0_2).

Variables	Model 1 Dependent variable = estate (pooled ordered probit regression)		Model 2 Dependent variable = estate_0_2 (pooled probit regression)	
	(1)	(2)	(3)	(4)
acc-rate		**−.2542***		**−.2644***
roa	.0070***	.0072***	.0078***	.0080***
sales	6.81e-07***	6.56e-07***	6.04e−07***	5.79e−07***
asset-turnover	−.0577**	−.0465*	−.0592**	−.0471*
/cut1	−.7289	−.7495		
/cut2	−.5985	−.5287		
cons			.6664***	.6869***
Goodness of fit	LR $Chi^2(3)$ = 45.41***	LR $Chi^2(4)$ = 48.55***	LR $Chi^2(3)$ = 39.59***	LR $Chi^2(4)$ = 42.41***
Pseudo R-squared	.0170	.0182	.0220	.0236
No. of observ.	1661	1661	1539	1539

***Significance level: $p < 0.1$; **Significance level: $p < 0.05$; *Significance level: $p < 0.01$
(1) (3) Baseline specification; (2) (4) Full model

Ordered probit regression is used for fitting the model 1, appropriate method to fit models with discrete dependent variables. For model 2 we run a probit regressions (pooled). STATA 13.1. has been used to fit our two models.

4 Discussion

The objective of this investigation is to improve the knowledge of the relationship between the accidents that the workers of a construction firm has suffered and the survival of their companies. In order to carry on this study, a hypothesis was suggested considering two different models. The difference between both models, is that in model 1 were included all the firms of the study, a total of 344 construction firms. Meanwhile, in the model 2 we excluded the companies that went into a process of insolvency during the period of our study, up to 25 firms of the total. Pooled ordered probit regression was used to fit the model 1 and pooled probit regression to fit the model 2. Results are reported in Table 3, where the column (1) shows the results of the baseline specification of the model 1 and the column (2) shows the results of the full model 1. The column (3) shows the baseline specification on the model 2 and the column (4) shows the results of the full model 2.

Looking at our results, more specifically at the likelihood-ratio tests, we can confirm that the goodness of fit of the both models is quite good. Observing the p-values < .01 shown in Table 3 for the economic performance variables, such us return on assets and net sales of the firm, we can affirm that there is a low but positive and significant impact of these variables on the state of the firm and its survival likelihood. It can be also seen in Table 3 that asset turnover (which indicates firm efficiency) has a negative and significant estimated effect on the firm survival likelihood. Although it seems not easy to explain this result, it is important to take into account that during part of the period of this study there was a great economic and financial crisis, and we are afraid that this result might be affected by the recession.

With the results reported in Table 3, and taking in account our initial idea of estimating the relationship between the survival of a construction firm and the rate of accidents, we can conclude that a high number of accidents has a negative effect on the ability of the company to survive. Thus, based on the evidence found, our hypothesis is confirmed. A construction firm that has an increase in the number of workplace accidents of its workers, will also see increased its probability of having survival problems as a company. We can see it when we regress in the analysis the variable "estate" on the independent variables. The effect of accident rate in self-preservation of the firm is negative and significant (*p-value* < .1), and it has similar behavior in the two different proposed models.

5 Conclusions

We address whether or not the accident rate of a construction company determines the firm survival likelihood. Since construction is one of the sectors with the highest accident rate, and potential losses in social and economic terms, it seems to be a suitable and relevant environment to test this hypothesis. We have pointed out above the lack of empirical papers in the literature where have been demonstrated the goodness of having safer organizations in order to reinforce a competitive advantage. It is obvious that an insolvent firm who is forced to shut down its operations, has lost the war for sustainable competitiveness.

We have found evidence that the accident rate of a construction firm has a significant effect on the likelihood of economic failure. If any construction firm experiences an increase in the number of accidents suffered, our results predict that is more probable this firm will have difficulties for continuing running the business, falling into an insolvency procedure, or going bankrupt. This is our main contribution.

As it cannot be other way, this paper present also some limitations which can inspire research agendas with more new questions related to occupational H&S. For instance: a) Develop in depth case studies to know the process through which firms move from an active to bankrupt state; b) Analyze if organizational design impact the probability of accidents occurrence; c) Study periods of year different to 2008–2016, where the great economic recession noise was huge.

Acknowledgments. This research has been partially funded by project ECO2017-86305-C4-1-R from the Spanish Ministry of Economy, Industry and Competitiveness, funds from the

Department of Business Economics at University of the Balearic Islands, and with the help of the Labor Authority of Balearic Islands Government.

References

1. Alexander, D., Hallowell, M., Gambatese, J.: Precursors of construction fatalities. II: predictive modeling and empirical validation. J. Constr. Eng. Manag. **143**(7), 04017024 (2017). https://doi.org/10.1061/(ASCE)CO.1943-7862.0001297
2. Bavafa, A., Mahdiyar, A., Marsono, A.K.: Identifying and assessing the critical factors for effective implementation of safety programs in construction projects. Saf. Sci. **106**, 47–56 (2018). https://doi.org/10.1016/j.ssci.2018.02.025
3. Jin, R., Zou, P.X.W., Piroozfar, P., Wood, H., Yang, Y., Yan, L., Han, Y.: A science mapping approach based review of construction safety research. Saf. Sci. **113**, 285–297 (2019). https://doi.org/10.1016/j.ssci.2018.12.006
4. Zhou, Z., Goh, Y.M., Li, Q.: Overview and analysis of safety management studies in the construction industry. Saf. Sci. **72**, 337–350 (2015). https://doi.org/10.1016/j.ssci.2014.10.006
5. Giang, D.T., Pheng, L.S.: Role of construction in economic development: review of key concepts in the past 40 years. Habitat Int. **35**(1), 118–125 (2011)
6. López-Alonso, M., Ibarrondo-Dávila, M.P., Rubio-Gámez, M.C., Munoz, T.G.: The impact of health and safety investment on construction company costs. Saf. Sci. **60**, 151–159 (2013). https://doi.org/10.1016/j.ssci.2013.06.013
7. Biddle, E., Ray, T., Owusu-Edusei, K., Camm, T.: Synthesis and recommendations of the economic evaluation of OHS interventions at the company level conference. J. Saf. Res. **36** (3), 261–267 (2005). https://doi.org/10.1016/j.jsr.2005.06.008
8. Breslin, C., Tompa, E., Zhao, R., Amick III, B.A., Pole, J.D., Smith, P., Hogg-Johnson, S.: Work disability absence among young workers with respect to earnings losses in the following year. Scand. J. Work Environ. Health **33**(3), 192–197 (2007)
9. Cheng, C.-W., Leu, S.-S., Lin, C.-C., Fan, C.: Characteristic analysis of occupational accidents at small construction enterprises. Saf. Sci. **48**(6), 698–707 (2010)
10. Crichton, S., Stillman, S., Hyslop, D.: Returning to work from injury: longitudinal evidence on employment and earnings. Ind. Labor Relat. Rev. **64**(4), 765–785 (2011)
11. Feng, Y., Zhang, S., Wu, P.: Factors influencing workplace accident costs of building projects. Saf. Sci. **72**, 97–104 (2015)
12. Gurcanli, G.E., Bilir, S., Sevim, M.: Activity based risk assessment and safety cost estimation for residential building construction projects. Saf. Sci. **80**, 1–12 (2015)
13. HSE. Costs to Britain of workplace fatalities and self-reported injuries and ill health, 2013/14 (2015). Accessed Sept. 2017. http://www.hse.gov.uk/statistics/pdf/cost-tobritain.pdf
14. Ibarrondo-Dávila, M.P., López-Alonso, M., Rubio-Gámez, M.C.: Managerial accounting for safety management. The case of a Spanish construction company. Saf. Sci. **79**, 116–125 (2015). https://doi.org/10.1016/j.ssci.2015.05.014
15. Labelle, J.E.: What do accidents truly cost? - ProQuest. Profess. Saf. **45**(4), 38–42 (2000)
16. Lim, S.J., Chung, W.J., Cho, W.H.: Economic burden of injuries in South Korea. Inj. Prev. **17**(5), 291–296 (2011)
17. Reville, R.T., Schoeni, R.F.: Disability from injuries at work: the effects on Earnings and Employment. RAND. Labor and Population Program. Working Paper 01–08 (2001). http://www.rand.org/content/dam/rand/pubs/drafts/2005/DRU2554.pdf. Accessed Sept 2017

18. Woock, C.: Earnings losses of injured men: reported and unreported injuries. Ind. Rel.: J. Econ. Soc. **48**(4), 610–628 (2009)
19. Barth, A., Winker, R., Ponocny-Seliger, E., Sögner, L.: Economic growth and the incidence of occupational injuries in Austria. Wien. Klin. Wochenschr. **119**(5–6), 158–163 (2007)
20. Mainardi, S.: Earnings and work accident risk: a panel data analysis on mining. Resour. Policy **30**, 156–167 (2005)
21. Argilés-Bosch, J.M., Martí, J., Monllau, T., Garcia-Blandón, J., Urgell, T.: Empirical analysis of the incidence of accidents in the workplace on firms' financial performance. Saf. Sci. **70**, 123–132 (2011)
22. Rechenthin, D.: Project safety as a sustainable competitive advantage. J. Saf. Res. **35**(3), 297–308 (2004)
23. Teo, E., Ling, F.: Developing a model to measure the effectiveness of safety management systems of construction sites. Build. Environ. **41**(11), 1584–1592 (2006)
24. Forteza, F.J., Carretero-Gómez, J.M., Sese, A.: Occupational risks, accidents on sites and economic performance of construction firms. Saf. Sci. **94**, 61–76 (2017). https://doi.org/10.1016/j.ssci.2017.01.003
25. Fairfield, P.M., Yohn, T.L.: Using asset turnover and profit margin to forecast changes in profitability. Rev. Acc. Stud. **6**(4), 371–385 (2001)
26. Tan, J., Wang, L.: Flexibility–efficiency tradeoff and performance implications among Chinese SOEs. J. Bus. Res. **63**(4), 356–362 (2010)

Influence of Improper Workload on Safety Consciousness and Safety Citizenship Behavior of Construction Workers

Xiangcheng Meng(✉) and Alan H. S. Chan

City University of Hong Kong, Tat Chee Avenue, Kowloon, Hong Kong SAR
xcmeng3-c@my.cityu.edu.hk, alan.chan@cityu.edu.hk

Abstract. The safety status of construction workers has been considered a promising research domain for the safety improvement of the construction industry. In this study, safety consciousness (SC) and safety citizenship behavior (SCB) of construction workers were tested in terms of the difference among various workloads, which was measured by "weekly working hours" in this study. Over 161 respondents were involved in a questionnaire survey. To explore the relationships of different workloads with SC and SCB statistically, analysis of variance (ANOVA) was conducted by using SPSS 24. For the results, a significant correlation was observed among different workloads in terms of SC and SCB. The improper workloads will lead to poor SC and SCB of construction worker. In addition, further recommendations were provided to optimize the reasonable working hour of construction workers.

Keywords: Organizational citizenship · Safety consciousness · Human safety · Risk cognition

1 Introduction

With the continuous development of social economy, the construction industry has become one of the most important manufacturing industries in the world. However, on account of the high intensity of physical labor and poor working conditions, safety accidents still frequently occur, causing serious damage to the lives of people and properties [1]. Take on-site construction workers in China as an example. An unreasonable workload will force them to be exposed directly to the strong sunshine and industrial dust for extended periods, thus negatively impacting the working efficiency and further increasing the potential risks of personal injury and danger [2].

Also, the unreasonable workload will cause work-related musculoskeletal disorders, which may create significant financial costs for medical treatment and lead to losses in productivity and higher workforce turnover [3, 4]. Skogstad et al. found that unreasonable working hours had a strong effect in leading to increased occupational injuries and illnesses [5]. Alameddine et al. found that unsuitable work hours can negatively affect work productivity, job satisfaction, worker health, and cause inconsistent job performance [6]. In addition, studies have shown that unaffordable workload is a risk factor for occupational injuries and occupational psychosocial hazards [7]. To

P. M. Arezes and R. L. Boring (Eds.): AHFE 2020, AISC 1204, pp. 111–116, 2020.
https://doi.org/10.1007/978-3-030-50946-0_16

measure this key factor affecting occupational safety, the working time has been considered as one of effective indicators [8, 9]. Therefore, in this study, the weekly working hour was used to reflect the working load of construction workers.

Furthermore, two research hotspots aimed at enhancing the working safety of construction workers have recently been discussed by several scholars as direct roles in improving the safety of construction personnel, namely, safety consciousness (SC) and safety citizenship behavior (SCB). In detail, SC is considered the perception and understanding of the safety of working circumstances [10]. Moreover, SCB is identified as a voluntary behavior focusing on improving group-level safety performance and mutual support [11, 12]. However, up to now, no quantitative research has been conducted to examine the performing differences of SC and SCB from different workloads. This research gap will lead to difficulty targeting the safety interventions of particular groups of construction workers with different working loads. Therefore, this study has two objectives:

(a) to verify whether a significant relationship exists between workload and SC.
(b) to determine whether the level of SCB of workers is affected by the workload of construction workers.

2 Methodology

2.1 Questionnaire Survey

To achieve the objectives of this research, an on-site questionnaire survey was applied to collect both the workload information and the SC and SCB of construction workers in China. A total of 161 construction workers were recruited as the respondents to participate in this study. All the respondents had normal visual acuities and healthy physical conditions. They did not have any history of neurological and mental diseases and any reading disability. In addition, the numbers of respondents with different personal characteristics such as genders and workloads were relatively equal to ensure the reliability of the research outcomes [13]. The questionnaire used in this study was adopted from Meng et al. [14]. All items were presented together with five-point Likert scale ranging from one to five to show the agreements of the respondents

2.2 Data Analysis

Preliminary analyses included computing frequencies and descriptive statistics for all demographical features, which were first conducted to examine the performing differences of SC and SCB in line with different weekly working hours. For the data analysis, the average scores of SC and SCB were calculated based on the classification of different workloads. Descriptive information was provided in the forms of tables and graphs, and further comparisons among groups with different workloads were made. The significance of any observed difference was further tested by ANOVA [15, 16].

3 Results

3.1 Descriptive Analysis

As depicted in Table 1, the results of descriptive statistics were provided by comparing the SC and SCB of construction workers with different genders. In this study, a total of 161 valid responses were collected from 86 male construction workers and 75 female construction workers. Based on the distribution of the data, the weekly workload of female workers was mostly less than 40 h, whereas most of their male counterparts had more than 55 working hours per week. Female workers had lower scores (3.82 for SC and 3.80 for SCB) on both SC and SCB compared with their male counterparts (3.92 for SC and 4.19 for SCB).

Table 1. Data distributions of SC and SCB in line with different weekly working hours: gender specific

Characteristic		Male (N = 86)			Female (N = 75)		
		N (%)	Average SC	Average SCB	N (%)	Average SC	Average SCB
Weekly working hours	<40	8 (9.30%)	4.39	4.08	21 (28.00%)	4.01	3.67
	40–45	10 (11.63%)	4.13	4.19	17 (22.67%)	3.86	3.86
	46–50	10 (11.63%)	3.94	4.22	15 (20.00%)	3.75	3.89
	51–55	37 (43.02%)	3.84	4.26	11 (14.67%)	3.70	3.93
	>55	21 (24.42%)	3.75	4.10	11 (14.67%)	3.63	3.71
Overall average			3.92	4.19		3.82	3.80

The SC of both male and female construction workers was observed to decrease as weekly workloads increased. However, the distinction of SCB was relatively sophisticated compared with that of SC. Both female and male workers were observed to have poorer SCB at the lower level of workload. As the weekly working hours increased, the SCB initially showed a rising trend and then declined as the working hours per week increased to more than 55 h. Further quantitative analysis would be conducted to clarify if the significant differences of SC and SCB existed among different subgroups of workloads.

3.2 Anova

As depicted in Table 2 the significance of distinctions of SC and SCB were verified among various weekly working hours given that all the significant levels were less than

0.05, which quantitatively verified the significant impacts of workload on SC and SCB for construction workers. The distinction was more pronounced for male construction workers due to the higher significance (P = 0). However, the significant levels of distinction for the SC and SCB of female workers were also acceptable because the levels were all less than 0.05. Furthermore, the change tendencies of SCB and SC along with the various workloads were also confirmed. For both male and female workers, the transition of SC was observed as a significant downward trend, whereas the SCB was verified to have a up and down tendency.

Table 2. ANOVA of SC and SCB for construction workers with different workloads

Characteristic				Quadratic sum	DOF	Mean square	F	Sig
Weekly working hours	Male	SC	Interclass	12.79	4	3.20	7.13	.000
			Intraclass	36.34	81	0.45		
			Total	49.13	85			
		SCB	Interclass	7.30	4	1.83	6.25	.000
			Intraclass	23.67	81	0.29		
			Total	30.97	85			
	Female	SC	Interclass	14.88	4	3.72	7.71	.000
			Intraclass	34.25	70	0.48		
			Total	49.13	74			
		SCB	Interclass	7.62	4	1.90	5.79	.003
			Intraclass	23.35	70	0.33		
			Total	30.97	74			

4 Discussion and Conclusion

For the proposed achievement, this study discovered the influence of different workloads (indicated by the weekly working hours of workers) on the SC and SCB of construction workers. The descriptive analysis showed that various degrees of workloads can cause the distinctions on the SC and SCB of both male and female construction workers. the SC was observed to show a decreasing transition tendency as the workload continued increasing, whereas the SCB first increased and then declined after the workload increased to more than 50 h per week. In this study, the different variation trends of SC and SCB were considered important and worth examining for further related research and industrial practice. In terms of the decreasing trend of SC, the increasing workload tends to negatively influence the SC of workers. The prolonged working time will make them exhausted and have difficulty concentrating; therefore, their attitude toward working safety will decline [17, 18].

However, the SCB first improved as the workload increased, which was mainly due to the increasing familiarity of the group members [19]. Given that SCB is an organization-based behavior, the degree of organizational familiarity will positively influence the SCB [20]. Nevertheless, conflicts will arise among group members after working together for a long time, in which the group-based safety level will be reduced [21]. To solve this problem, this study suggests rescheduling the workload of construction workers. The government and the management need to reschedule the working hours properly to optimize the SCB.

For the practical implications, SC needs to be reinforced regularly. Safety education and training should be regularly conducted to promote risk perception and recognition. A popular way of conducting safety training is by recalling and analyzing the actual accident scenes to provide direct experience for risk identification [22, 23]. For the enhancement of SCB, adjusting the internal structure of the working organization regularly is suggested to avoid the storming stage of the working group, which may cause conflicts and contradictions among group members. In particular, the construction group should maintain post adjustment and personnel change regularly to renew the structure of the group composition to optimize the organizational-level performance of working safety of construction workers [24].

References

1. Pinto, A., Nunes, I.L., Ribeiro, R.A.: Occupational risk assessment in construction industry - overview and reflection. Saf. Sci. **49**, 616–624 (2011)
2. Zou, P.X.W., Zhang, G., Wang, J.: Understanding the key risks in construction projects in China. Int. J. Proj. Manag. **25**, 601–614 (2007)
3. Goggins, R.W., Spielholz, P., Nothstein, G.L.: Estimating the effectiveness of ergonomics interventions through case studies: implications for predictive cost-benefit analysis. J. Saf. Res. **39**, 334–339 (2008)
4. Leigh, J.P.: Economic burden of occupational injury and illness in the United States. Milbank Q. **89**, 728e772 (2011)
5. Skogstad, M., Mamen, A., Lunde, L.K., Ulvestad, B., Matre, D., Aass, H.C.D., Øvstebø, R., Nielsen, P., Samuelsen, K.N., Skare, Ø., et al.: Shift work including night work and long working hours in industrial plants increases the risk of atherosclerosis. Int. J. Environ. Res. Public Health **16**, 521 (2019)
6. Alameddine, M., Otterbach, S., Rafii, B., Sousa-Poza, A.: Work hour constraints in the German nursing workforce: a quarter of a century in review. Health Policy (New York) **122**, 1101–1108 (2018)
7. Swaen, G.M., van Amelsvoort, L.P., Bultmann, U., Slangen, J.J., Kant, I.J.: Psychosocial work characteristics as risk factors for being injured in an occupational accident. J. Occup. Environ. Med. **46**, 521–527 (2004)
8. Åkerstedt, T., Knutsson, A., Westerholm, P., Theorell, T., Alfredsson, L., Kecklund, G.: Sleep disturbances, work stress and work hours a cross-sectional study. J. Psychosom. Res. **53**, 741–748 (2002)
9. Caplan, R.D., Cobb, S., French, J.R.P.: White collar work load and cortisol: disruption of a circadian rhythm by job stress? J. Psychosom. Res. **23**, 181–192 (1979)
10. Conrad, P., Bradshaw, Y.S., Lamsudin, R., Kasniyah, N., Costello, C.: Helmets, injuries and cultural definitions: motorcycle injury in urban Indonesia. Accid. Anal. Prev. **28**, 193–200 (1996)
11. Hofmann, D.A., Morgeson, F.P.: Safety-related behaviour as a social exchange: the role of perceived organizational support and leader-member exchange. J. Appl. Psychol. **84**, 286–296 (1999)

12. Didla, S., Mearns, K., Flin, R.: Safety citizenship behaviour: a proactive approach to risk management. J. Risk Res. **12**, 475–483 (2009)
13. Swiger, L.A., Harvey, W.R., Everson, D.O., Gregory, K.E.: The variance of intraclass correlation involving groups with one observation. Biometrics **20**, 818–826 (1964)
14. Meng, X., Zhai, H., Chan, A.H.S.: Development of scales to measure and analyse the relationship of safety consciousness and safety citizenship behaviour of construction workers: an empirical study in China. Int. J. Environ. Res. Public Health **16**, 1411 (2019)
15. Shaw, R.: Anova for unbalanced data: an overview. Ecology **76**, 2000 (1995)
16. Levin, J.R., Donahue, B., Keselman, J.C., Keselman, H.J., Petoskey, M.D., Olejnik, S., Lix, L.M., Lowman, L.L., Huberty, C.J., Cribbie, R.A., Kowalchuk, R.K.: Statistical Practices of Educational Researchers: An Analysis of their ANOVA, MANOVA, and ANCOVA Analyses. Rev. Educ. Res. **68**, 350–386 (1998)
17. Rundmo, T., Hestad, H., Ulleberg, P.: Organisational factors, safety attitudes and workload among offshore oil personnel. Saf. Sci. **29**, 75–87 (1998)
18. Kruglanski, A.W., Thompson, E.P., Higgins, E.T., Atash, M.N., Pierro, A., Shah, J.Y., Spiegel, S.: To "do the right thing" or to "just do it": locomotion and assessment as distinct self-regulatory imperatives. J. Pers. Soc. Psychol. **79**, 793–815 (2000)
19. Leyden, D.P., Goodman, P.S.: Familiarity and group productivity. J. Appl. Psychol. **76**, 578–586 (1991)
20. Koulaei, A., Bharadwaj, S.G., van Oest, R.: Team familiarity mitigates the effect of team-based reward on functional conflict and citizenship behavior (2019). SSRN 3397513
21. Tuckman, B.W.: Developmental sequence in small groups. Psychol. Bull. **63**, 384–399 (1965)
22. Eiris, R., Gheisari, M., Esmaeili, B.: PARS: using augmented 360-degree panoramas of reality for construction safety training. Int. J. Environ. Res. Public Health **15**, 2452 (2018)
23. Wang, P., Wu, P., Wang, J., Chi, H.L., Wang, X.: A critical review of the use of virtual reality in construction engineering education and training. Int. J. Environ. Res. Public Health. **15**, 1204 (2018)
24. Monk, T.H., Folkarc, S., Wedderburn, A.I.: Maintaining safety and high performance on shiftwork. Appl. Ergon. **27**, 17–23 (1996)

A Mediation Analysis on the Relationship between Safety Climate and Work Abilities of Hong Kong Construction Workers

Jacky Yu Ki Ng(✉)

Department of Systems Engineering and Engineering Management, City University of Hong Kong, Kowloon Tong, Hong Kong, China
jacky.ngyk@my.cityu.edu.hk

Abstract. Construction is one of the most dangerous industries in the world. The incident rate in the construction industry is comparatively high in Hong Kong. To avoid occupational injuries and accidents involving construction workers, safety climate and work abilities should be addressed and promoted. In this way, work abilities and shared perceptions of workers in the Hong Kong construction industry can focus on the importance of safety conduct in their occupational behavior.

Advancement in technologies in the construction industry have improved work abilities and productivity, reduced costs, and enhanced product quality and work environment safety.

However, only few studies have examined the technology adoption in construction firms. Specifically, the case of Hong Kong construction workers' adoption of latest technologies in their daily work to promote safety culture and work abilities and avoid further incidents in their immediate work environment.

This study aims to assess the technology adoption in the Hong Kong construction industry and identify the effects and relationships between safety climate and work abilities of construction workers.

The findings of this study can contribute to widen the literature and provide evidence of the relationship among technology adoption, safety climate, and work abilities of construction workers in Hong Kong.

Recommendations are also proposed to further improve the safety culture and work abilities of construction workers.

Keywords: Construction safety · Safety performance · Work Ability · Construction worker

1 Introduction

Safety climate is defined as the shared perceptions of employees on the relative importance of safety conduct in their occupational behavior [1]. Safety climate is used to predict safety performance (SP) and applied to different industries, such as manufacturing and construction. A positive safety culture within an organization enhances the safety knowledge of employees by increasing the frequency of their safety behavior

P. M. Arezes and R. L. Boring (Eds.): AHFE 2020, AISC 1204, pp. 117–120, 2020.
https://doi.org/10.1007/978-3-030-50946-0_17

at their immediate work environment. Existing studies have shown that low safety climate scores indicate poor SP and high accident rates within an organization [2–5].

Technological advancement has increased the use of innovative machineries in the construction industry and improved productivity, reduced construction costs, and enhanced work quality and work environment safety [6]. Technology acceptance model has been widely used for explaining and predicting usage intentions and adoption behavior for decades [7]. However, the relation between construction technology adoption and safety improvement has not been explored to date in construction management, despite the frequent interaction of construction workers with a wide variety of construction technologies in construction sites.

When people face new technology, their decision on whether to use such technology will depend on several factors, such as perceived convenience and usefulness. Perceived convenience refers to the extent to which individuals believe that using a particular technology would free them from additional effort of their routine tasks for finishing a job. Perceived usefulness refers to the extent to which individuals believe that using a particular technology would enhance their job performance [8].

To quantify work ability and provide a work capacity indicator for current workers, a set of questions was developed to form the Work Ability Index (WAI) [17]. The WAI has been widely used for assessing work ability in workplace health promotion and occupational health reintegration. Work ability can be improved continuously if the right measures are taken, regardless the age of workers [18, 19]. Determining work ability can facilitate better management of working lives. Thus, quality of work and life and productivity can be improved on the basis of the interplay between personal factors and work demands. Accordingly, this proposed study aims to use work ability measurement to quantify work ability and identify possible factors affecting safety climate at work environment; wherein workers can be further efficient and productive, which in turn, extend their working lives and promote high-quality retirement [20].

In this proposed study, construction technology adoption and work abilities will be assessed to identify their possible effects on SP of construction workers in Hong Kong [21]. In particular, this study aims to explore the relationship among work ability, safety climate, and SP of Hong Kong construction workers with technology adoption as a mediator.

2 Method

Accordingly, the following hypotheses are formulated:

H1. Safety climate index (SCI) has a positive effect on perceived usefulness.
H2. SCI has a positive effect on perceived convenience.
H3. SCI has a positive effect on SP.
H4. SCI has a positive effect on work ability.
H5. Work ability has a positive effect on SP.

3 Measurement and Discussion

Measurement: SCI questionnaire survey developed by the Occupational Safety and Health Council of Hong Kong [9] will be used. The measures of perceived convenience and usefulness will be adopted from a previous study [10]. SP will be measured by using the scale developed by Siu et al. [11]. WAI will also be measured [17].

Discussion: From the theoretical implications perspective, our study will contribute to the safety climate literature by investigating the relationships among work ability, safety climate, and perceived convenience and usefulness of construction technologies.

Evidence supporting the influence of safety climate on SP mediated by perceived usefulness is found. Hence, perceived usefulness of construction technologies is a mediator that explains the underlying mechanism of the indirect relationship among work abilities, safety climate of an organization, and the SP of construction workers.

4 Conclusion

First, the data were collected from self-reported questionnaires survey, thereby resulting in common method biases.

Second, the sample size may be insufficient for representing the entire construction industry in Hong Kong.

These limitations may be addressed by collecting longitudinal data for 6 months or 12 months for follow-up studies and recruiting a large number of workers to join the interview for wider representation of the construction job category. Further research is expected to help generalize our findings and discussion.

References

1. Zohar, D.: Safety climate in industrial organizations: theoretical and applied implications. J. Appl. Psychol. **65**(1), 96–102 (1980)
2. Clarke, S.: The relationship between safety climate and safety performance: a meta-analytic review. J. Occup. Health Psychol. **11**(4), 315–327 (2016)
3. Teo, E.A.L., Feng, Y.: The role of safety climate in predicting safety culture on construction sites. Architect. Sci. Rev. **52**(1), 5–16 (2009)
4. Vinodkumar, M.N., Bhasi, M.: Safety management practices and safety behaviour: assessing the mediating role of safety knowledge and motivation. Accid. Anal. Prev. **42**(6), 2082–2093 (2010)
5. Hon, C.K.H., Chan, A.P.C., Yam, M.C.H.: Relationships between safety climate and safety performance of building repair, maintenance, minor alteration, and addition (RMAA) works. Saf. Sci. **65**, 10–19 (2014)
6. Fung, K., Rowlinson, S.: The adoption of advanced technologies in the Hong Kong construction industry. Survey. Built Environ. **17**, 15–26 (2006)
7. Lucas, H.C., Spitler, V.K.: Technology use and performance: a field study of broker workstations. Decis. Sci. **30**(2), 291–311 (1999)

8. Davis, F.D., Bagozzi, R.P., Warshaw, P.R.: User acceptance of computer technology: a comparison of two theoretical models. Manag. Sci. **35**(8), 982–1003 (1989)
9. Occupational Safety & Heal Council: Construction Industry Safety Climate Index Software. http://www.oshc.org.hk/oshc_data/files/trgkit/2016/SCI_book_eng.pdf (2008)
10. Wu, L., Chen, J.L.: An extension of trust and TAM model with TPB in the initial adoption of on-line tax: an empirical study. Int. J. Hum. Comput. Stud. **62**(6), 784–808 (2005)
11. Siu, O., Phillips, D.R., Leung, T.: Safety climate and safety performance among construction workers in Hong Kong: the role of psychological strains as mediators. Accid. Anal. Prev. **36** (3), 359–366 (2004)
12. Hair, J.F., Black, W.C., Babin, B.J., Anderson, R.E., Tatham, R.L.: Multivariate Data Analysis, vol. 5, no. 3. Prentice Hall, Upper Saddle River (1998)
13. Churchill Jr., G.A.: A paradigm for developing better measures of marketing constructs. J. Mark. Res. **16**, 64–73 (1979)
14. Fornell, C., Larcker, D.F.: Evaluating structural equation models with unobservable variables and measurement error. J. Mark. Res. **18**, 39–50 (1981)
15. Nunnally, J.C.: Psychometric Theory. McGraw Hill, New York (1967)
16. Kline, R.B.: Principles and Practice of Structural Equation Modeling, 4th edn. Guilford Press, New York (2015)
17. Tuomi, K., Ilmarinen, J., Martikainen, R., Aalto, L., Klockars, M.: Aging, work, life-style and work ability among Finnish municipal workers in 1981–1992. Scand. J. Work Environ. Health **23**(1), 58–65 (1997)
18. Illmarinen, J.: Towards a longer worklife! Ageing and the quality of worklife in the European Union Helsinki: Finnish Institute of Occupational Health (2005)
19. Ilmarinen, J.: 30 years of work ability and 20 years of age management in Finland. In: 4th Symposium on Work Ability, Tampere, Finland, June (2010)
20. Maltby, T.: Extending working lives? Employability, work ability and better quality working lives. Soc. Policy Soc. **10**(3), 299–308 (2011)
21. Man S.S., Ng, J.Y.K., Chan A.H.S.: A review of the risk perception of construction workers in construction safety. In: Ahram, T., Karwowski, W., Pickl, S., Taiar, R. (eds.) Human Systems Engineering and Design II. IHSED 2019. Advances in Intelligent Systems and Computing, vol 1026. Springer, Cham (2020)

Application of the PRECEDE-PROCEED Model to Design a Program for Prevention of Low Back Pain in a Thai Community Hospital

Chuliporn Sopajareeya[1(✉)], Chompunut Sopajaree[2], and OiSaeng Hong[3]

[1] The Department of Community Health Nursing, Faculty of Nursing, Thammasat University, Pathumthani, Thailand
csopajareeya@gmail.com, juraipor@tu.ac.th
[2] School of Nursing, Mae Fah Luang University, Chiang Rai, Thailand
chompunut.sop@mfu.ac.th
[3] School of Nursing, University of California San Francisco (UCSF), San Francisco, CA, USA
oisaeng.hong@ucsf.edu

Abstract. Although low back pain (LBP) is one of the major health problems among nursing personnel in Thai hospitals, very few comprehensive interventions have been designed and there is minimal application of a planning model to design a theory-based program. The aims of this research were to design, implement, and evaluate a program for the prevention of LBP among nursing personnel in a community hospital in Thailand. The samples were selected using purposive sampling and they were divided into a program developer group and an attendant group. Quantitative and qualitative data were collected using research instruments. The program was designed, implemented, and evaluated based upon the PRECEDE-PROCEED model and consisted of three activities: 1) regular education sessions on safe patient lifting techniques and good working postures; 2) training on lifting overweight patients by using lifting devices and a lifting team; and 3) training with back muscle exercises. The results revealed that the participants' knowledge, attitude, and health behavior scores increased at three months after the intervention ($p < .05$). This model can be applied as a conceptual framework and demonstrates the possibility of the use of the model to plan interventions for community hospital nursing personnel.

Keywords: PRECEDE-PROCEED model · Low back pain · Program for prevention · Community hospital · Nursing personnel

1 Introduction

Nursing personnel are known to be a high-risk group for low back pain (LBP) [1]. The prevalence of LBP during the previous 12 months and at a point of time among hospital nurses in many countries was high [2], [3], presenting 36.2% [2] and 76.5% [3], and in Thailand the prevalence during the last 12 months was 61.5% [4]. With the high

P. M. Arezes and R. L. Boring (Eds.): AHFE 2020, AISC 1204, pp. 121–129, 2020.
https://doi.org/10.1007/978-3-030-50946-0_18

prevalence of LBP, nursing personnel suffer from acute and chronic LBP during direct and indirect nursing care activities, such as manual lifting, transferring, and repositioning patients, bending and twisting their postures, and standing or walking for a long time. These activities lead to the occurrence of LBP affecting their health and ability to work.

There are several causes of LBP among nursing personnel, including inappropriate knowledge of back care [5], moving patients without assistance, lack of back muscle exercise [4], frequent standing on duty, and awkward working postures [3]. Therefore, resolving the LBP problem should include several intervention strategies. In many countries, intervention methods have been proposed to resolve this problem among nursing personnel, including providing education, training, and a prevention program. Very few comprehensive interventions however have applied a planning model to design a theory-based program in these countries. Planning, designing, and evaluating interventions are important processes to complete a program for the prevention of LBP.

In this study, the researchers used the PRECEDE-PROCEED model (PPM) [6] as a framework to design, implement, and evaluate a program for the prevention LBP among nursing personnel. The PPM, a nine-step model, was originally developed by Green and Kreuter [6], and it is composed of the PRECEDE (the first five phases) and the PROCEED (the second four phases) frameworks. The PRECEDE framework can be applied for planning and designing a program for LBP prevention by assessing the LBP problem and the impacts of LBP, assessing epidemiological data, behaviors and environments, the potential factors affecting health behaviors, and administration and policies. The PROCEED is a framework for implementing and evaluating the program and consists of implementation, process evaluation, impact evaluation, and outcome evaluation.

This model has been used widely in the field of medicine, public health, education, and community health education [7]. There has been minimal application of a planning model however to design a theory-based program in Thai hospitals. Therefore, the main objectives of this study were to design, implement, and evaluate a program for the prevention of LBP among nursing personnel at a community hospital in Thailand.

2 Materials and Methods

The PPM was used as the conceptual framework to design, implement, and evaluate a program for the prevention of LBP among nursing personnel at Takhli Hospital, a community hospital in Thailand. The samples were selected using purposive sampling and they were divided into two groups. The first group, the program developers, consisted of a director of the hospital, the director of nurses, three heads of nursing units, a physiotherapist, an occupational health nurse, and 10 nursing personnel altogether comprising 17 people. The second group was composed of program attendants, including 81 nursing personnel.

The research instruments consisted of 1) a self-reported questionnaire on LBP problems, knowledge, attitudes, and health behaviors; 2) a program for the prevention of LBP among nursing personnel; 3) an in-depth interview guideline; 4) a focus group guideline; and 5) record forms. The questionnaire was divided into four parts:

demographic characteristics (7 items), questions on LBP prevalence (4 items), knowledge (10 items), attitudes (10 items), and health behaviors for LBP prevention (15 items). In the section concerning knowledge (four multiple choices), the reliability test of Kuder-Richardson (KR-20) was 0.74. The questionnaire on attitudes about LBP and prevention was on a five-point rating scale, including strongly agree (5), agree (4), uncertain (3), disagree (2), and strongly disagree (1). A four-point Likert scale was used to indicate the frequency of performing health behaviors during working tasks from never (1), to occasionally (2), to frequently (3), and to very frequently (4). The reliability of the questionnaire on attitudes and health behaviors which was tested using Cronbach's alpha, was 0.81 and 0.83, respectively. The interview guideline and the focus group guideline were confirmed for quality by examining content validity by three experts in occupational health nursing, qualitative research, and in program planning and evaluation.

Descriptive statistics were used to describe the characteristics of the study participants and the study variables. Paired t-test was applied to compare the knowledge, attitude, and health behavior scores before and at three months after the intervention. The contents of the in-depth interviews and the focus group discussion were transcribed and analyzed. By summarizing the content analysis [8], all of the recordings were explored and categorized as important themes. The level of statistical significance was set at 0.05.

Before conducting this study, the protocol was approved by the Institutional Review Board (IRB) at Mae Fah Luang University (Number REH-5955) in Thailand. All of the invited participants were provided consent forms and the study information sheets.

2.1 Designing the Program for the Prevention of LBP Based on the PRECEDE Framework

The design of this study was action research. The PRECEDE framework of the PPM was applied to design the program for the prevention of LBP among nursing personnel. In the first five phases in Table 1, we assessed the perspective of the program developers on the impact of LBP and the needs for LBP prevention, identified the prevalence of LBP among the nursing personnel, assessed the risk factors of LBP, the predisposing, reinforcing, and enabling factors affecting the health behaviors of the nursing personnel, and the organizational policies, resources, and management that could facilitate the implementation of the prevention of LBP in the hospital. We identified the prevalence of LBP among nursing personnel by distributing the questionnaire on LBP prevalence, knowledge, attitudes, and health behaviors for LBP prevention to 81 nursing personnel. In order to assess the perspective of the program developers on the impact of LBP and the needs for LBP prevention, the risk factors of LBP, the predisposing, reinforcing, and enabling factors affecting the health behaviors of the nursing personnel, and the organizational policies, resources, and management that could facilitate the implementation of the prevention of LBP in the hospital, in-depth interviews and a focus group discussion were used as the methods for collecting the data.

The in-depth interviews were conducted at different workplaces for seven program developers, consisting of the director of the hospital, the director of nurses, three heads of nursing units, a physiotherapist, and an occupational health nurse. The questions in

Table 1. Application of the PRECEDE framework to design the program for the prevention of LBP among nursing personnel

Phases	Application	Study participants	Methodology
Phase 1: social assessment	- Assessed the impact of LBP and needs for LBP prevention	- 17 program developers	- In-depth interviews - A focus group discussion
Phase 2: epidemiological assessment	- Identified the prevalence of LBP among nursing personnel	- 81 nursing personnel	- Distribution of a questionnaire on LBP prevalence, knowledge, attitudes about LBP and prevention, and health behaviors for LBP prevention
Phase 3: behavioral and environmental assessment	- Assessed risk factors of LBP among nursing personnel	- 17 program developers	- In-depth interviews - A focus group discussion
Phase 4: educational and ecological assessment	- Assessed predisposing, reinforcing, and enabling factors affecting health behaviors of the personnel	- 17 program developers	- In-depth interviews - A focus group discussion
Phase 5: administrative and policy assessment	- Assessed the organizational policies, resources, and management that could facilitate the implementation	- 17 program developers	- In-depth interviews - A focus group discussion

the in-depth interview guideline asked about the impact of LBP and the needs for LBP prevention, the risk factors of LBP, the factors affecting health behaviors, and the policies, resources, and management that could facilitate the implementation of the prevention of LBP, and how to solve this problem.

The focus group discussion was conducted at a meeting room for 10 nursing personnel, including nursing personnel that had LBP and never had LBP. The focus group guideline questions asked about the impact of LBP and the needs for LBP prevention, the risk factors of LBP, the factors affecting health behaviors, and how to solve this problem.

2.2 Implementing and Evaluating the Program for the Prevention of LBP Based on the PROCEED Framework

The PROCEED framework was used for implementing and evaluating the program. In the second three phases in Table 2, we implemented the program for the prevention of LBP, evaluated the process by which the programs are being implemented, and

Table 2. Application of the PROCEED framework to implement and evaluate the program for the prevention of LBP among nursing personnel

Phases	Application	Study participants	Methodology
Phase 6: implementation	- Implemented the program for prevention of LBP	- 81 program attendants	
Phase 7: process evaluation	- Evaluated the process by which the programs are being implemented	- 81 program attendants	- Distribution of evaluation forms on overall satisfaction of the program
Phase 8: impact evaluation	- Evaluated the program effectiveness, including knowledge, attitude, and health behavior scores	- 81 program attendants	- Knowledge, attitude, and health behavior scores were measured before and at three month after the intervention

evaluated the program effectiveness, including knowledge, attitude, and health behavior scores.

The program for the prevention of LBP among nursing personnel was designed and developed based on the integration of the results of the in-depth interviews, the focus group, and the questionnaire which came from the five phases of the PRECEDE framework. The program for the prevention of LBP consisted of three activities; 1) regular education sessions on safe patient lifting techniques and good working postures; 2) training on lifting overweight patients by using lifting devices and a lifting team; and 3) training with back muscle exercise. We conducted the three activities within three days in the afternoon in March 2016 in a large meeting room in the hospital. Meanwhile, we evaluated the process of the three projects by distributing evaluation forms on overall satisfaction with the program. In terms of evaluating the program effectiveness, we measured the knowledge, attitude, and health behavior scores before and at three months after the intervention.

3 Results

3.1 Results from Applying the PRECEDE Framework to Design the Program for Prevention of LBP Among Nursing Personnel

Participant characteristics from in-depth interviews and the focus group (17 program developers). Most of the participants were female (94.12%) and the ages ranged from 32 to 54, with a mean age of 40.3 (S.D. = 10.36). More than a half of them (64.70%) graduated with a master's degree and worked as a staff member (70.59%). Their years of work experience ranged from 5 to 32, with a mean of 18.9 (S.D. = 12.39) years.

After conducting the in-depth interviews and the focus group discussion, we analyzed all of the data and categorized the important themes.

Results from the In-depth Interviews (Seven Program Developers). All of the participants recognized that LBP has been a great problem among nurses in the hospital. According to the impact of LBP, two themes emerged from the interviews, consisting of the impact on administrators' management and the hospital. They considered that LBP among nursing personnel has impacts on their administration in terms of staffing and scheduling.

"When a nurse in my unit has moderate to severe LBP symptom, she cannot work and has a sick leave from work. Therefore, it has an impact on my management because I have to reschedule a new one to replace the old one for work. It is quite a difficult problem for me."

The participants also considered that LBP among nursing personnel has impacts on the hospital in terms of the quality of nursing care and the economic aspect.

"Although there is no claim about the quality of nursing care, it has an impact on the quality of nursing care because nurses who have LBP symptoms cannot work or turn patients effectively.

In the case of the program developers' views regarding the needs for LBP prevention, three themes emerged from the interviews; a) education session on safe patient lifting and good work postures; b) back muscle exercises; and c) training on lifting patients by using lifting devices and a lifting team.

Regarding organizational policies, resources, and management that could facilitate the implementation of the prevention of LBP in the hospital, four themes emerged from the interviews: man, money, materials, and management. Man includes the readiness and participation of nursing personnel and other staff members to assist in the program. Money is defined as the support in budgeting for conducting the program. Materials include the facilities for implementation of the program. Management includes existing policies or systems that support the prevention of LBP, such as policies to promote health and safety in the workplaces of the hospital.

"Because the policies of hospital accreditation (HA) and the health promoting hospital (HPH) evaluations from the Ministry of Public Health have been used as strategies to promote health and safety in this hospital, we have a risk assessment and management system, the environment and safety committee for the hospital, and therefore we have a policy to promote health and safety in the workplace."

Results from the Focus Group Discussion (10 Program Developers). According to the impact of LBP, two themes emerged from the focus group, including the impact on work ability, both at home and in the workplace, and scheduling and staffing. In terms of the needs for LBP prevention, three themes emerged from the focus group: a) having a lifting team; b) education and training in LBP prevention during work; and c) adjusting work postures.

According to the risk factors of LBP, two themes emerged from the focus group discussion, consisting of behaviors and work environment. The behaviors were heavy lifting, frequent lifting, handling patients without assistance, prolonged standing, and awkward postures during work. *"During working in the Inpatient Care Unit, I usually turn patients by myself every two hours because most of the patients are semi-conscious."*

Concerning the organizational policies, resources, and management that could facilitate the implementation of the prevention of LBP in the hospital, three themes emerged from the interviews, consisting of man, money, and management.

Characteristics of the Program Attendants (81 Nursing Personnel). Most of the participants were female (95.1%). The participants' ages ranged from 23 to 59, with a mean age of 41.1 (S.D. = 10.06). The nursing experience ranged from 1 to 31 years, with mean years of 8.4 (S.D. = 7.97). None (100.00%) of the subjects had ever smoked cigarettes. Most of the participants (97.5%) had never had previous diseases related to LBP and most of them (87.7%) had never experienced back trauma from work. Most of them (88.9%) had never attended any LBP prevention training.

Results from the Questionnaire. The result from the 81 nursing personnel responding to the questionnaires revealed that the prevalence of LBP during the previous seven days and during the previous 12 months was 39% and 63%, respectively.

Before the intervention, the mean score regarding knowledge about LBP and prevention was 4.30 (S.D. = 5.63, Min = 2 Max = 7), and the mean score for attitudes about LBP and prevention was 32.56 (S.D. = 10.52, Min = 22 Max = 42), while the mean score on health behaviors for LBP prevention was 38.57 (S.D. = 9.63, Min = 21 Max = 40).

3.2 Results from Applying the PROCEED Framework to Implement and Evaluate the Program for the Prevention of LBP Among Nursing Personnel

The results from the 80 evaluation forms on the overall satisfaction with the program revealed that the overall satisfaction scores was 4.09 (S.D. = 0.53) from the total score of 5. An additional suggestion was that it should be an important policy for nurses, especially back muscle exercises, before beginning work. All of the participants attended the program. At three months after the intervention, the mean score for knowledge about LBP and prevention was 9.20 (S.D. = 2.48, Min = 5 Max = 10); the mean score for attitudes about LBP and prevention was 42.36 (S.D. = 8.60, Min = 32 Max = 50); and the mean score for health behaviors for LBP prevention was 46.23 (S. D. = 8.63, Min = 33 Max = 56). A paired t-test was applied to compare the knowledge, attitude, and health behavior scores before and at three months after the intervention. The results revealed that there was a statistically significant difference in mean scores for the knowledge, attitude, and health behaviors before the intervention and at three months after the intervention ($p < .05$).

4 Discussion

This study demonstrated that applying the PPM to design, implement, and evaluate a program for the prevention of LBP among nursing personnel was successful. The program was designed by involving the target population, both administrators and nursing personnel, to meet their needs. Following this model step by step provided greater understanding of LBP problems and the processes involved in resolving these problems.

Compared with the contents of the program for LBP prevention in other countries, it was found that the contents of this program were similar to those in some previous studies. In the studies by Gray and colleagues [9], Langerstrom and Hagberg [10], Fanello and colleagues [11], and Karahan and Bayraktar [12], for example, the contents of the program in manual lifting techniques included both theoretical and practical parts. The theoretical part consisted of patient manual lifting techniques, patients' experiences of comfort and safety, the quality of care from transfer, and the causes of LBP and risk factors. The practical part consisted of practice in patient handling and transfers based on the assessment of different kinds of situations. However, there were few differences regarding the place for practice sessions. In this study, practice sessions were organized at the venue of the hospital, whereas it was organized in the nursing units in a study by Gray and colleagues [9]. This could be due to the different time spent in skill training on lifting and transferring patients. The practice sessions in this study were organized at the venue of the hospital, where a nursing unit was simulated, with a patient bed, a wheelchair, and a stretcher. In this way, this method did not disturb the patients in nursing units.

In this study, the effectiveness of the program was assessed according to the knowledge, attitude, and health behavior scores. The results revealed that the program had effects at three months after the intervention. This achievement may have come from the contents of the three activities: 1) regular education sessions on safe patient lifting techniques and good working postures; 2) training on lifting overweight patients by using lifting devices and a lifting team; and 3) training with back muscle exercises, which focused on academic knowledge and skill training on lifting and transferring patients and establishing a lifting team in each nursing unit. The finding of this study was also supported by the study of Nelson and colleagues [13], which found that a multifaceted ergonomics program decreased the number of unsafe patient handling practices performed daily among nursing staff. The findings of this study also demonstrated that the program has an effect on attitude by increasing the mean scores from before the intervention to three months after the intervention. This success may have come from the content and activities of this program, which focused on providing information on health and safety, particularly regarding the causes of LBP and LBP prevention via back muscle exercise practice. The results of this study were also confirmed by the finding from Buchbinder and colleagues [14], which found that a back-pain campaign increased the participants' positive attitude about back pain between baseline and the second and third survey.

5 Conclusion

The PRECEDE-PROCEED model was seen to be beneficial and can be applied as a conceptual framework and demonstrates the possibility of use as a model to plan interventions for community hospital nursing personnel.

6 Implications/Recommendations

Hospital administrators can use the findings of this study to promote organizational policies on LBP prevention. The sustainability of the LBP prevention program needs to be tested by expanding the duration of the evaluation to and 12 months after the intervention.

References

1. Chan, S.G.: Factors associated with low back pain among nurses in critical care units, hospital Universiti Sains Malaysia. Biomed. J. Sci. Tech. Res. **1**(7), 2025–2030 (2017)
2. Sanjoy, S.S., Ahsan, G.U., Nabi, H., Joy, F., Hossain, A.: Occupational factors and low back pain: a cross-sectional study of Bangladeshi female nurses. BMC Res. Notes **10**, 173 (2017)
3. Ibrahim, M.I., Zuair, I.U., Yaacob, N.M., Ahmad, M.I., Shafei, M.N.: Low back pain and its associated factors among nurses in public hospitals of Penang, Malaysia. Int. J. Environ. Res. Public Health **16**, 4254 (2019)
4. Sopajareeya, C., Viwatwongkasem, C., Lapwongwatana, P., Hong, O., Kalampakorn, S.: Prevalence and risk factors of low back pain among nurses in a Thai public hospital. J. Med. Assoc. Thai. **92**(Suppl.7), s93-9 (2009)
5. Sikiru, L., Hanifa, S.: Prevalence and risk factors of low back pain among nurses in a typical Nigerian hospital. Afr. Health Sci. **10**(1), 26–30 (2010)
6. Green, L.W., Kreuter, M.W.: Health Promotion Planning: An Educational and Ecological Approach, 3rd edn. Mayfield, California (1999)
7. Haling, P.H., Sullivan, P.E., Chaiyawat, P.: Application of PRECEDE-PROCEED planning model in transforming the clinical decision making behavior of physical therapists in Myanmar. Front. Public Health **7**, 114 (2019)
8. Flick, U.: An Introduction to Qualitative Research, 3rd edn. Sage Publications, London (2006)
9. Gray, J., Cass, J., Harper, D.W., O'Hara, P.A.: A controlled evaluation of lifts and transfer educational program for nurses. Geriatr. Nurs. **17**(2), 81–85 (1996)
10. Langerstrom, M., Hagberg, M.: Evaluation of a 3 year education and training program: for nursing personnel at a swedish hospital. AAOHN J. **45**(2), 83–92 (1997)
11. Fanello, S., Jousset, N., Roquelaure, Y., Chotard-Frampas, V., Delbos, V.: Evaluation of a training program for the prevention of lower back pain among hospital employees. Nurs. Health Sci. **4**, 51–54 (2002)
12. Karahan, A., Bayrakta, N.: Effectiveness of an education program to prevent nurses' low back pain. Workplace Health Saf. **61**(2), 73–78 (2013)
13. Nelson, A., Matz, M., Chen, F., Siddharthan, K., Lloyd, J., Fragala, G.: Development and evaluation of a multifaceted ergonomics program to prevent injuries associated with patient handling tasks. Int. J. Nurs. Stud. **43**, 717–733 (2006)
14. Buchbinder, R., Jolley, D., Wyatt, M.: Population based intervention to change back pain beliefs and disability: three part evaluation. BMJ **322**, 1520 (2001)

The Role of Human Factors and Ergonomics Professionals on Sustainable Development

Eduardo Ferro dos Santos[1(✉)], Karine Borges de Oliveira[2], Gustavo Aristides Santana Martinez[1], and Messias Borges Silva[1]

[1] Department of Basic and Environmental Sciences, USP, University of São Paulo, Estrada do Campinho s/n, Lorena, Brazil
{eduardo.ferro,gustavo.martinez,messias.silva}@usp.br
[2] Production Department, UNESP, São Paulo State University, Av. Ariberto Pereira da Cunha, 333, Guaratinguetá, Brazil
karine.borges@unesp.br

Abstract. Sustainable development goals (SDG) have been proposed by Agenda 2030 (UN), and challenges relate to the application of the triple bottom line (TBL) - people, planet, and profit. Human factors and ergonomics (HFE) professionals develop actions integrating an equal TBL – health (people), safety (planet), and performance (profit). The issue of sustainable development has become one of the main priorities all over the world, and to establish projects by HFE professionals to collaborate for SDG goals is essential. The SDG is a relatively young area of research, and the main purpose of this study is to identify actions, projects, gaps, and trends in the literature integrating HFE and SGD. A bibliometric analysis explored the Scopus database, examining documents from 1999 to 2019. The results suggest that HFE professionals still have a long way to produce more research about SGD. We must act to raise awareness among HFE professionals. These actions will be established new researches and projects related to sustainability and collaborate with Agenda 2030.

Keywords: Sustainability · Sustainable development · Agenda 2030

1 Introduction

The sustainability evaluation criteria comprise the triple bottom line (TBL) sustainability categories, i.e., environmental, economic, and social [1]. For this paper, it is most appropriate to define sustainability as the potential to develop high-quality interventions that are integrated into local communities and services and maintained over time [2].

Environmental issues point to growing concerns about the impact on human factors and the planet [3]. Policy and economic factors have economic issues. The intense exploitation of natural resources for economic growth is one of the known issues sometimes with a negative impact and unsustainable environmental degradation, contributing a potentially catastrophic climate change. It remains an open problem and

P. M. Arezes and R. L. Boring (Eds.): AHFE 2020, AISC 1204, pp. 130–138, 2020.
https://doi.org/10.1007/978-3-030-50946-0_19

intense exploitation of non-renewable resources. In practice, there exists no exact solution.

Balance economic expectation, environmental, and social responsibility is a big challenge. Companies have intensely focused on production, profit, and the intense exploitation of non-renewable resources. The consumer desire for new products or services results in the unsustainability of many production systems [4]. The world is directly affected by social inequalities, unequal distribution of wealth and healthcare, unfair distribution of income, lack of access to education, low quality of life [5].

To tackle this problem, the United Nations introduces the SGD - Sustainable Development Goals, by Agenda 2030 [6], as an urgent call for action to integrate all countries in a global partnership to comply with a series of processes and practices that involve strategic actions to improve sustainability [7].

Human Factors and Ergonomics (HFE) as science can address global human problems. HFE projects can act through the use of solutions for sustainable development [8]. In the future, all HFE actions will consider the impact on sustainability [9]. However, there are gaps and trends for HFE professionals to stay more involved with sustainability. Most of the theories of HFE and SDG are recently, and it is necessary to increase the research and actions in the field of HFE and sustainability, that relate the actions of Agenda 2030. There have been several studies and documents examining green ergonomics products [8, 10–16], with ecology, natural design, and relations comfort with environmental work. There does not seem to exist any acceptable solution for the issue of the relationship between TBL and HFE.

HFE professionals must consolidate actions in sustainability, and declare the importance and relevance to work and research in this integrated subjects. It is necessary to apply the existing set of knowledge in HFE to the design of interventions and improvements, contributing to the SDG. Agenda 2030 became a field that is relevant and important for all nations. It is essential to understand the full extent of SGD to guide the role of HFE professionals.

2 Methodology

In this study, we use a comprehensive bibliometric analysis to trace and obtain the global trends on HFE and Sustainability. We collected data using the PRISMA model [17].

There are two related questions here: "how can be linked HFE professionals to SDG projects?" and "what the gaps and trends in the HFE and SDG relationships?".

A research tool utilized by bibliometric analysis includes the use of Scopus, a bibliographic database containing the abstracts and citations for academic journal articles. Collected data ranged from 1999 to 2019, including all published documents from the Scopus database.

The search term for original and review articles was TITLE-ABS-KEY (("ergonomic" OR "human factors") AND ("sustainability" OR "development sustainable" OR "Agenda 2030" OR "triple bottom line")). This combination increases the chances for the combination of sustainability and HFE subjects.

However, we consider only open access has. The results of this study showed the most cited papers, authors, countries, institutions, journals. The five most cited papers and five indicated as the most important papers by the filter of Scopus were identified and further analyzed.

3 Results

This section outlines the results of the study. Our systematic search yielded 1114 papers. However, 202 papers in open access were included in the systematic review. The search strategy and inclusion criteria are in Fig. 1 – PRISMA Model flowchart.

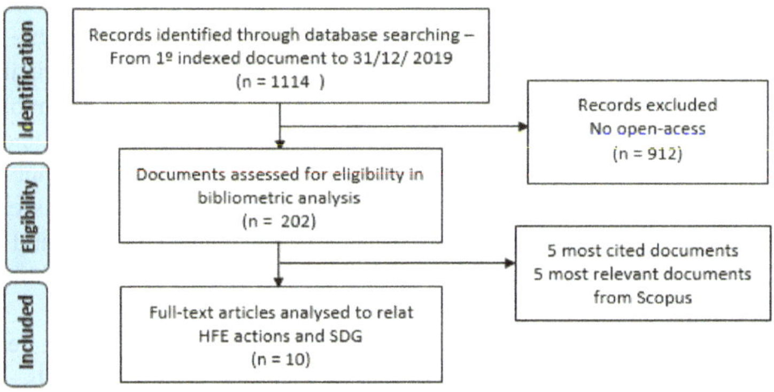

Fig. 1. Flow chart adopted by Prisma model

The present study was designed to move beyond these initial, preliminary observations. This overall work advances the state-of-the-art, including more HFE actions and projects for achieving SDG goals. This approach will provide a framework for HFE professionals in future documents.

4 Discussion

According to the evolution of documents (Fig. 2), the first publication indexed in Scopus was in 1999 [18]. It is clear from the analysis the exponential growth.

The results were analyzed by source. During this period, the Sustainability journal (MDPI) had the largest number of publications. Amarria Dila Sari and Muhammad Ragil Suryoputro, from Universitas Islam Indonesia, are the authors identified as having the largest number of publications (4 papers). However, they were written in collaboration. Antonio Marano, from the University G. d'Annunzio Chieti and Pescara, has 3 papers. Analyzing the Marano papers [19–21], it is evident that HFE and sustainability are linked. Sari and Suryoputro describing the benefits of ergonomics regarding organizational and human factors, superficially including sustainability [22–25].

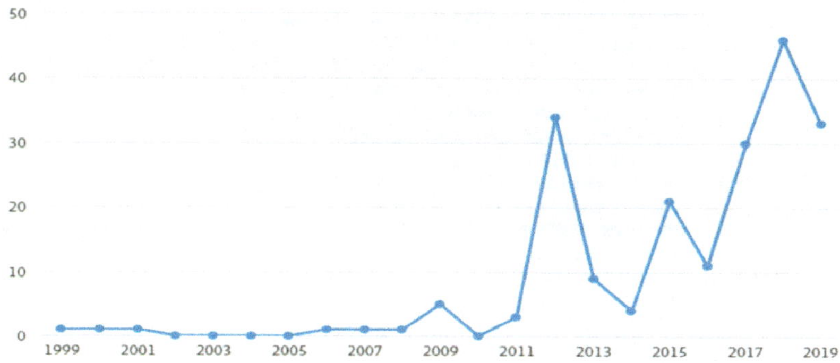

Fig. 2. Number of publications per year

Universitas Islam Indonesia is the institution with the highest number of documents published (in a total of 8 - Fig. 3). The Federal University of Rio Grande do Norte, from Brazil, is in the second position, followed by the University of Naples Federico II, from Italy.

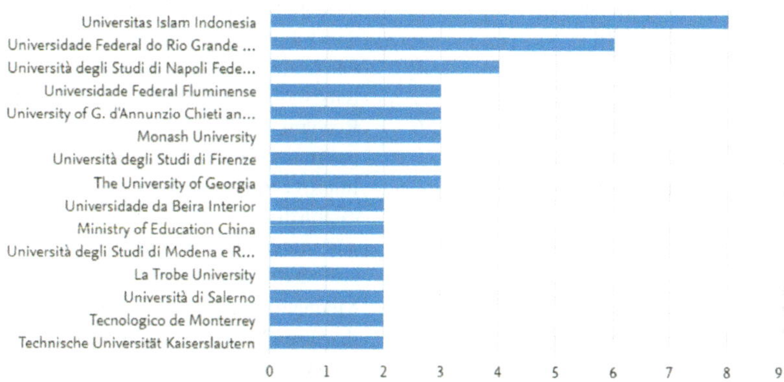

Fig. 3. Documents by affiliation

Italy and the United States are leaders in papers published (24 each - Fig. 4).

The Sustainability journal (MDPI) published 26 papers of the 106. The conference paper (proceedings) includes Work (IOS Press) that published 30 documents. Work is an influential journal; however, it published conference papers in 2012 from 18th World Congress on Ergonomics (IEA), totaling 30 papers cited in only one volume.

We select 5 papers defined as more important by Scopus (Table 1), and the 5 most cited (Table 2) to analyze interrelation between HFE and SDG.

SDG 4 refers to quality education, and SDG 17 refers to partnerships for the achievement of the goals. These are the two SDG most present in the most cited documents. Models are present in 3 documents; however, it needs to be tested to ensure improvement and maturity. The special edition in the Ergonomics journal published 17

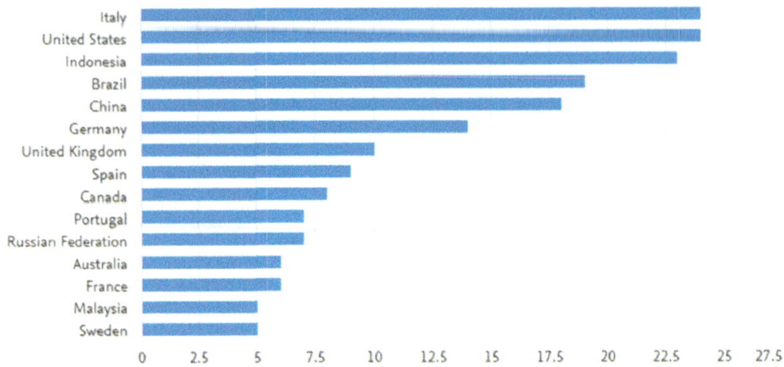

Fig. 4. Documents by country

Table 1. Most cited documents in Scopus

Document	Cite	Approach	SDG
The conceptual framework for integrating workplace health promotion and occupational ergonomics programs [26]	90	Models of participatory structures for the establishment of health goals, design, and development programs, to increase the effectiveness and long-term sustainability	17
Workplace health protection and promotion through participatory ergonomics: an integrated approach [27]	56	Recognized the workplace programs are designed to improve the health of employees, establish employee partnerships, and integrate work and sustainability in an organization	4, 17
Living systematic reviews 2 Combining human and machine effort [28]	46	Models to address the evidence that uses human effort and automation in an integrated manner to improve the viability and sustainability of systematic reviews	4, 9
Environmental and health perceptions before and after relocation to the green building [11]	38	Analysis of the green buildings impacts human health, better comfort for occupants, and better well-being	3, 7, 9, 11, 12
Ergonomics and Sustainability [9] - Editorial	38	A special issue of the journal Ergonomics showing how sustainability reflects in the field of HFE, with 17 papers	4, 9, 11, 12, 17

papers, and the editorial has 38 citations. Meanwhile, it could not guarantee the same performance in papers inside.

Table 2. More important by Scopus

Document	Cite	Approach	SDG
Early variability in the conceptualization of "sustainable development and human factors [10]	13	Definitions about sustainable development are discussed concerning future challenges and opportunities for future theoretical and empirical work	4, 17
The broader consideration of human Factor to Enhance sustainable building design [29]	8	HFE contribute to the design of sustainable buildings, where methodologies and ergonomic tools improve the performance and sustainability during all phases of the building lifecycle	3, 7, 9, 11, 12
Defining elements of sustainable work systems - The system-oriented approach [30]	11	Conceptual model of a sustainable system of work, with definitions about ergonomics and TBL	4, 8, 9, 11, 12, 17
The importance of ergonomics to sustainability throughout a building's life cycle [31]	6	Discuss ergonomics integrated into the building lifecycle to promote sustainability targets on a university campus	3, 7, 9, 11, 12
The hybrid HFACS-BN model for analysis of Mongolian aviation professionals' awareness of human factors related to aviation safety [32]	4	A model to evaluate how security is crucial to ensure the sustainability of aviation, through the study of variables and influences of human factors in aviation	8, 9

We found the most relevant document by Scopus presenting models of HFE and sustainability actions, with predominance to green ergonomics in building architectures. In general. There is considerable uncertainty in the exact results. We not satisfied. The research will continue to run in other databases. However, the results are exciting, and it will be possible to conclude and move on to new challenges.

5 Conclusions

Sustainability is a global challenge, with the individual attention of the nations in achieving SDG. We have the hypothesis that HFE professionals are disconnected from SGD and must show more results, involving the TBL. In conclusion, the relationship between HFE and SDG needs to be explored more. It is an opportunity to become opportunities to produce change, new researches, propose models, and given the opportunity to show HFE professionals skills. A brief conclusion to the advantages and shortcomings is that we have a lot to do, not limited to show the benefits of green buildings. There are 17 SDG to be related to HFE actions. The role of HFE professionals to collaborate with society to achieve SDG is extremely necessary and relevant. New projects, research, actions, ways to modify people's behavior to be more sustainable, opportunities in education, develop new products and optimize processes.

HFE and SDG goals provide safe, healthy, and productive environments. It is the main propose of ergonomic science and practice. Maybe the triple bottom line is still connected with HFE theoretical propose. The papers analyzed pointed to HFE solutions for sustainability, but the practice needs to be more widespread and embedded in business and society, from the educational role of the HFE professionals and the search for partnerships, both present in the SDG 4 and 17, but not limited to this. There are new possibilities in all SDG. The knowledge in HFE is essential to guide society and ensure efficient and effective results in any area. HFE professionals are required to contribute to innovative interventions and must focus on the future, including sustainability.

A more concise understanding of this concept is required to provide additional information about the necessity of HFE professionals exploring SDG goals in their actions. With a few exceptions, sustainability seems to be a new concept for HFE professionals. One limitation of this work naturally includes the necessity of new approaches, and it is the first part of a long way to develop our research group, included in our agenda. The Scopus has restricted data, and it is necessary to include more databases. It is necessary to expand the arguments and relate collaborative networks, expand the state of the art, and consider to be a significant step forward to support projects for HFE professionals. This theoretical study should provide further benefits to propose a framework for future research.

References

1. Birkel, H.S., Veile, J.W., Müller, J.M., Hartmann, E., Voigt, K.I.: Development of a risk framework for Industry 4.0 in the context of sustainability for established manufacturers. Sustainability **11**, 1–27 (2019)
2. Sartori, S., Latrônico, F., Campos, L.M.S.: Sustainability and sustainable development: a taxonomy in the field of literature. Ambient. Soc. **17**, 01–22 (2011)
3. Kim, S., Osmond, P.: Analyzing green building rating tools for healthcare buildings from the building user's perspective. Indoor Built Environ. **23**, 757–766 (2014)
4. Bonilla, S.H., Silva, H.R.O., da Silva, M.T., Gonçalves, R.F., Sacomano, J.B.: Industry 4.0 and sustainability implications: a scenario-based analysis of the impacts and challenges. Sustainability **10**, 1–24 (2018)
5. Boström, M.: A missing pillar? Challenges in theorizing and practicing social sustainability: introduction to the special issue. Sustain. Sci. Pract. Policy **8**, 3–14 (2012)
6. Colglazier, W.: Sustainable development agenda: 2030 (2015)
7. Feil, A.A., de Quevedo, D.M., Schreiber, D.: An analysis of the sustainability index of micro- and small-sized furniture industries. Clean Technol. Environ. Policy **19**(7), 1883–1896 (2017). https://doi.org/10.1007/s10098-017-1372-7
8. Thatcher, A., Waterson, P., Todd, A., Moray, N.: State of science: ergonomics and global issues (2018)
9. Haslam, R., Waterson, P.: Ergonomics and sustainability. Ergonomics **56**, 343–347 (2013)
10. Thatcher, A.: Early variability in the conceptualisation of "sustainable development and human factors". Work **41**(1), 3892–3899 (2012)

11. MacNaughton, P., Spengler, J., Vallarino, J., Santanam, S., Satish, U., Allen, J.: Environmental perceptions and health before and after relocation to a green building. Build. Environ. **104**, 138–144 (2016)
12. Hanson, M.A.: Green ergonomics: challenges and opportunities. Ergonomics **56**(3), 399–408 (2013)
13. Poon, W.C., Herath, G., Sarker, A., Masuda, T., Kada, R.: River and fish pollution in Malaysia: a green ergonomics perspective. Appl. Ergon. **57**, 80–93 (2016)
14. Thatcher, A., Milner, K.: Green ergonomics and green buildings. Ergon. Des. **22**(2), 5–12 (2014)
15. Pilczuk, D., Barefield, K.: Green ergonomics: combining sustainability and ergonomics. Work **49**(3), 357–361 (2014)
16. Hedge, A., Dorsey, J.A.: Green buildings need good ergonomics. Ergonomics **56**(3), 492–506 (2013)
17. Moher, D., Liberati, A., Tetzlaff, J., Altman, D.G., et al.: Preferred reporting items for systematic reviews and meta-analyses: The PRISMA statement **6**(7), e1000097 (2009)
18. Kawakami, T., Batino, J.M., Khai, T.T.: Ergonomic strategies for improving working conditions in some developing countries in Asia. Ind. Health **37**, 187–198 (1999)
19. Marano, A., Di Bucchianico, G., Rossi, E.: Strategies and arguments of ergonomic design for sustainability. Work **41**, 3869–3873 (2012)
20. Di Bucchianico, G., Marano, A., Rossi, E.: Toward a transdisciplinary approach of ergonomic design for sustainability. Work **41**, 3874–3877 (2012)
21. Marano, A., Di Nicolantonio, M.: Ergonomic design in eHealthcare: a study case of eHealth technology system. Procedia Manuf. **3**, 272–279 (2015)
22. Suryoputro, M.R., Sari, A.D., Burhanudin, R., Sugarindra, M.: Lean production design using value stream mapping and ergonomics approach for waste elimination on buffing panel upright process. In: IOP Conference Series: Materials Science and Engineering (2017)
23. Sari, A.D., Suryoputro, M.R., Rahmillah, F.I.: A study of 6S workplace improvement in Ergonomic Laboratory. In: IOP Conference Series: Materials Science and Engineering (2017)
24. Suryoputro, M.R., Wildani, K., Sari, A.D.: Analysis of manual material handling activity to increase work productivity (case study: manufacturing company). In: MATEC Web of Conferences (2018)
25. Sari, A.D., Anwar, A.R., Suryoputro, M.R.: Work postural analysis and musculoskeletal injury risk in critical working station at XYZ Ceramics Yogyakarta. In: MATEC Web of Conferences (2018)
26. Punnett, L., Cherniack, M., Henning, R., Morse, T., Faghri, P.: A conceptual framework for integrating workplace health promotion and occupational ergonomics programs. Public Health Rep. **124**(Suppl), 16–25 (2009)
27. Henning, R., Warren, N., Robertson, M., Faghri, P., Cherniack, M.: Workplace health protection and promotion through participatory ergonomics: an integrated approach. Public Health Rep. **124**(Suppl), 26–35 (2009)
28. Thomas, J., Noel-Storr, A., Marshall, I., Wallace, B., McDonald, S., Mavergames, C., Glasziou, P., Shemilt, I., Synnot, A., Turner, T., Hill, S., Pearson, L.: Living systematic reviews: 2. Combining human and machine effort. J. Clin. Epidemiol. **91**, 31–37 (2017)
29. Attaianese, E.: A broader consideration of human factor to enhance sustainable building design. Work **41**, 2155–2159 (2012)

30. Fischer, K., Zink, K.J.: Defining elements of sustainable work systems - a system-oriented approach. Work **41**, 3900–3905 (2012)
31. Miller, L., Dorsey, J., Jacobs, K.: The importance of ergonomics to sustainability throughout a building's life cycle. Work **21**, 2129–2132 (2012)
32. Zhou, T., Zhang, J., Baasansuren, D.: A hybrid HFACS-BN model for analysis of Mongolian aviation professionals' awareness of human factors related to aviation safety. Sustainability **10**(12), 4522 (2018)

Human Factors Impact on Smart Cities Construction: The Case of Lisbon City and Dwelling Suburbs

Cristina Caramelo Gomes[1,2](✉)

[1] CIAUD, Faculdade de Arquitectura, Universidade de Lisboa, Lisbon, Portugal
Cris_caramelo@netcabo.pt
[2] CITAD, Faculdade de Arquitectura e Artes, Universidade Lusiada de Lisboa, Lisbon, Portugal

Abstract. The aim of this paper is to highlight the need to humanise smart urban environment by improving human interactions to a qualified user experience. Numerous municipalities of Portugal aim for smart and sustainable solutions, making available equipment and services and encouraging people appropriation of the urban space. Lisbon emerge as a smart place, displaying samples of sustainable, economic, social and cultural dimensions. Municipalities near important city centres, preserve a traditional model featured by careless urban design solutions, where commuting possibilities and proximity of areas of interest are the main assets. Despite the innovative and creative solutions supported by ICT, the traditional model with its intrinsic fragilities is still a common practice, framing technical developments and possibilities, the sense of place identity, user belonging and user qualified experience. A new approach to urban planning is needed to empower the place, the context and the users, where creative solutions support smart, sustainable and inclusive communities.

Keywords: Human factors interaction · Smart cities · Lisbon · Users · Suburbs dwelling areas

1 Introduction

In the present time, cities are witnessing dramatic changes due to the migration of people searching for better quality of life and access to social equipment. Statistical data supports the concentration of individuals in metropolitan areas. In the meantime, the advancements in ICT (Information and Communication Technologies) invaded our ways of living and interacting, promoting new models of work, dwell, entertainment and interaction. ICT incursion in our daily routines surpassed our individuality and took over the city and its dynamic. The problems raised by the dichotomy model city centre and suburbs, with the split between dwelling and work, commuter movements, waste of time and energy, and the lack of place identity and human sense of belonging boost the need of smart and humanized cities.

Smart cities have technical solutions to problematic contemporaneous realities; however, it is not evident how much and which users are benefiting from such implemented solutions. People are struggling with deficient conditions of dwelling,

P. M. Arezes and R. L. Boring (Eds.): AHFE 2020, AISC 1204, pp. 139–146, 2020.
https://doi.org/10.1007/978-3-030-50946-0_20

unhealthy conditions of workplaces, and insecure and undesirable urban spaces. Urban environments are planned in function of its location, real estate speculation, economic pressure disregarding human requirements and desires.

If main cities, by their importance, are running some changes to raise reputation and investment attraction, suburban areas are repeating the traditional model, where the mono-functionality of dwelling competes with the offer of work environments reiterating the traditional model characterised by the lack of green areas, public spaces to encourage human physical interaction and commuting hours. Suburban areas are dependent of main cities, featured by lack of identity, deficient social and cultural equipment and spatial humanised solutions. Smart built environ requires a different approach, detached from hi-technical developments and trend solutions, and focused on human requirements and ambitions. Changing suburbs' reality means humanise areas and breaking traditional models that support unsustainable urban solutions. For that, Human factors offer a significant contribute oriented to different areas of expertise: from public to private space, human factors contribute to the conception and management of space to efficient uses and qualified interactions with users.

Lisbon as a capital city reveals the changes due to the political intentions to transform it into a smart city. Some solutions are technology oriented, but some others featured qualified urban solutions that encourage and qualify user experiences. Nevertheless, suburban areas less aware of the possibilities offered by smart concept, or by political attention, highlight questionable solutions, unaware of users' needs although some of which still show some solutions related with e-mobility as well as remote access to services. Urban spaces have the private car as the focus, showing solutions that disregard human security, safety, quality of life and above all inclusion.

2 Smart Cities: A Buzz Trendy Expression

There is a common awareness that the 21st century will be oriented to the development of cities namely the ones that by their dimension aggregates a metropolitan area with which develop a reciprocal relationship. The city is focused on services provision - regardless their economic, social and cultural nature – while adjacent areas are focused on dwelling users' requirements. This model shaped by the growth of public and private transportation accentuates the monofunctional character of each area. Regardless of ICT evolution, with multidisciplinary functions and users' acceptance, a long way is needed for efficient exchange of information instead of people. As a result, urban environment is planned to facilitate cars access, movement and parking whilst public spaces lack functional, pleasant and inclusive solutions. Built environment shows an image of massive building to meet housing needs, characterized by construction processes harmful to the environment and finishing as well as layout solutions influenced by (economic, political and aesthetical) trends neglecting geographical, cultural and daily living local requirements [1]. The smartness of urban areas emerges as a response to the challenges that cities are facing, such as concentration of population and unsustainable ways of living. The different approaches to the smart city's definitions show a common pattern to the sustainable use of natural resources and the promotion of humanized user centred built environments [2]. Smart cities are the ones that seek for sustainable economic growth supported by human and social capital and communications infrastructures - such as transport and ICT – with strong participatory policies to handle with natural resources, to achieve a higher quality of life [3].

A city becomes a place as a result of the quality of the experience offered to the user, which is formed by the way urban planning is responsive to user's needs, the encouraged interactions (physical and digital) and consequently the sense of belonging.

Notwithstanding the closeness between the smart city concept and technology, a city deserves the smart label if it respects the sense of place, in other words, the identity of the place, the one built up with its social and cultural heritage, granting features that make it unmatched and pleasing to its inhabitants and visitors (regardless the purpose of the visit). Smart city initiatives ought to be more than technical innovations display - grounded on technical progress, global movements and public/private interests - so often unconscious of contextual realities and users' requirements.

The smart city should encourage the co-design of its urban fabric, in a close collaboration between academy and industry as well as experts and users. A more humanised place requires the reality observation of daily uses and practices to support the conceptualisation and implementation of creative and innovative solutions, in a human centred methodology, in a close measurement result. Users, despite of their role while experiencing the city, emerge as the crucial providers of information and the cluster that better understands the impact of the delivered solutions. The aim of the smart city is to boost users' quality of living supported by smart technology [4].

3 Portugal: A Collection of Smart … Cities

Portugal contributes to the objectives of European Union pursuing Smart cities. For that it created RENER – living lab [5], a network of municipalities, which aims for the renovation of the cities to fulfil the principles that support the smart city concept. A study by University Nova [6] found that almost 60% of the Portuguese municipalities have a strategy towards smart city but, due to low investment, has little implementation. According to Opalka [7] most of the projects concern energy, mobility and governance; however, it is notorious the lack of integration/ connectivity between the different municipalities and/or projects. This observation shows in the inefficiency of institutional websites, or websites that present the smart cities subject and initiatives - most of them framed by meetings and conferences with a broad(?) audience – but without any information about its implementation, follow-up and evaluation. Public participation is framed by the lack of opportunities, deficient communication, inaccessible presential meetings and websites inhibitors of a user-friendly experience, driving users away.

According to ONU (2018) [8] the urban population is expected to almost double by the year of 2050, making urbanization one of the most transformative trends of the 21st century. Populations, economic activities, social and cultural interactions, as well as environmental and humanitarian impacts, are more and more focused in cities, boosting challenges to sustainability related with housing, infrastructure, basic services, food security, health, education, jobs decent security and resources among others. To all these areas, smart concept can be an important factor of improvement. A city without strategical spatial planning becomes less attractive and increases social and environmental problems while decreases the sense of belonging and endangers the identity of the place. The innovative and creative projects concerning smart cities, are not always able to respond to the problems that cities and their users are facing nowadays. The absence of an efficient evaluation of implemented solutions allows the perpetuation of these realities.

4 Lisbon: A Smart City Scenario

Lisbon as the capital of Portugal benefits from more involvement by the stakeholders and more investment to implement smart solutions. Although the information by the city council is framed by municipality smart objectives and related projects - information on the implemented solutions and related evaluation process is absent - the city presents significant urban rehabilitation and a more sustainable, inclusive and smart urban environment. The municipality of Lisbon is an active part of European Smart Cities which aims the use of new technologies of information and communication to improve the quality of life of its users. The objective of Lisbon city [9] is to attract individuals and organisations throughout energy-efficient constructions, e-mobility, smart living and smart citizens solutions, with attention to the inclusion of human heterogeneity namely elderly, which has a strong representation on local demography. Social cohesion is supported by agendas for civic participation, rehabilitation of the urban environment and the boost of efficient urban services and systems. Thus, the city of Lisbon was awarded with The European Green Award 2020, a city with 550,000 residents and closely 40,000 daily commuters [10].

The spread of smart and sustainable services - supported by real-time data collection and sharing - are easily adopted by users with technical literacy. Regardless of the ICT positive impact in the city and consequence acceptance by the users, the urban responses to ageing, accessibility and inclusion to a qualified urban spatial experience, arise as the main contribution to the physical and digital interactions to ensure a positive and significant user experience.

Solutions towards e-mobility are smoothly accepted; accessible and inclusive practices offer flexibility of use to individuals with physical, functional and sensorial constraints. The public space is enhanced by new green areas and resting spaces, along with coffee houses and commercial spots. Traditional neighbourhoods, economic and social debilitated, are suffering significant rehabilitation processes to attract multicultural patterns of uses and users. The encouraged interactions, between the city and the users, are multiple, flexible regarding physical equipment and digital applications accessible throughout mobile devices. The interventions on built urban environment follow the Lisbon Strategic Charter 2010–2024 [11]. Notwithstanding the theoretical and generalist message of the document, Lisbon offers the sense of security, humanised urban solutions as well as the preservation of its own identity, allows the sense of welcome and belonging to individuals, qualifying the experience of each user.

Lisbon shows a strong effort to deserve the label of smart city, implementing coordinated solutions in the areas of urban, social and cultural environment, economy, energy management and inclusion improving the quality of life of its citizens.

However, the implementation of such innovative and smart solutions demands also, a long and intense process of gentrification [12] (Fig. 1).

Fig. 1. Urban space rehabilitated to a positive user experience. The smart concept of the city is extended to humanised and inclusive solutions. (Photos from the author)

5 Suburban Dwelling Areas: The Smartness of Intentions in an Unhumanised Place

The urban environment where individuals interact with space, equipment, services and people is the scenario to understand the needs and expectations of users and consequently create innovative, responsive and smart solutions. When technical and advanced solutions are available, but the urban environment does not promote security, inclusion and quality of life, smartness fades in the blind pursuit of short-lived fashions. From metropolitan municipalities to interiors dwelling suburbs, the majority aims to be labelled smart. A survey throughout municipal websites shows a common intention to provide innovative and technological solutions; most of them are focused on projects related with environment, namely the reduction of the CO2 emissions. Regardless the objective, the social and cultural patterns of society continue to be focused on the need of commuting people rather than information. The will of reduction of Co2 by the use of new ways of mobility are not supported by incentives to develop telework policies, and the services that could be accessed by digital means are framed by platforms featured by insufficient information, inefficient writing and graphical communication and a non-user friendly experience. Some municipalities, near to main metropolitan cities centres, have natural areas of interest, good services and the effort to attract people and organisations to diversify the area functionalities. They demonstrate bright smart innovations intentions, stimulating interventions and digital platforms to engage users on its usage; nevertheless, the lack of survey and evaluation of urban environment experienced by users, can reduce the possibilities of interaction, with a significant impact in the user's sense of belonging and the quality of user's experience. Massive construction, mono-functional areas, urban space oriented to public and private transportation, poor urban equipment, questioning human heterogeneity security, comfort and interactions with the place. These finding don't undercut the value of the

Portuguese municipalities nor unqualify the smart solutions implemented, just highlights the need to look at urban environment and the impact of irrational solutions design. It is important to highlight the built environment features to encourage interactions (physical and digital) in a secure and pleasant way (Fig. 2).

Fig. 2. Playground set between roads in Vila do Conde (photo from Pedro Correia/Global Imagens). Roundabout with green area, fountain and sculpture (photo from the author).

6 Discussion

The city as a dynamic and complex organism demands for strategical planning. A strategical planning requires a critical analysis of the context to support innovative and creative solutions. Regardless the objectives of a municipality for its environment domain, it is urgent to understand the requirements and expectations of new ways of living and this includes organisations to provide services to users as well as users as performers with strong interactions with built environment. The need for sustainable solutions boosts the smartness of built environment. Cities more than a geometric space area are a place featured by individuals' interactions with the spatial layout, equipment, products and people to build up social and cultural experiences to improve human quality of life. A city labelled as a smart one, is a city responsive to users' needs and desires, regardless the physical or digital nature of the response. Technology set in as a facilitator to strength combined actions among individuals, individuals' and organisations, organisations and organisations with the aim to improve the community´s quality of life. Human nature requires the sense of belonging so well expressed by literature, and its construction depends on the interactions encouraged by city environment which are connected with the place identity and its ability to grant public participation (in a community and municipality level) and produce the sense of community answering by physical and digital structures to human demands.

Most metropolitan cities aim for the adoption of smart solutions oriented mostly to minimise the carbon footprint. Thus, some municipalities are committed to the promotion of debates as well as the emergence of creative and innovative ideals. Yet, experiencing the urban environment, emerge the experience of the traditional model, oriented to functionalism with a strong dependence on public and private transport. Real estate development and speculation is a consequence of gentrification processes of the areas closest to the centre, public transport or areas of interest. Commuting movements, traffic congestion and waste of time are still a reality, and urban environment layout highlights these key issues and related requirements more than sustainable and users' requirements. It is not enough anymore to promote debates and creative ideas, it is more than ever important to contextualise the problems, imagine creative solutions, implement them and evaluate the results to understand what is vital and what is collateral in community daily living. This traditional model grounded on the public and private transportation to overcome great distances encourage de number of cars, the associated costs of parking which empties payed parking and overloads with disordered parking the residential as well as quieter functional mixed areas. Solutions to minimise the mono-functionality of the place results in a displacement of the fragilities of the traditional model, displacing commuting, traffic congestion and parking problems. This reality affects the quality of the air, and the security for drivers and pedestrians, even if some municipalities are measuring the number of cars and air quality, as a strategy to deserve the label of smart, a smartness questioned by thoughtless implemented solutions, unaccompanied and evaluated. A new planning paradigm is needed: conscious of the transformations boosted by technology, with a profound understanding of the context, to favour wider, consistent and humanised solutions, while granting for the built environment sustainability.

Acknowledgments. CITAD FCT Fundação para a Ciência e a Tecnologia MINISTÉRIO DA CIÊNCIA, TECNOLOGIA E ENSINO SUPERIOR Portugal This work is financed by national funds by FCT - Foundation for Science and Technology, under the Project UID/AUR/04026/2019.

References

1. Caramelo Gomes, C.: A conceptual model to introduce telework in Lisbon. Universidade Lusíada Editora Lisboa (2004)
2. Albino, V., Berardi, U., Dangelico, R.M.: Smart cities: definitions, dimensions, performance, and initiatives. J. Urban Technol. **22**(1), 3–21 (2015)
3. Thuzar, M.: Urbanization in South East Asia: Developing Smart Cities for the Future? Regional Outlook, pp. 96–100 (2011)
4. Alawadhi, S., et al.: Building understanding of smart city initiatives. LNCS, vol. 7443, pp. 40–53 (2012)
5. RENER: Rede Portuguesa de Cidades Inteligentes (2016). http://www.forumdascidades.pt/content/rede-portuguesa-de-cidades-inteligentes
6. Ferreira, R.: Como Portugal está a transformar-se num país de cidades inteligentes, Dn-Insider (2018). https://insider.dn.pt/featured/portugal-smart-cities-agueda-seixal/4918/
7. Opalka, W.: Smart buildings and smart cities to take center stage, Pike Research (2014). https://www.environmentalleader.com/2014/01/smart-buildings-and-smart-cities-to-take-center-stage/
8. ONU: The World's Cities in 2018 (2018). https://www.un.org/en/events/citiesday/assets/pdf/the_worlds_cities_in_2018_data_booklet.pdf
9. Sharing Cities: Building smart cities together - common solutions for shared challenges (2016). http://www.sharingcities.eu/sharingcities/smartcities

10. Smart Cities Brussels: Smart City Lisbon (2018). https://smartcity.brussels/news-598-smart-city-lisbon
11. Câmara Municipal de Lisboa: Carta Estratégica de Lisboa 2010 - 2024 - um compromisso para o futuro da cidade (2014). http://www.cmlisboa.pt/fileadmin/MUNICIPIO/Camara_Municipal/Carta_Estrategica/A_Carta_Estrategica_Lisboa_Comissariado.pdf
12. Caramelo Gomes, C., Almeida, C.D.: Urban design: colour impact in the human sense of comfort and well-being. In: Proceedings of the International Colour Association (AIC) 2018 – Colour and Human Comfort, pp. 117–123. International Colour Association (2018)

Characterization of Occupational Health and Safety Management in Companies of Bogotá – Colombia

Luis Gabriel Gutiérrez Bernal(✉) and Wilder Alfonso Hernández Duarte(✉)

Virtual and Distance Education, UVD, Corporación Universitaria Minuto de Dios, Bogotá, Colombia
{lgutie55,whernandezd}@uniminuto.edu.co

Abstract. This research presents the characterization of the occupational safety and health management system - OSHMS, for this purpose a quantitative study and descriptive scope was carried out, with a sample of 150 companies in Bogota - Colombia, assessing the level of compliance with the minimum standards of the OSHMS, later the perception of entrepreneurs was subsequently identified as compared to their development in companies.

It was observed that: 70% of companies have a critical level of development meeting less than 60% of the requirements, and only 12% reach the acceptable level with compliance between 85% and 100%. The perception of entrepreneurs shows that the OSHMS is implemented with a mainly legal motivation, the allocation of resources and the knowledge about the OSHMS implementation is insufficient, and there are factors recognized as challenges to overcome like change regulation and the safety culture.

Keywords: Risk management · Safety and health at work · Colombia · Management system

1 Introduction

For the World Health Organization – WHO, workplace health promotion regards a set of workplace policies and activities designed to help employers and workers at all hierarchical levels to increase control and improve their health, promoting the productivity and competitiveness of enterprises and contributing to the economic and social development of countries [1]. However, despite the benefits of a healthy workplace, it kills more people than wars even though work should not be a dangerous task [2].

According to the International Labor Organization – ILO every year, some 374 million non-fatal accident-related injuries occur, and more than 2.78 million people lose their lives to occupational diseases and work-related accidents, these events generate losses of about 4% of each year's global gross domestic product [3]. The IberoAmerican social security organization –OISS reports that in Latin America by 2016, 30 million workplace accidents were caused with 240,000 deaths from work-

P. M. Arezes and R. L. Boring (Eds.): AHFE 2020, AISC 1204, pp. 147–153, 2020.
https://doi.org/10.1007/978-3-030-50946-0_21

related causes, meaning that in the region die about nearly 1,090 people for work-related causes [4].

This situation is no different in Colombia, where according to the Federation of Colombian Insurers - FASECOLDA for 2018, 645.119 work accidents occurred, leaving a total of 567 deaths and 456 people with loss of work capacity over 50%, which means; equally, 10.450 occupational diseases were rated. Particularly in the country's capital, city that hosts the companies under study, there were 189.112 work accidents, 95 deaths from work accidents and 3.686 qualified illnesses [5].

Given this situation, it is necessary to identify the legal and administrative strategies and actions that the Colombian's government has been developing to deal with this serious situation. In the year of 2012, the 1562 law was approved, which modified the structure of the general occupational risk system, giving an important advance in the restructuring of occupational risks in the country by turning the area of occupational health, into an occupational safety and health management system; developed through the Deming cycle [6] and framed in continuous improvement.

Subsequently in 2015 the 1072 decree compiled the regulations of health and safety at work, becoming the road map of the OSHMS. Finally, in 2017 the Resolution 1111 is issued, which was modified by the resolution 0312 of 2019, setting the minimum standards of compliance for companies, based on their economic activity, size and level of risk, establishing for the first time the specific responsibilities that small and medium-sized factories must develop in relation to the prevention of occupational risks. These actions highlight the concern that the Colombian government has been addressing in terms of occupational risks and becomes a way forward for companies to enter the dynamics of culture towards the prevention of work accidents and occupational diseases. Remarking that it has been a slow process, and that still there are many challenges for both, the government and the enterprises [7].

2 Materials and Methods

The study was based on the assessment of the development level on OSHMS implementation, in a group of 150 companies from different economic sectors in the city of Bogota selected non-probabilistically for convenience, in which an instrument was applied proposed and validated by the Colombian Ministry of Labor, which contains the minimum requirements that organizations must meet according to their economic activity, the level of risk and size, and it also raises three levels of compliance: Critical from 0 to 60%, moderately acceptable from 61% to 85% and acceptable from 86% to 100% [8].

On the other hand, through a semi-structured interview, it was sought to identify the perception of entrepreneurs facing the challenges of the implementation of this management system, taking into account the short time with a new structure that this regulation appeared, and the specific requirements it contains.

3 Results

The implementation of the management system is defined in four phases of the PHVA cycle, to plan, to do, to verify and to act, through which activities related to the promotion and prevention of occupational risks take place; it was found companies assessed in this study showing the phase where the highest level of development was to plan with 66% (Fig. 1) including actions defining the policy and objectives of the OSHMS, technical and financial human resources, applicable legal requirements, the training plan and the annual work plan, among others; it was also evidenced that as the stages progressed, the percentage of compliance decreases.

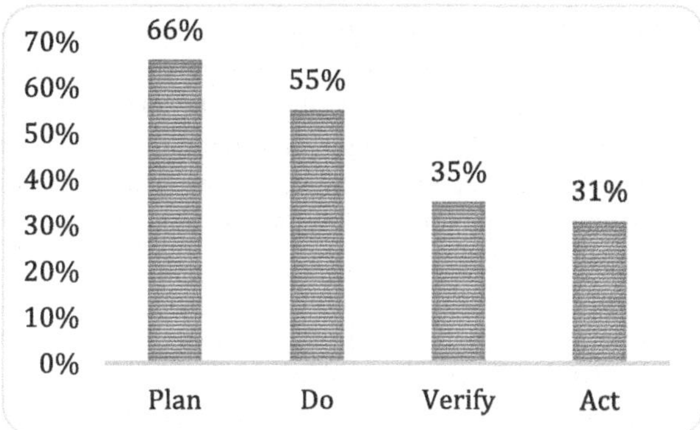

Fig. 1. Level of compliance with the minimum standards

Even though it is truth the success of companies is directly related to a good planning process [9], the development of the planned activities is even more, a situation that is not reflected in the companies under study, considering the fact that the phase called to do, in which activities aimed at managing the workers' health, the threats, the hazards and the risks are carried out, meaning actions on working conditions themselves [10] is not fully implemented (55%), even though this phase represents the largest amount in the assessment of compliance with the minimum requirements of the OSHMS with 66% of the total [8].

The phase to verifying, that evaluates the effectiveness of the intervention measures proposed and carried out through the auditing processes and the analysis of indicators; as well as the phase to act, where preventive, corrective and improvement measures are taken based on the results of the OSHMS, allowing to close the cycle of continuous improvement, however, these phases are the ones with the least progress (35% and 31% respectively).

It was found 74% of companies below 60% of the minimum standards compliance level, which is equal to a critical level. The 20% presented a moderately acceptable level of development, and only 6% of companies achieved an acceptable level of compliance. In the same way, the implementation of OSHMS has a direct relationship with the size of the company, considering that large companies have an average level of compliance of 71%, median companies 55% and the small ones 52%. Situation related to the few resources organizations allocate for the development of management systems [8], with the assignment and level of training of those responsible for leading OSHMSs in companies, and also with the fact that without an appropriate training for all subjects involved in prevention, the OSHMS can hardly be effectively implemented [11].

The interview applied to employers made it possible to establish that one of the main motivations for implementing the OSHMS is legal compliance (62%), in order to prevent penalties of supervisory entities and legal action by workers, as well as to be able to meet contractual requirements in tendering or contracting processes. Just 27% of employers implement OSHMS to generate safe and healthy workspaces, protecting their workers from accidents and diseases related to working conditions [12].

The limited knowledge of employers facing the minimum requirements of the OSHMS, the benefits of its implementation and the technical aspects of its development, all play a role against prevention of occupational risks, because employers consider that hiring a worker in charge of leading the OSHMS is enough for this purpose, without understanding that by regulation they are responsible for safety and health in their organization, for defining and allocating staff, financial and technical resources for design, implementation, review, evaluation and improvement of prevention and control measures [13].

53% of entrepreneurs consider the resources allocated for the development of the OSHMS to be insufficient (Fig. 2), however, they attribute this situation due to the fact of not being able to evidence the profit obtained or the return of investment in occupational safety and health issues, trying to generate the least impact on the overall budget of the organization [14], developing interventions only to deal correctively with contingencies arise from the occurrence of work accidents or occupational diseases.

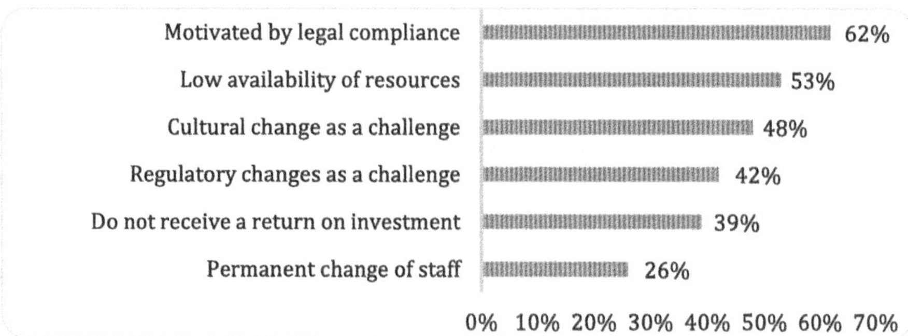

Fig. 2. Businessmen perception of the implementation of OSHMS

The regulatory changes that the Colombian government has been establishing in recent years represent a clear impetus and concern for the improvement of safety and health at work in the country, however, for entrepreneurs it is evidenced as a challenge, because it substantially modified what for more than 30 years has been running as an occupational health program [15], to become a systematic process of management of occupational safety and health [16]. This normative evolution also implies a cultural and awareness transformation not only in the managers of organizations, but in workers of all levels, reflecting their well-being and quality of life [17].

4 Discussion

This research was intended to characterize the management of occupational safety and health in Bogotá-Colombia companies, from the point of view of the entrepreneurs who lead them and the fulfillment of the minimum requirements proposed by the Ministry of protection [8], finding that most companies have a critical level of development, which confirms the established in the II National Survey of Occupational Safety and Health Conditions in the General Risk System [18].

Health and safety seek to build a culture of healthy working environments, safe and decent work, which is why the Colombian state promotes public safety policies based on recommendations from international agencies, some of which are framed in the challenges of the millennium [16], however, a process of promotion, divulgation and technical support by the state and the Labor Risk Administrators - ARL [13], is needed in order to eliminate the gaps between the legal requirements and the reality of business contexts [19], enabling the development of actions focused at promoting healthy work and life styles, safe work environments, and prevention of accidents and diseases in workers [16].

One of the most important challenges shown by the entrepreneurs is the insufficient assignment of an specific budget to do activities related to OSHMS, this is due to the economic situation of the country and of the companies in general, but also due to the inability of occupational safety and health leaders to demonstrate the impact of this process on the productivity of the organizations (cost-benefit) [16] and due to the lack of planning, as most companies do not have specific occupational safety and health objectives, a training plan, or an annual work plan, generating a reactive system focused on contingencies, losing the preventive orientation as well as the continuous improvement of management systems [20].

From the results it can be established that success in the implementation of the OSHMS depends to a large extent on the commitment´s managers, not only by allocating the resources necessary for its implementation, but also by establishing the culture of prevention [21], which must be developed at all levels of the organization, motivating the participation of workers not only as an object of intervention, but as an active part of the solution, considering their experience, and the relationship with the risks according to the conditions in which they are working [2]. Given the above, it is additionally required that the health and safety at work leader is located in the strategic or tactical level of the company, having the possibility to make decisions and to have an easy communication and integration with the other functional areas [13].

One of the limitations of this research and also in the employment general situation in Colombia, is the high rate of business informality, according to data of the National Administrative Department of Statistics (DANE), in the second half of 2019, the informal employment represented 46.2% [22], which involves that almost half of domestic production is not being regulated, and therefore it cannot be assured that employees of these informal companies are protected with safe environments at work and under health care conditions [19].

References

1. Organización mundial de la salud – OMS: Estrategia de Promoción de la salud en los lugares de trabajo de América Latina y el Caribe: Anexo Nº 6 - Documento de Trabajo. Ginebra, Suiza (2000)
2. Gutiérrez, L., Malagón, M.: Estrategias de Éxito para la Implementación de los Sistemas Gestión en Seguridad y Salud en Trabajo, en Investigación en Administración y su Impacto en Comunidades Académicas Internacionales, pp. 2148–2166 (2017)
3. Organización Internacional del Trabajo – OIT. https://www.ilo.org/global/topics/safety-and-health-at-work/lang–es/index.htm
4. Organización iberoamericana de seguridad Social – OISS: Declaración Prevencia. IX congreso de prevención de Riesgos laborales en Iberoamerica, Colombia. http://www.oiss.org/prevencia2016/libponencias/DeclaracionPrevencia2016.pdf
5. Federación de Aseguradores Colombianos – FASECOLDA: Estadísticas del sector. http://www.fasecolda.com/index.php/fasecolda/estadisticas-del-sector/
6. Deming, W.E.: Calidad, Productividad y Competitividad: la salida de la crisis. Ediciones Díaz (1989)
7. Gutiérrez, L., Ortiz, M., Malagón, M.: Gestión de seguridad y salud en el trabajo en PYMES colombianas, en Estrategias regionales de innovación y universidades. Visiones y herramientas para el desarrollo en Iberoamérica, pp. 28 a 42 (2019)
8. Ministerio del Trabajo: Resolución 1111. Por la cual se definen los Estándares Mínimos del Sistema de Gestión de Seguridad y Salud en el Trabajo para empleadores y contratantes (2017)
9. Porter, M.E.: On Competition. Harvard Business School Publishing, Boston (1998)
10. Adolfo, L.: Seguridad e higiene en el trabajo. Marcombo, pp. 13 y 20 (2003)
11. Palatsí, B.: La importancia de la formación y la información para la integración de la prevención, p. 5 (2015)
12. Ruiz Frutos, C., et al.: Salud laboral: conceptos y técnicas para la prevención de riesgos laborales, pp. 327–328 (2007)
13. Ministerio del trabajo: Decreto 1072. Por medio del cual se expide el Decreto Único Reglamentario del Sector Trabajo (2015)
14. Vega, N.: Nivel de implementación del Programa de Seguridad y Salud en el Trabajo en empresas de Colombia del territorio Antioqueño. Cuadernos de Saúde Pública 33 (2017)
15. Ministerio de Trabajo y Seguridad Social: Resolución 1016. Por la cual se reglamenta la organización, funcionamiento y forma de los Programas de Salud Ocupacional que deben desarrollar los patronos o empleadores en el país (1989)
16. Torres, S., Álvarez, H., Riaño, M.: La política pública de seguridad y salud en el trabajo: el caso colombiano. Gerencia y Políticas de Salud, pp. 17–35 (2018)
17. Ministerio de Trabajo: Plan Nacional de Seguridad y Salud en el Trabajo 2013–2021, Bogotá (2013)
18. Ministerio del Trabajo: Informe ejecutivo. II Encuesta Nacional de Condiciones de Seguridad y Salud en el Trabajo en el Sistema General de Riesgos. Grafiq Editores, Bogotá (2013)
19. Alarcón, J., Ortega, A.: Importancia de la seguridad de los trabajadores en el cumplimiento de procesos, procedimientos y funciones. Academia & Derecho, pp. 155–175 (2017)

20. Gutiérrez, L., Malagón, M., Ortiz, M: "Desafíos en la Implementación del Sistema de Gestión de Seguridad y Salud en el Trabajo en Empresas de Bogotá" la Empresa Comunica: Protocolo y lenguaje Organizacional. En: España ed: GEDISA, pp. 169–184 (2018)
21. Sánchez, J: Cultura organizacional en la seguridad y salud en el trabajo. Bachelor's thesis, Universidad Militar Nueva Granada (2013)
22. Departamento Administrativo Nacional de Estadística – DANE: Boletín técnico medición del empleo informal y seguridad social (2020)

Distribution of Occupational Accidents in Coal Fired Thermal Power Plant Using HFACS Technique

Akide Cerci Ogmen(✉) and Ismail Ekmekci(✉)

Istanbul Commerce University, Istanbul, Turkey
akidecerci@gmail.com, iekmekci@ticaret.edu.tr

Abstract. Energy is one of the most important and essential requirements in the social and economic development of countries. Therefore, in order for the energy to be uninterrupted, the necessary measures must be taken before and after. In this study, 50 accident reports occurring in X coal fired power plant between 2016 and 2019 were examined. The causal factors of these accidents were statistically analyzed and classified using the human factor analysis and classification system (HFACS) method. According to the results of the analysis, the inadequate management factor that contributes primarily to human errors in all accidents examined, the other factors are the routine violations and organizational process. The findings of this study will give an idea people in academia of the factors that contribute to accidents in the thermal power plants.

Keywords: Human factors analysis and classification system (HFACS) · Supervision · Organization · Occupational accidents · Coal Fired Thermal Power Plant

1 Introduction

Human error is seen as an inappropriate or unacceptable human decision or action that reduces efficiency, security, and system performance [3]. When the accidents that occur and the sectors they occur are examined, it is difficult to find out what caused this situation if the accidents are caused by the human factor. In accident analysis, if the cause of the accident is not a mechanical failure, the analysis to be performed will be intuitive. However, advanced tests are sufficient to detect mechanical faults. There is the HFACS Method used to create the accident database for the detection of human error in the accidents that occur. The HFACS Method was developed by Shappell and Wiegmann. HFACS is derived from Reason's Swiss Cheese Model [2]. HFACS is a very useful method used in aviation, maritime, health and many other sectors. In the study, the accidents that occurred in the coal-fired thermal power plant were examined. The causes of the accidents were determined using the HFACS method and an accident database was tried to be created.

P. M. Arezes and R. L. Boring (Eds.): AHFE 2020, AISC 1204, pp. 154–159, 2020.
https://doi.org/10.1007/978-3-030-50946-0_22

1.1 Swiss Cheese Model

It is an approach recommended by Reason (1990) that draws particular attention to the occurrence of human error. Reason (1990) explains that accidents can be examined up to four levels of failure. These are Unsafe actions, Prerequisites for unsafe actions are Unsafe control and Organizational effects. The system is similar to the Swiss cheese slice, as shown in the Fig. 1. Cheese represents barriers and protects against failure. Holes also show remaining errors. When holes and errors are lined up at every level in the system, the system is prone to an accident [3]. The Swiss Cheese Model shows the reasons for the occurrence of errors. The model explains that this is a series of events, not necessarily an event that caused an error [7].

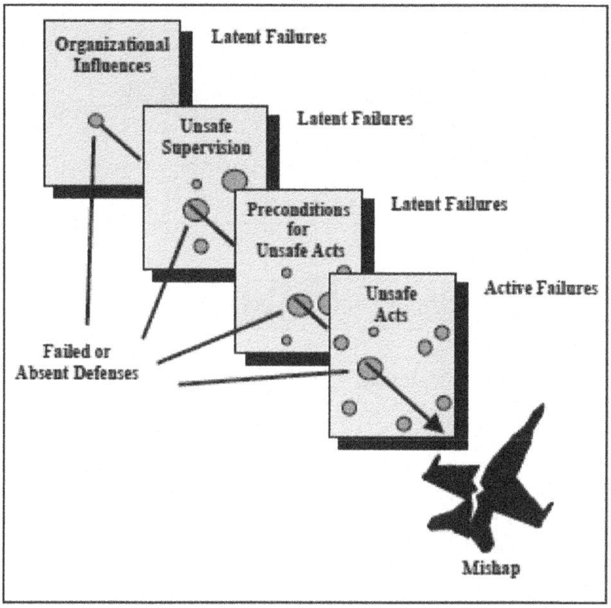

Fig. 1. The "Swiss Cheese" Model of human error cautions [8].

1.2 Human Factors Analysis and Classification Systems (HFACS)

The human factor analysis and classification system (HFACS) was introduced in 2000 by Shappell and Wiegmann, and Reason (1990) is a method derived from the Swiss Cheese Model. It is the most widely used and reliable method to examine and classify the causal factors of accidents and incidents in various industries such as medicine, construction, marine, mining industry, oil, aviation, railway [5].

The human factor analysis and classification system is a model designed to find the root causes of human errors. HFACS has defined the human factor at four levels, namely Unsafe Acts, Preconditions for Unsafe Acts, Unsafe Supervision and Organization Influences [1]. Within these four levels, there are many subcategories that further explain the contributing human factor [4, 6]. The HFACS framework is shown in Fig. 2 [6].

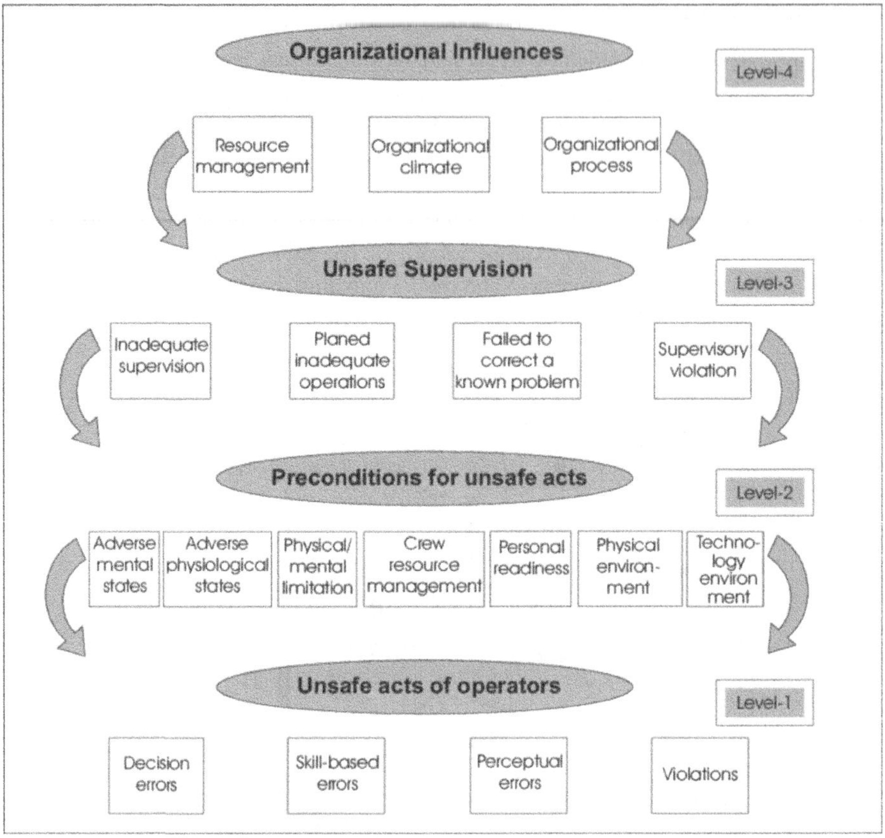

Fig. 2. HFACS framework, each upper level would affect downward level, proposed by Wiegmann and Shappel [6].

HFACS examines human error at four levels. Each higher level affects the next downward level in HFACS framework.

Level-1 Unsafe acts of operators: This level is where the majority of causes of accidents are focused. Such causes can be classified into the two basic categories of errors and violation.
Level-2 Preconditions for unsafe acts: This level tackles the hidden failures within the causal sequence of events as well as more obvious active failures. It also identifies the context of substandard conditions of operators and the substandard practices they adopt.
Level-3 Unsafe supervision: This level traces the causal chain of events producing unsafe acts up to the front-line supervisors.
Level-4 Organizational influences: This level involves the hardest to get of these latent failures, fallible decisions of upper levels of management, which directly affect supervisory practices, as well as the conditions and actions of front-line operators [6].

2 Methodology

2.1 Data Sources

This study is based on the analysis of accidents that occurred in X Coal Fired Thermal Power Plant between 2016 and 2019. Accidents will be classified according to the HFACS method and an accident database will be occurred.

2.2 Analysis of the Data

50 serious accidents occurred between 2016 and 2019 at the X coal fired thermal power plant. These accidents have occurred due to falling from height, fracture, falling object, fire, chemical contact. The causes and percentages of the accidents are given in Table 1. According to Table 1, the highest accident causality is object crash with 24%. Incision shows the lowest cause of accident with a rate of 2%.

Table 1. Accident causality and percentages

Accident causality	Percentage
Crushed	16%
Object crash	24%
Fall from height	18%
Broken	6%
Burnt	10%
Chemical contact	4%
Hand tools use	4%
Incision	2%
Object drop	12%
Explosion	4%

In order to classify the factors underlying 50 boilers in the X coal fired thermal power plant, 280 categories have been assigned to each of the four HFACS categories. The obtained statistics are analyzed and shown separately in the tables. 50 thermal power plant accidents are shown in tables with HFACS codes (Tables 2, 3, 4 and 5).

Table 2. The frequency and percentage of 50 accidents by HFACS Unsafe Acts of Operators Categories

Unsafe Acts of Operators Categories	Frequency	% of all accidents
Decision error	18	36
Skill-based error	20	40
Perceptual errors	2	4
Routine violations	36	72
Exceptional violations	3	6

Table 3. The frequency and percentage of 50 accidents by HFACS Preconditions for Unsafe Acts Categories

Preconditions for Unsafe Acts Categories	Frequency	% of all accidents
Adverse Mental States	30	60
Adverse Physiological States	3	6
Physical/Mental Limitation		
Crew Resource Management	26	52
Personal Readiness		
Physical Environment	10	20
Technology Environment	11	22

Table 4. The frequency and percentage of 50 accidents by HFACS Unsafe Supervision Categories

Unsafe Supervision Categories	Frequency	% of all accidents
Inadequate Supervision	49	98
Planed Inadequate Operations	5	10
Failed to Correct a Know Problem		
Supervisory Violation	23	46

Table 5. The frequency and percentage of 50 accidents by HFACS Organizational Influences Categories

Organizational Influences Categories	Frequency	% of all accidents
Resource Management	3	6
Organizational Climate	5	10
Organizational Process	35	70

3 Result and Discussion

When the tables are analyzed, it is seen that the most important contributing factor to the accidents is the inadequate management factor. Inadequate management factor indicates the primary cause of accidents that cause accidents in all types of operations in thermal power plants. It is undeniably important to carry out the necessary supervision and to provide the necessary training to the staff. The data also showed that accidents are errors that can also be caused by organizational processes and routine violations. Since accidents are caused by inadequate management, management should implement more innovative approaches and conduct regular inspections.

50 accidents occurring in the thermal power plant were examined and the human factors contributing to the accidents were classified correctly. With the help of the HFACS method, it was determined that accidents occurring in the thermal power plant were caused by inadequate management, routine violations and human errors caused by the organizational process. The data showed that the criteria in conditions that

preconditions for unsafe acts had the lowest percentage of accidents. As a result, thc management unit is more important to reduce the probability of human error in accidents.

References

1. Bakar, Y.: Evaluating the contribution of Human Factors Analysis and Classification System (HFACS) the accidents investigation model to flight safety, III. National Aviation Technology and Applications Congress, İzmir, 23–24 (2016)
2. Dönmez, K., Uslu, S.: Insan Faktörleri Analiz ve Sınıflandırma Sistemi'nin (HFACS) Literatürde Yaygın Kullanımının Değerlendirilmesi. J. Aviat. **2**(2), 156–176 (2018). https://doi.org/10.30518/jav.463607. e-ISSN 2587-1676
3. Ergai, A.: Assessment of the Human Factors Analysis and Classification System (HFACS): intra-rater and inter-rater reliability. Clemson University (2013). http://tigerprints.clemson.edu/all.dissertations/1231
4. Inglis, M., Sutton, J., McRandle, B.: Human Factors Analysis of Australian Aviation Accidents and Comparison with the United States (2007). ISBN 1.92109259.9
5. Kılıc, B.: HFACS analysis for investigating human errors in flight training accidents. J. Aviat. **3**(1), 28–37 (2019). https://doi.org/10.30518/jav.553315. e-ISSN 2587-1676
6. Li, W., Harris, D.: HFACS analysis of roc air force aviation accidents: reliability analysis and cross-cultural comparison. Int. J. Appl. Aviat. Stud. **5**(1), 65–81 (2005)
7. Semeryuk, N.: The Swiss Cheese Model. HCM 370 Quality and Risk Management in Healthcare Colorado State University (2015)
8. Shappell, S., Wiegmann, D.A.: The Human Factors Analysis and Classification System – HFACS. U.S. Department of Transportation Federal Aviation Administration, Washington (2000)
9. Shappell, S., Wiegmann, D.A.: A Human Error Analysis of Commercial Aviation Accidents Using the Human Factors Analysis and Classification System (HFACS). U.S. Department of Transportation Federal Aviation Administration (2001)

A Fuzzy Decision Making Method for Preventing the Loss of Knowledge in Nuclear Organizations

Jaqueline Vianna[1,2], Paulo V. R. Carvalho[1,2], Carlos A. N. Cosenza[2], and Claudio H. S. Grecco[1,2(✉)]

[1] Instituto de Engenharia Nuclear, Comissão Nacional de Energia Nuclear, Rio de Janeiro, Brazil
jaqueline.vianna@bolsista.ien.gov.br, {paulov,grecco}@ien.gov.br
[2] Laboratório de Lógica e Matemática Fuzzy (LabFuzzy), COPPE/UFRJ, Rio de Janeiro, Brazil
cosenzacoppe@gmail.com

Abstract. Appropriate technical expertise and experience, along with a strong safety culture, must be developed and kept available throughout the nuclear technology life cycle. For organizations using nuclear technology, the ability to take safe decisions and actions can be affected by knowledge gaps or knowledge loss. For preventing the loss of knowledge in nuclear organizations, it is important to identify the barriers or critical factors that affect the success of the knowledge management (KM) process. From the perspective of the nuclear organizations, no systematic structure exists on characterizing a set of critical success factors (CSFs) for KM. Furthermore, there are no appropriate models that can satisfactorily assess the CSFs and fully solve the subjectivity of KM assessment. In this context, this work presents a fuzzy decision making method for preventing the loss of knowledge in nuclear organizations using CSFs able to identify the barriers that affect the KM.

Keywords: Knowledge Management · Fuzzy logic · Human factors · Safety culture

1 Introduction

The knowledge of workers constitute as valuable resources, as they enable organizations to perform their functions successfully. Nuclear projects usually last for many years or decades, and can be divided into numerous phases involving different stakeholders. Nuclear equipment, installations and facilities may have long life cycles with changing operational conditions. The safe use of licensed nuclear facilities and technologies is dependent on the ongoing availability and maintenance of suitable knowledge and expertise, and an adequate understanding of related safety issues.

There is an evident risk that in the absence of knowledge transfer plans, essential knowledge can be lost between different phases of a nuclear facility lifetime, for a variety of reasons [1]. The most evident risk is that for every phase of a nuclear project,

P. M. Arezes and R. L. Boring (Eds.): AHFE 2020, AISC 1204, pp. 160–167, 2020.
https://doi.org/10.1007/978-3-030-50946-0_23

a different workforce is employed. Some competences needed during the design phase may not be required during operation, but at the operation stage, new competences will be needed. When moving from one phase of a nuclear facility lifetime to another, independently of other risk factors, knowledge gaps, such as personnel attrition, diminishing job tenures, decreasing availability of skills on the market for the nuclear sector and reducing knowledge transfer between generations, can be created [1]. Organizational and workforce changes are not the only risks related to knowledge preservation in a nuclear organization. Some projects related to IT and data management might implement different information systems, with different media storage formats that require constant upgrades [2].

Furthermore, there are others conditions that favor the loss of this knowledge in organizations, as for example, the natural aging of workers and consequently the retirement and staff turnover. Then, it becomes important for organization to seek the preservation these knowledge. For a successful implementation of Knowledge Management (KM), it is important to identify the barriers or critical factors that affect the success of the KM process.

From the perspective of the nuclear organizations, no systematic structure exists on characterizing a set of critical success factors (CSFs) for implementing KM. Additionally, the CSFs assessment deals with imprecision of human judgments. In this context, this paper presents a method for preventing the loss of knowledge in nuclear organizations, based on: 1) the use of critical success factors in order to be able to monitor the success of a knowledge management initiative; 2) the approach of knowledge management in the development of critical success factors, which are on seven themes: top-level commitment, organizational culture, organizational structure, human resources management practices, measuring and results, information technology and learning culture; 3) the use of properties of decision making and fuzzy set theory to model of critical success factors. Fuzzy theory is essentially used in mapping quantitative models for decision making and representation methods in imprecise and uncertain environments; 4) the prioritization of CSFs of a given organization to serve as a standard for the KM process assessments.

2 Critical Success Factors

The set of critical success factors can act as a list of items for organizations to address when adopting knowledge management. This helps to ensure that the essential issues and factors are covered during design and implementation phase. For academics, it provides a common language for them to discuss and study the factors crucial for the success of knowledge management program in an organization.

For a successful implementation of knowledge management, it is important to identify the barriers or critical factors that affect the success of the KM process. Interest associated with barriers and critical success factors rose strongly after 2003. This was also because the basic models of KM already existed, which, despite not providing a solution to the failures of KM projects, created a strong basis to build basic concepts and identification and description of the process steps of knowledge management. Many empirical studies are focused of them, for example, much has been stated about culture, information technology and leadership as important considerations for its accomplishment [2].

However, no systematic work exists on characterizing a collective set of CSFs for implementing KM in nuclear organizations. CSFs are critical areas of managerial planning and action that must be practiced in order to achieve effectiveness. In terms of KM, CSFs can be viewed as those activities and practices that should be addressed in order to ensure its successful implementation. These practices would either need to be nurtured if they already existed or be developed if they were still not in place [1].

Our research aims to propose a CSFs framework based on seven themes to identify barriers or critical factors that affect the success of the knowledge management process.

3 Fuzzy Decision Making for CSFs Modeling

Fuzzy decision making is a powerful tool for dealing with human expert knowledge when one is designing an appropriate logical-mathematical. This tool handles with problems with such characteristics [2–4]: (1) it allows the computation of variables linguistics; (2) it allows the modeling of the heuristic knowledge that cannot be described by traditional mathematical equations and; (3) it deals with uncertainty and imprecision of situations and reasoning processes.

Fuzzy set theory (FST) is an outgrowth of the classical set theory where elements have degrees of membership. Let X be the universe of discourse and x be its elements. According to the classical set theory, crisp set B of X is defined by the characteristic function $f_B(x)$ of set B. A fuzzy set B of X is defined by its membership function $\mu_B(x)$. The $\mu_B(x)$ is the membership grade x in B. The membership function associates to each element x of X, a real number $\mu_B(x)$, in the interval [0, 1]. A fuzzy number is a quantity whose value is imprecise, rather than exact as is the case with "ordinary" (single-valued) numbers. Its membership function is a continuous mapping from R (real line) to a closed interval [0, 1]. A triangular fuzzy number B can be denoted by (a, b, c) (Fig. 1) and its membership function is described in Eq. 1.

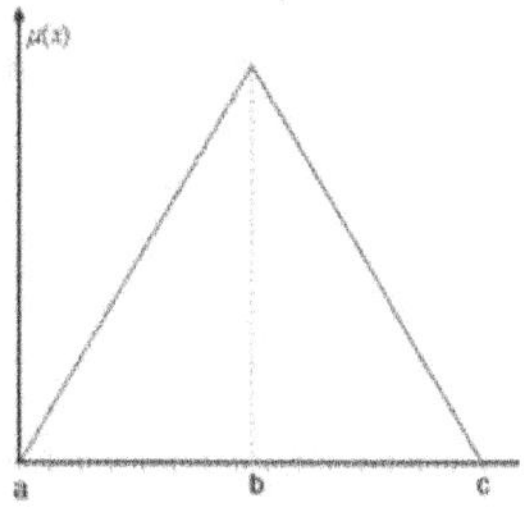

Fig. 1. Triangular fuzzy number

$$\mu_B(x) = \begin{cases} \frac{x-a}{b-a}, & if\ a \leq x \leq b \\ \frac{c-x}{c-b}, & if\ b \leq x \leq c \\ 0, & otherwise \end{cases} \tag{1}$$

An important concept in fuzzy set theory is the concept of linguistic variables. A linguistic variable is a variable whose values are words or sentences in natural language. We can consider a CSF as a linguistic variable represented by set of four linguistic terms (Unimportant, Little Important, Important and Very Important) which satisfy to the importance degrees used to assess the weight of this CFS by experts.

4 Fuzzy Decision Making Method for Preventing the Loss of Knowledge

The method developed in this paper was structured according to the following steps: (1) determination of the CSFs framework; (2) prioritization of CSFs.

4.1 Determination of the CSFs Framework

The list of CSFs was developed in seven themes based on the literature [1, 2, 4–7]. Table 1 list the themes and exemplify some CSFs and the metrics used to calculate their values.

4.2 Prioritization of CSFs

The second step of this fuzzy framework is to obtain from experts on KM the degree of importance of each CSF, so that the implementation of KM in organization can be considered good. The relative importance of the expert will be calculated on the basis of subjective attributes (experience, knowledge of KM). We will use a questionnaire (Q) to identify the profile. Each questionnaire will contain information of a single expert. The relative importance of expert (RIEx) Ex_i (i = 1, 2, 3, … , k) will be a subset μ_i (k) Є [0, 1] defined by Eq. 2. According to the Eq. 2, *tsQi*, will be the total score of expert *i*.

$$RIEx_i = \frac{tsQ_i}{\sum_{i=1}^{n} tsQ_i} \tag{2}$$

Each CSF can be seen as a linguistic variable, related to a linguistic terms set associated with membership functions. These linguistic terms will be represented by triangular fuzzy numbers to represent the importance degree of each CSF (Fig. 2). The linguistic terms will be: U (Unimportant), LI (Little Important), I (Important) and VI (Very Important) to evaluate the importance of each CSF.

The similarity aggregation method [8] will be used to combine the experts' opinions which are represented by triangular fuzzy numbers. The agreement degree

Table 1. Themes, CSFs and metrics

Themes	CSFs	Metric used
1 Top-level commitment	1.1 Mission and values 1.2 … 1.3 …	1.1 …clear definition of the mission and values of the institution …
2 Organizational culture	2.1 Organizational climate 2.2 … 2.3 …	2.1 …positive environment, encouraging the knowledge sharing …
3 Organizational Structures	3.1 Multidisciplinary groups 3.2 … 3.3 …	3.1 …groups are multidisciplinary and have autonomy in decisions …
4 Human resources management practices	4.1 Training 4.2 … 4.3 …	4.1 …training is often offered and encouraged by managers …
5 Measurement of results	5.1 Investment monitoring 5.2 … 5.3 …	5.1 …adequate monitoring of results of investments in training …
6 Information Technology	6.1 Technological structure 6.2 … 6.3 …	6.1 …technological structure for the storage of knowledge… …
7 Learning culture	7.1 Context changes 7.2 … 7.3 …	7.1 …adequate learning to accept new practices… …

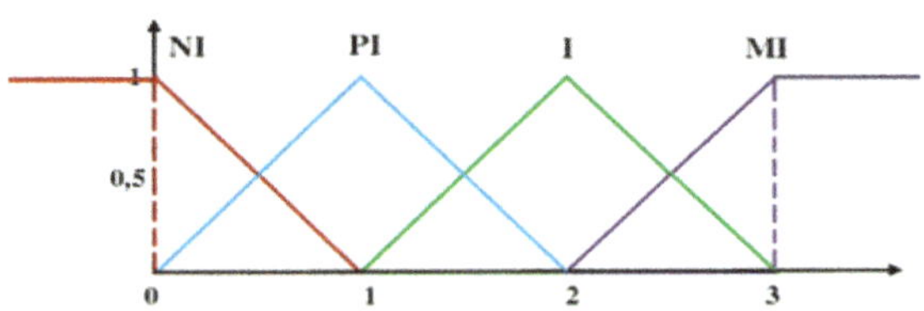

Fig. 2. Membership functions

(A) between expert Ex_i and expert Ex_j will be determined by the proportion of intersection area to total area of the membership functions. The agreement degree (A) is defined by Eq. 3.

$$A = \frac{\int_x \left(\min\left\{\mu_{\tilde{N}i}(x), \mu_{\tilde{N}j}(x)\right\}\right)dx}{\int_x \left(\max\left\{\mu_{\tilde{N}i}(x), \mu_{\tilde{N}j}(x)\right\}\right)dx} \tag{3}$$

If two experts have the same estimates, the agreement degree between them will be one. If two experts have completely different estimates, the agreement degree will be zero. The higher the percentage of overlap, the higher the agreement degree. After all the agreement degrees between the experts calculated, we will construct an agreement matrix (AMX), which will give us insight into the agreement between the experts.

$$AMX = \begin{bmatrix} 1 & A_{12} & \cdots & A_{1j} & \cdots & A_{1n} \\ \vdots & \vdots & & \vdots & \vdots & \\ A_{i1} & A_{i2} & \cdots & A_{ij} & \cdots & A_{in} \\ \vdots & \vdots & & \vdots & \vdots & \\ A_{n1} & A_{n2} & \cdots & A_{nj} & \cdots & 1 \end{bmatrix}$$

The relative agreement of expert (RAEx) Ex_i (i = 1, 2, 3, … , k) will be given by Eq. 4.

$$RAEx_i = \sqrt{\frac{1}{n-1} \cdot \sum_{j=1}^{n} (A_{ij})^2} \tag{4}$$

The calculate of the relative agreement degree of expert (RADEx) Ex_i (i = 1, 2, 3, … , k) by Eq. 5 and the consensus coefficient of expert (CCEx) Ex_i (i = 1, 2, 3, … , k) will be given by Eq. 6.

$$RADEx_i = \frac{RAEx_i}{\sum_{i=1}^{n} RAEx_i} \tag{5}$$

$$CCEx_i = \frac{RADEx_i \cdot RIEx_i}{\sum_{i=1}^{n} (RADEx_i \cdot RIEx_i)} \tag{6}$$

Let W be a fuzzy number for combining expert's opinions. W is the fuzzy value of each CSF which is also triangular fuzzy number. By definition of the consensus coefficient of expert (CCEx) Ex_i (i = 1, 2, 3, … , n), W can be defined by Eq. 7.

According to the Eq. 7, w_i, is the triangular fuzzy number relating to the linguistic terms, U (Unimportant), LI (Little Important), I (Important) and VI (Very Important).

$$W = \sum_{i=1}^{n} (CCEx_i \cdot w_i) \tag{7}$$

The prioritization of CSF as a standard for assessing the CFSs is established by calculating the normalized importance degree (NID) of each CSF that make up each property relevant to prevent the loss of knowledge. The normalized importance degree (NID) of each CSF will be given by defuzzification of its triangular fuzzy number $W\,(a_i,\, b_i,\, c_i)$, where b_i represents the importance degree. Then, NID will be defined by Eq. 8.

$$NID_i = \frac{NID_i}{\text{the largest numerical value of } b_i} \tag{8}$$

5 Research Status and Results

To exemplify the use of the fuzzy decision making method we will performed an exploratory case study on the complex systems engineering department located in a research institute of the Brazilian Nuclear Energy Commission. This research institute contributes to the scientific community by facilitating experiments in nuclear engineering. The complex systems engineering department aims to perform studies on human factors engineering, human reliability analysis and artificial intelligence techniques. At present, we selected the team of experts for determination of the degree of importance of each CSF, so that the implementation of KM in complex systems engineering department can be considered good. The team of experts is comprised fourteen researchers with experience and, knowledge of KM. Afterwards, the relative importance score assigned to each expert will be determined by a questionnaire with ten questions, whose items were associated with a score.

6 Conclusions

In this paper, we described a fuzzy decision making method for preventing the loss of knowledge in nuclear organizations using CSFs able to identify the barriers that affect the KM. We develop a prioritization of CSFs using a similarity aggregation method to aggregate fuzzy individual opinions, considering the difference of importance of each expert. This prioritization is important to evaluate the CSFs that will influence the efficiency of the implementation of KM in nuclear organization. This means that this fuzzy approach is a proactive tool to provide a basis for checking the CFSs in organization. As the result, the organizations that use this method will be able to proactively evaluate and manage knowledge.

References

1. IAEA, Nuclear Energy Series NG-T-6.11: Knowledge loss risk management in nuclear organizations, International Atomic Energy Agency, Vienna (2007)
2. Wong, K.Y.: Critical success factors for implementing knowledge management in small and medium enterprises. Ind. Manag. Data Syst. **105**(3), 261–279 (2005)
3. Zadeh, L.A.: Fuzzy logic - computing with words. IEEE Trans. Fuzzy Syst. **4**, 103–111 (1996)
4. Takeuchi, H., Nonaka, I.: Gestão do Conhecimento, 1st edn. Porto Alegre, Brasil (2008)
5. Terra, J.A.: Gestão do Conhecimento: O Grande Desafio Empresarial, 5th edn. Rio de Janeiro, Brasil (2005)
6. Wang, J., Peters, H.P., Guan, J.: Factors influencing knowledge productivity in German research groups: lessons for developing countries. J. Knowl. Manag. **10**(4), 113–126 (2006)
7. Figueiredo, S.: Gestão do Conhecimento – Estratégia Competitiva para a Criação e Mobilização do Conhecimento na Empresa, 4th edn. São Paulo, Brasil (2005)
8. Hsu, H.M., Chen, C.T.: Aggregation of fuzzy opinions under group decision making. Fuzzy Sets Syst. **79**, 279–285 (1996)

Assessing Risks

Approaches to Human Performance Modeling of Electric Grids Operators

Ruixuan Li(✉) and Katya Le Blanc

Idaho National Laboratory, Idaho Falls, ID, USA
{Ruixuan.Li,Katya.LeBlanc}@Inl.gov

Abstract. Due to the integration of digital technology, electric grids have become more complex and vulnerable to cyber-attacks. A worst-case scenario cyber-attack could result in a blackout, leading to significant financial loss or life-threatening events. For prioritizing limited investment, cascading failure analysis with simulation is one way to understand impacts and to provide insights into how a blackout might happen. Human operators play an important role in mitigating risk or maintaining the system state to be considered in the simulation. In this research, we review the cognitive architecture model, task analysis approach, and human reliability analysis in the context of characterizing humans in electric grid cascading failure analysis. Recommendations for an operator modeling method are made to address the scalability.

Keywords: Human performance modeling · Cyber-attack · Cascading failure · Risk analysis · Power systems reliability

1 Introduction

Along with the development of information technology, the power industry has started to take advantage of digital systems and communication networks for meeting the increasing complexity of electric grids in the 21st century. Two-way communication between the utility and customers, remote monitoring and checking, and distributed power generation have become common in the industry to minimize the impact of outages [1]. It allows operators to respond to equipment failures and unplanned events to prevent power system collapse. However, unlike analog systems, which typically can only be compromised through physical access, the digital-analog system is more complex and vulnerable to cyber-attack. One or more major security vulnerabilities have been detected in 96% of the analyzed digital applications in 2013 [2]. Though those data are not necessarily representative of the electric utility industry, it highlights that vulnerabilities likely exist in the power industry due to the fast evolution of technology and attack capabilities.

The Ukraine power outage caused by cyber-attack in 2015 is an example of how cyber security can impact national security [3]. Attackers could compromise the confidentiality and integrity if the software and network architecture vulnerabilities had not been minimized, or if there were not enough cyber protections (e.g., firewalls, intrusion detection system) in place. The availability and controllability of critical components could also be threatened through the communication network remotely. However, with

P. M. Arezes and R. L. Boring (Eds.): AHFE 2020, AISC 1204, pp. 171–177, 2020.
https://doi.org/10.1007/978-3-030-50946-0_24

the limited resources for investment in security controls and technology, it is necessary to analyze the potential impacts of the component or functional failures, which can result from cyber-attacks.

Cascading failure analysis and simulation are one way to analyze the expected and potential impacts of cyber-attacks. The event failure paths of simulation scenarios provide system evolution insights in response to unexpected failures. Operators play an essential role in cascading failure events, and the insufficient situation awareness of operators was one of the factors in multiple disturbances [4]. Operators can also reduce the risk of a system by taking appropriate mitigating actions and prevent widespread outages. Modeling operators' performance is essential for understanding impacts and which steps or actions might disturb the power system further.

Ideally, the human performance modeling method in cascading failure analysis needs to have the following characteristics:

1. The model needs to be complex enough to model operators' effects on electric grids. Operators' behaviors and actions (e.g., re-dispatch generation, re-route power, or shed load), which might change system state, are essential.
2. The model needs to be simple enough to be replicated and expanded to a larger-scale electric grid network. If the modeling methods are overly complicated, it could not be implemented in complicated settings.
3. The operator model needs to have a response or reaction time for each action to fit in the cascading failure simulation. During event propagation, operators respond to the system states based on the operating procedures and their interpretations of what might go wrong. The response time might include the time that it takes operators to detect and identify problems, to make decisions, to communicate, and to carry out actions.
4. The model needs to have the probabilities of operator actions built in to consider uncertainties. Operators might make different decisions, especially in ambiguous situations. The deviations can be due to various factors (e.g., experience, training, stress), which is hard to capture and quantify in the working context. Having the probabilities of operator actions in the model is to simplify the problem without losing individual differences.
5. The operator model needs to include omission and commission errors. Omission errors are relatively straightforward to model, but most errors resulting in severe consequences are commissions due to the lack of situation awareness [5]. Commission errors, which are hard to model, are also important to be considered in this context.
6. The model construction should have some data to support its validity but not rely on large data samples because of the limited time and data resources that we have. Ideally, collecting electric grid operational data would be the best way to model operator actions. However, the time and complexity of collecting large scale operational data for modeling actions is prohibitive for the current research.

In this paper, we review cognitive architecture modeling methods, task analysis approaches, and human reliability analysis methods in human performance modeling based on the characteristics of the human performance modeling objectives described above. These modeling approaches are not mutually exclusive and can be used in combination. The overall goal of the current paper is to provide insights and research directions for modeling operators in cascading failure simulation.

2 Cognitive Architectures

Cognitive architectures generally have a set of assumptions about information processing mechanisms (e.g., knowledge representations and beliefs, goals, and perception and motor modules) [6]. The layouts and interconnecting relationship between modules are slightly different to represent various aspects of human cognition. Cognitive architectures help synthesize cognitive psychology theories and findings and provide a bigger picture of how cognitive, perception, and motor modules work together. They also play an important role in interconnecting cognitive psychology to artificial intelligence (AI). Cognitive architectures have the capabilities to simulate human perception and actions, which satisfy the first need of human performance modeling methods in cascading failure simulation context.

Cognitive architectures often have a set of "if-then" rules in the production systems and can adapt to new environments. The architecture can be seen as hardware, whereas the rule-based production functions are software that runs on the hardware. Cognitive architectures often contain two types of memory: short-term and long-term memory. Short-term memory stores dynamic information that changes over time, while long-term memory has persistent content, rules, and knowledge during the system execution period. Adaptive Control of Thought-Rational (ACT-R) and Executive-Process/Interactive Control (EPIC) models have similar construction of modules that send inputs and outputs to the external world. In ACT-R, production rules are in serial [7]; the probability of the next action has been used to model the reinforcement learning process; researchers have used the predictions from ACT-R to locate neural processes in the brain [8]. In EPIC, production rules can fire in parallel [9]; it does not consider the probability of next actions; this mechanism lacks network interactions, and it does not explain the sequential neural network activation [10]. Though ACT-R has fewer disadvantages than EPIC, the serial and parallel neural network processing challenges the assumptions which these two models have. To better accommodate neuroscience and psychological findings, the queueing network-model human processor (QN-MHP) was developed with both serial and parallel neural network structures by integrating queueing network operations research theories [11]. QN-MHP has been decomposed to QN servers under perceptual, cognitive, and motor subnetworks. Because the cognitive QN servers might have more than one route to move forward, the reaction time in the reinforcement learning process can be described by the probability of choosing an information processing route. Each QN server has productional procedures which are represented by the goal, operator, method, and selection (GOMS) task analysis. Except for the fact that QN-MHP can have an accurate estimation of reaction time, fixation duration, and the number of fixations, human behaviors generated by production rules in QN-MHP is not much different from a traditional GOMS, which is discussed in the following session.

Cognitive architectures can describe the operators' actions. Researchers often use experimental data to support model construction, especially defining the response time or reaction time based on the cognitive processes. The possibility of operator actions can be described by the reinforcement process or adding the probability of actions in the model. The omission (possibility of not taking actions) and commission errors (possibility of taking an incorrect action) can be covered. However, building a cognitive architecture model is not simple. It requires a good understanding of inferential cognitive processes in electric grid operation. The simplicity criterion has not been satisfied because different system layouts can result in different cognitive processes, and the operator model for each system must be custom built based on experiments with the cognitive architectural approaches.

3 Task Analysis Approaches

Task analysis methods, which describe users' activities, decision making, and problem-solving strategies, analyze what tasks users try to do [12]. These methods sometimes focus on users' actions on an interface and sometimes emphasize the users' reasoning, inferential, and heuristic process. These two approaches are not mutually exclusive, but researchers often emphasize one over the other dependent on their needs. They are equally important in the context of simulating electric power grid operators' actions.

Some methods (e.g., keystroke-level model (KLM), GOMS) were constructed for analyzing interface design and human-machine interactions through simulations so that no participant is needed during the initial protocol design phase. KLM links time to various keystroke and mouse movements, and the time to complete a task is the sum of each typing and mouse moving time for the task [13]. GOMS was introduced to predict user's behaviors by considering goals, operators, and the constraint of the task [14]. In GOMS, operators are elementary perceptual, motor, or cognitive acts, whose exertion is necessary to change any aspect of the user's mental state or to affect the task environment. The structure of the goals is in the hierarchy that the overall goal can be decomposed to goals and sub-goals, and there are selection rules to choose the proper method to go forward. It can predict the execution time of a task and the operator sequence. These approaches are designed to model experts' execution of tasks that the sequence of operation is fixed given a system state. However, as it is known that expert judgments and elicitations deviate considerably, combining the probability with GOMS can overcome the difficulty of describing uncertainties.

Other common methods (e.g., decision ladder (DL) [15], hierarchical task analysis (HTA) [16]) represent static processes. They were built for analyzing what information is important to support accomplishing goals and decision making in a complex work environment. Common data collection methods include interviews, self-reports, observations, and automated capture, operating procedures, performance records [12]. DL is often implemented at the second stage of work domain analysis after the abstraction hierarchy. It describes possible cognitive activities, cognitive states, and actions. The decision ladder structure, with information processing activities and states at the left side and the decision-making states and actions at the right side, allows researchers to discover possible shortcuts between actions and states. HTA is a goal-directed method and can be understood through the hierarchical structure. HTA and GOMS share some principles that the meaning of goal is the same. The operation in

HTA is specified by a goal, the conditions under which the goal is activated (or the input), the actions by which the goal is achieved, and the feedback. Actions are like the *operators*; the input is like "if-then" rules starting condition; the feedback is explicit while implicit in GOMS. In HTA, researchers sometimes emphasize the hierarchical goal structure over the ultimate stop rule, which is "stop when you have all the information you need to meet the purposes of the analysis." It can be the same as the stop rule in the risk assessment process to "stop when the product of the probability of failure and the cost of failure is judged acceptable" [16]. Like human reliability analysis (HRA), it asks not only what should happen but what might happen with estimate probability and cost of failure. However, it does not include response time intentionally due to the complexity of nonrepetitive tasks for conducting workload and error assessment, designing a new system, and determining training content or method. Along with the improvement of computing speed, simulating complex tasks with response time becomes applicable.

Task analysis approaches are complex enough to model operators' actions. The data collection process is simple when ignoring the *operators* and describing tasks at a higher level because the probability of actions and response time distribution can be obtained by interviews and questionnaires. The possibility of operators' actions can be added in task analysis approaches, so omission and commission errors can be included.

4 Human Reliability Analysis (HRA)

HRA aims to estimate and reduce the probabilities of human errors to an acceptable minimum [17]. Most commonly implemented HRA methods (e.g., Technique for Human Error Rate Prediction (THERP), A Technique for Human Error Analysis (ATHEANA), Standardized Plant Analysis Risk-Human Reliability Analysis (SPAR-H), Cognitive Reliability Error Analysis Method (CREAM)) are static [18]. The basic human error probability (HEP) can be calculated based on expert estimation or the use of performance shaping factors (PSFs). PSFs often include available time, stress and stressors, complexity, experience and training, procedures, ergonomics and human-machine interface, fitness for duty, and work processes. HRA methods have different weight multipliers towards the PSFs, and the effect of PSFs toward the HEP is not always the same. In SPAR-H, the PSFs have both positive and negative effects on performance. In CREAM, the positive effects include the influence of time, training, and work processes on task performance reliability.

Probability risk assessment (PRA) (e.g., fault tree analysis (FTA) and event tree analysis (ETA)) can be used to determine the probability of a specific outcome. ATHEANA considers the error-forcing context (EFC) and creates human failure events (HFEs) in PRA. THERP considers both the error of omission and the error of commission, which includes the error of selection, error of sequence, error of timing, and error of quality, but it does not consider errors during decision making, which can result in severe impacts of the system. HRA methods satisfy most criteria in developing operator models except having some data to support the validity of the operator model. One serious problem of HRA is that it often does not collect performance data, which leads to low accuracy in quantifying human error probabilities in a complex system [17]. It can introduce uncertainty and threaten the validity in cascading failure analysis.

Dynamic HRA is a more recent approach in analyzing human errors. One of the goals of conducting dynamic HRA is to calculate the contextual HEPs through dynamic Monte Carlo simulations. GOMS-HRA has been developed to link subtasks with HEPs, which is adjusted by PSFs changing over time [19]. The corresponding time of completing a subtask can be from the operators' response time probabilistic distribution. This approach requires the decomposition of tasks to perceptual, cognitive, and motor levels in which the operators' actions are described. It is complicated and maybe unnecessary to describe the operational environment when the deviation of tasks execution and completion time is higher than each operator's response time. It does not fulfill the simplicity criterion.

5 Conclusions

In this paper, we discuss some human performance models in the context of electric grid cascading failure analysis. Cognitive architectural approaches often require a good understanding of tasks at the information processing level, which makes the simulation complex. Task analysis approaches are more appropriate for the complicated environment, especially the HTA with decomposition stop rule, which excludes unnecessary details. Limited performance data is a problem in HRA, so whether including human error probability in the simulation would be dependent on research purposes. In the context of cyber-attack, the omission error probability, which is relatively small, might be ignored due to the high computational cost. Observations and interviews can be used to decompose tasks. And questionnaires can be used in collecting the probability distribution of decisions and response time, which can be calibrated by Cooke's classical method [20]. Overall, we recommend combining the task analysis approach, response time, and the probability of actions to build the operator model.

References

1. Farhangi, H.: The path of the smart grid. IEEE Power Energy Mag. **8**(1), 18–28 (2009)
2. Bendovschi, A.: Cyber-attacks–trends, patterns and security countermeasures. Proc. Econ. Finance **28**, 24–31 (2015)
3. Venkatachary, S.K., Prasad, J., Samikannu, R.: Economic impacts of cyber security in energy sector: a review. Int. J. Energy Econ. Policy **7**(5), 250–262 (2017)
4. Bialek, J., Ciapessoni, E., Cirio, D., Cotilla-Sanchez, E., Dent, C., Dobson, I., Henneaux, P., Hines, P., Jardim, J., Miller, S., Panteli, M.: Benchmarking and validation of cascading failure analysis tools. IEEE Trans. Power Syst. **31**(6), 4887–4900 (2016)
5. Maas, G.A., Bial, M., Fijalkowski, J.: Final report-system disturbance on 4 November 2006. Union for the Coordination of Transmission of Electricity in Europe, Technical report (2007)
6. Langley, P., Laird, J.E., Rogers, S.: Cognitive architectures: research issues and challenges. Cogn. Syst. Res. **10**(2), 141–160 (2009)
7. Anderson, J.R.: The Architecture of Cognition. Psychology Press, London (2013)

8. Borst, J.P., Anderson, J.R.: A step-by-step tutorial on using the cognitive architecture ACT-R in combination with fMRI data. J. Math. Psychol. **76**, 94–103 (2017)
9. Kieras, D.E.: A summary of the EPIC cognitive architecture. In: The Oxford Handbook of Cognitive Science, vol. 1, p. 24 (2016)
10. Ku, Y., Hong, B., Zhou, W., Bodner, M., Zhou, Y.D.: Sequential neural processes in abacus mental addition: an EEG and FMRI case study. PLoS ONE **7**(5), e36410 (2012)
11. Liu, Y., Feyen, R., Tsimhoni, O.: Queueing Network-Model Human Processor (QN-MHP): a computational architecture for multitask performance in human-machine systems. ACM Trans. Comput.-Hum. Interact. (TOCHI) **13**(1), 37–70 (2006)
12. Crandall, B., Klein, G., Klein, G.A., Hoffman, R.R.: Working Minds: A Practitioner's Guide to Cognitive Task Analysis. MIT Press, Cambridge (2006)
13. Card, S.K., Moran, T.P., Newell, A.: The keystroke-level model for user performance time with interactive systems. Commun. ACM **23**(7), 396–410 (1980)
14. Card, S.K.: The Psychology of Human-Computer Interaction. CRC Press, Boca Raton (2018)
15. Burns, C.M., Vicente, K.J.: Model-based approaches for analyzing cognitive work: a comparison of abstraction hierarchy, multilevel flow modeling, and decision ladder modeling. Int. J. Cogn. Ergon. **5**(3), 357–366 (2001)
16. Annett, J.: Hierarchical task analysis. In: Handbook of Cognitive Task Design, vol. 2, pp. 17–35 (2003)
17. Swain, A.D.: Human reliability analysis: need, status, trends and limitations. Reliab. Eng. Syst. Saf. **29**(3), 301–313 (1990)
18. Boring, R.L.: Fifty years of THERP and human reliability analysis (No. INL/CON-12-25623). Idaho National Laboratory (INL) (2012)
19. Boring, R.L., Rasmussen, M.: GOMS-HRA: a method for treating subtasks in dynamic human reliability analysis. In: Proceedings of the 2016 European Safety and Reliability Conference, pp. 956–963 (2016)
20. Ryan, J.J., Mazzuchi, T.A., Ryan, D.J., De la Cruz, J.L., Cooke, R.: Quantifying information security risks using expert judgment elicitation. Comput. Oper. Res. **39**(4), 774–784 (2012)

Evidence of the Use of Fuzzy Techniques in Occupational Safety

Celina P. Leão(✉) and Susana P. Costa

ALGORITMI Centre, Department of Production and System, School of Engineering, University of Minho, Guimarães, Portugal
{cpl, susana.costa}@dps.uminho.pt

Abstract. Risk assessment (RA) and Risk management (RM) are compulsory valuable tools for helping companies to set strategies in order to achieve goals through informed decision-making. The techniques for RA can be selected according to data type and data availability. On balance, semi-quantitative RA techniques seem to be a good compromise in risk estimation. For applying these techniques, it is necessary to build the hierarchy scale of Probability, Severity and Risk Index. The Fuzzy Logic (FL) approach is useful when dealing quantitatively with imprecision and has been applied in a wide range of fields, being a valuable tool for professionals who deal with Human Factors and human contingencies, and relevant in both RA and decision-making process of which workers' safety and health rely on. This work aims at evidencing the pertinence of the use of FL theory to semi-quantitative RA techniques by a systematic literature review (SLR). The early results highlight the pertinence of the use of FL approach to analyze RA and the need for continuity.

Keywords: Risk assessment · Risk management · Safety · Fuzzy logic theory · Collaborative qualitative analysis

1 Introduction

The World Health Organization (WHO) defines Occupational Safety and Health (OSH) as "occupational health deals with all aspects of health and safety in the workplace and has a strong focus on primary prevention of hazards." [1]. Safety is considered a strategic target for companies, for the gains and advantages that certification provides to them by improving their productivity, organization, credibility and efficiency. In occupational settings, the number of accidents' occurrence is of utmost importance. Apparently, it is expected that the compliance with the certification norms has a direct bearing in the reduction of the accident's occurrence. However, this link is not as straightforward as it would seem. Leão et al. [2], in their work, found out that regardless of the decreasing tendency of the mortal accidents, between the period of 2008–2015 in Portugal, in average, there was no statistically significant difference between Economic activity (EA) in mean accident incidence rate difference. The EA "Construction" was identified as the activity sector with the highest, increasing incidence rates for accidents, coherent with literature [3]. With regard to non-fatal

P. M. Arezes and R. L. Boring (Eds.): AHFE 2020, AISC 1204, pp. 178–184, 2020.
https://doi.org/10.1007/978-3-030-50946-0_25

accidents, some increases with increasing company size was observed for the same period [2]. This pattern was more apparent in the companies of size ≥ 500 employees.

Extend understanding of oscillate behaviour in the number of accidents' occurrence, it was possible to identify that through education it has been possible to reduce occupational accidents [4].

The present study goes one step further to understand the importance and impact that risk assessment and risk management have in the occupational safety. Before to start, a contextualization of the problem will be done in order to understand the main objective that this work is proposing to attain.

1.1 Contextualization

Risk assessment (RA) is compulsory in Portuguese companies [5, 6]. Risk assessment consists of risk analysis, which comprises identification of hazards and their extension (given, for instance, the number of people exposed) and risk evaluation (by obtaining a risk value and comparing it with reference values of acceptability). Risk management (RM) stands for the global process of risk assessment and risk control, whereby measures are taken to monitor or lower the risk levels to acceptable values [7]. Hence, besides being compulsory, RM is a valuable tool for companies, providing a range of valuable data for companies.

Safety Issues (Risk Analysis). RM is a continuous, iterative task that helps companies to set strategies in order to achieve goals through informed decision-making [8]. There are several techniques available for risk assessment [9], which can be selected according to the data available (e.g., the companies' available history of data); the type of data available – quantitative, qualitative and hybrid [10]; and the quantity of data available. On the one hand, quantitative RA techniques may be very laborious (even more so in large companies), expensive and may not be feasible when the quantity of the data available is scarce. On the other hand, qualitative RA may be very expeditious, but not provide enough detail on the risks that the workers are exposed to, so not as rich in information and, hence, not as serviceable to the decision-making process as the quantitative RA techniques. So, a good compromise seems to be the semi-quantitative RA techniques, whereby risk indices are obtained. Basically, semi-quantitative RA techniques estimate the numerical value of the magnitude of the occupational risk (R), from the product between the probability (P) estimate that the risk materializes and the expected severity (S) of the lesions. Therefore, for applying these techniques, it is necessary to construct the hierarchy scale of Probability, Severity and Risk Index. The number of classes of each variable may vary between different techniques, but the same logic imposes for instance, Probability may range from Impossible to Very Likely in three or more well-defined classes. But are these classes' boundaries really well-defined as crisp? Can one company rigorously compare their RA results to other companies' RA results, even when the same RA technique is used, with guaranteed comparability? Can one ensure that, if a different technician performs the same semi-quantitative RA technique his/hers to the same worker at the same time, results will be the same?

Advantages for the Use of Fuzzy Techniques. According to [11], fuzzy approach is useful when dealing quantitatively with imprecision. Fuzzy logic (FL) is a mathematical technique that enables computational models of real systems to act on vagueness, enabling the analyst to preserve coherence in his evaluation by making judgements using imprecise information [12, 13]. FL has been applied in a wide range of science fields such as engineering, management and medicine, by virtue of its ability of translating qualities to mathematics [10, 14].

Hence, fuzzy logic theory is a recognized valuable tool for professionals who deal with human factors and human contingencies and relevant in both RA and decision-making process of which workers' safety and health rely on.

Therefore, fuzzy logic seems to be a very important tool for the very important task of risk assessment, so relevant to the decision-making process of which workers′ safety and health greatly rely on.

1.2 Objective

This work aims at evidencing the pertinence of the use of fuzzy logic theory to semi-quantitative RA techniques. To accomplish this purpose a systematic literature review was conducted (SLR).

2 Methodology

This present study is the first stage of a larger project whose main objective is to assess workers' risk exposure as previously mentioned and is built on an earlier study [15].

A SLR on documents dealing with Workers' Perceptions on Safety and Risk Assessment in Occupational Environment, and Questionnaires and Fuzzy topics, was performed to help to understand what has been done and to identify possible gaps. The SLR methodology implemented was based on the following steps: (1) planning the review and research question formulation, (2) locating studies, (3) study selection and evaluations, (4) analysis and synthesis, and (5) reporting the results, as described [16, 17].

Following objective definition, the databases used were selected. Scopus and Web of Science (WoS) databases were considered with a combination of keywords. By considering the keywords combination that correspond to the main fields under study (["safety" and "risk assessment"] AND ["fuzzy" and "Likert"]) 1 (or 0) scientific work was found. However, by simply change the AND Boolean operator by OR, many more than 10,000 works were found (see Table 1). This result was expected since it shows the large number of works in the two main fields of RA and FL and not necessarily jointly, testifying the relevance of each matters. So, a second set of keywords was considered without including "safety" and "risk assessment".

According to the database, different filters types were considered: Topic, Keywords, Abstract, and All Fields. No constraints to the reference period of time was defined, allowing to find evidence and trends and if and how works are related, with the latest valuation performed at the 31st of October 2019. Only English-written documents were contemplated. When referring to documents, is meant as papers in journals (articles), conference papers, review and book chapters.

Table 1. Database search results by keywords combination considered by database and filter type considered.

No	Filter type	Database	Keywords combination	Number
#1	Topic	WoS	(Safety AND "Risk Assessment")	1 (0)
#2	All Fields		AND (Fuzzy AND Likert)	1 (0)
#3	Keywords	Scopus	(Safety AND "Risk Assessment")	0
#4	Abstract		AND	1
#5	All Fields		(Fuzzy AND Likert)	65
#6	Topic	WoS	(Safety AND "Risk Assessment")	10,801
#7	All Fields		OR (Fuzzy AND Likert)	19,241
#8	Keywords	Scopus	(Safety AND "Risk Assessment")	45,299
#9	Abstract		OR	9,594
#10	All Fields		(Fuzzy AND Likert)	215,258
#11	Topic	WoS	(Fuzzy AND Likert)	**115 (85)**
#12	All Fields			124
#13	Keywords	Scopus	(Fuzzy AND Likert)	**42 (33)**
#14	Abstract			120
#15	All Fields			1,064

As inclusion criteria, studies that highlighted the relationship between Workers' Perceptions on Safety and Risk Assessment in Occupational Environment and Fuzzy techniques in data analysis. As exclusion criteria the studies that are not full available and studies with low relevance regarding the main objective.

3 Discoveries

The database search results, Table 1, gives the base topics for the discussion. Based on the results obtained using the Scopus database, 33 of the 42 results, search results #13, were considered for analysis (9 documents were not considered since it were not full available). Notice that, from these 33 results, 24 are common to those 85 obtained using WoS database (see result #11).

3.1 Descriptive Analysis

All the 33 documents are written in English. The publication years range from 2005 and 2019 (Fig. 1a), presenting some variability, with the year 2017 with the highest obtained value (9), representing 27% of publications, and two years with no publication, 2008 and 2011. The majority (79%) come from periodic article (journals) and the remaining from conference proceedings (Fig. 1b). Regarding the research areas of the publication of these studies, four areas emerge: Computer Science (31%), Mathematics (22%), Engineering (15%) and Social Sciences (11%) (Fig. 1c). As expected the most used words were “Fuzzy”, “scale”, “data”, “Likert”, “analysis”, among others (Fig. 1d).

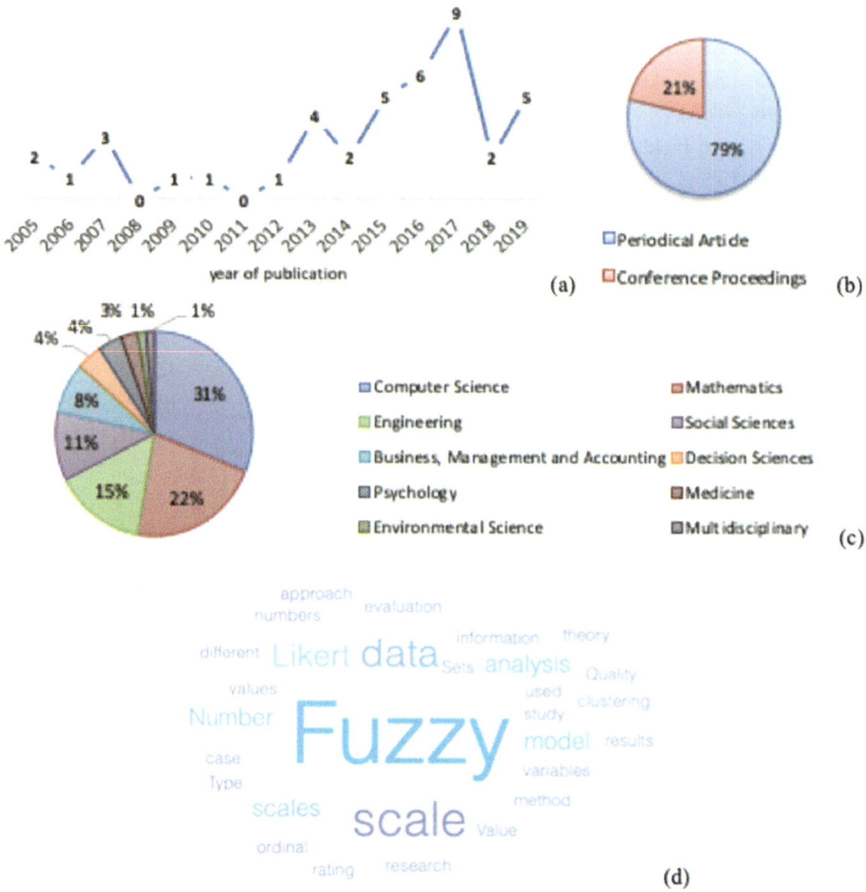

Fig. 1. Descriptive data, from Scopus, by: (a) year of publication, (b) Document type, (c) Document subject area, (d) the most frequent words.

3.2 Analysis and Interpretation of Results

The SLR was accomplished through a digital tool support, webQDA [18]. webQDA is a software that supports the analysis of qualitative data in a collaborative and distributed environment, helping on the organization of data and on the reliability of the results produced.

Two works were selected that, somehow, fulfill the main objective: Yazdi, Daneshvar, and Setareh (2017) [19] and Dapari, et al. (2017) [20].

The work 1 [19] uses Fuzzy Sets theory to make an adaption to the Failure Mode and Effects Analysis (FMEA) with an application for an aircraft landing system. The authors conclude that their model, the Fuzzy Developed FMEA (FDFMEA) yields a more reliable analysis result.

The work 2 [20] is an example of work that uses the methodology Fuzzy applied to the Delphi method to enhance the validation of a questionnaire which addresses a disease risk factors among metal industry workers. Hence, the scope of work 2 is slightly deviated from the focus in discussion in this paper.

4 Conclusions

The early results highlight the pertinence of the use of FL approach to analyze RA and the need for continuity, despite the different approaches that exist and are used namely focused on traditional RA. It should be noted that the technique chosen for the assessment of risks varies greatly between companies and that some techniques rely, on a great scale, on the subjective analysis of the evaluator. In general, companies are free to choose the methodologies for their RA and their commitment to the adequacy of the methodologies depends mostly on the companies' safety culture maturity, their mission and values.

Given that scientific disciplines bound to occupational matters may benefit greatly from FL, because they allow Health and Safety technicians to deal with evaluations' uncertainty in a coherent manner, these early results highlight the pertinence of the use of FL approach to analyze RA, and the need for continuity.

Acknowledgements. The authors acknowledge support from FCT – Fundação para a Ciência e Tecnologia, within the R&D Units Project Scope: UIDB/00319/2020.

References

1. WPRO—Occupational Health. www.wpro.who.int. Accessed 30 Jan 2020
2. Leão, C.P., Costa, S., Costa, N., Arezes, P.: Capturing the ups and downs of accidents' figures – the Portuguese case study. In: Ahram, T., Karwowski, W., Taiar, R. (eds.) Human Systems Engineering and Design, IHSED 2018. Advances in Intelligent Systems and Computing, vol. 876, pp. 675–681. Springer, Cham (2019). https://doi.org/10.1007/978-3-030-02053-8_103
3. Salguero-Caparros, F., Suarez-Cebador, M., Rubio-Romero, J.C.: Analysis of investigation reports on occupational accidents. Saf. Sci. **72**, 329–336 (2015)
4. Leão, C.P., Costa, S.: Safety training and occupational accidents – is there a link? In: Goossens, R., Murata, A. (eds.) Advances in Social and Occupational Ergonomics, AHFE 2019. Advances in Intelligent Systems and Computing, vol. 970, pp. 536–543. Springer, Cham (2020). https://doi.org/10.1007/978-3-030-20145-6_53
5. Law n.º 3/2014: Diário da República, 1ª série, Nº 9, 28 de janeiro de 2014, 554 (2014)
6. Law n.º 7/2009: Diário da República, 1ª série, Nº 30, 12 de fevereiro de 2009, 926 (2009)
7. Miguel, A.S.S.: Manual de higiene e segurança do trabalho (2010). (in Portuguese)
8. ISO 31000:2018: Risk management, International Standard Organisation, Genéve (2018)

9. ISO 31010:2019: Risk management – Risk assessment techniques, International Standard Organisation, Genéve (2019)
10. Gul, M.: A review of occupational health and safety risk assessment approaches based on multi-criteria decision-making methods and their fuzzy versions. Hum. Ecol. Risk Assess. **24** (7), 1723–1760 (2018). https://doi.org/10.1080/10807039.2018.1424531
11. Varela, M.L.R., Barbosa, R., Costa, S.: P2P web service based system for supporting decision-making in cellular manufacturing scheduling. In: Computational Intelligence and Decision Making, pp. 155–165. Springer, The Netherlands (2013)
12. Rodrigues, M.A., Leão, C.P., Nunes, E., Sousa, S., Arezes, P.: A fuzzy logic approach in the definition of risk acceptance boundaries in occupational safety and health. ASME J. Risk Uncertain. Part B **2**(4) (2016). https://doi.org/10.1115/1.4032923
13. Costa, S., Arezes, P.: Fuzzy logic theory applications on the occupational risk prevention domain. In: 10th International Conference on Occupational Risk Prevention, ORP 2012, Bilbao, Spain (2012)
14. Costa, S.R.P.: Contribution of the evaluation strategy and the use of hearing protection for the uncertainty associated with occupational exposure to noise. Ph.D. dissertation. University of Minho (2015)
15. Leão, C.P., Costa, S.P.: Fuzzy techniques for assessing workers' risk exposure–acknowledging the need. In: World Conference on Qualitative Research, vol. 1, pp. 179–180 (2019)
16. Briner, R.B., Denyer, D.: Systematic review and evidence synthesis as a practice and scholarship Tool. In: Handbook of Evidence-based Management: Companies, Classrooms and Research, pp. 112–129. Oxford University Press, New York (2012)
17. Denyer, D., Tranfield, D.: Producing a systematic review (chap. 39). In: Buchanan, D., Bryman, A. (eds.) The Sage Handbook of Organizational Research Methods, pp. 671–689. Sage Publications Ltd., London (2009)
18. Fornari, F., Pinho, I., de Almeida, C.A., Costa, A.P.: Literature review with support of digital tools. In: 9th IEEE Symposium on Computer Applications & Industrial Electronics, ISCAIE 2019, Kota Kinabalu, Malaysia (2019)
19. Yazdi, M., Daneshvar, S., Setareh, H.: An extension to Fuzzy Developed Failure Mode and Effects Analysis (FDFMEA) application for aircraft landing system. Saf. Sci. **98**, 113–123 (2017)
20. Dapari, R., Ismail, H., Ismail, R., Ismail, N.H.: Application of fuzzy Delphi in the selection of COPD risk factors among steel industry workers. Tanaffos **16**(1), 46–52 (2017)

Safety Analysis of an Industrial System Using Markov Reliability Diagram with Repair

Tony Venditti[1(✉)], Nguyen Duy Phuong Tran[2], and Anh Dung Ngo[1]

[1] Department of Mechanical Engineering, École de Technologie Superiéure, 1100 Notre-Dame West, Montréal, QC H3C1K3, Canada
tvenditti@asfetm.com, anh-dung.ngo@etsmtl.ca

[2] The Faculty of Mechanical Engineering, HCM City University of Technology, 268 Ly Thuong Kiet Street, Ward 10, HCMC, Vietnam
tnduyphuong@yahoo.com

Abstract. In many industrial systems, accidents result from equipment failures and human errors. The latter can thus be viewed as a form of system failure which must be identified and analyzed. In this paper, an industrial system made up of a brake press operated by a single operator is analyzed. A reliability Markov diagram including repair states is drawn to model the system. The probability of a work accident (failed state of the system) is calculated using fuzzy numbers. An innovative aspect of this study is that repair states for human failures are also defined and their repair rates are estimated. The equations of state of the system are then derived and solved. Thus, the probability of the system being in a failed state is calculated.

Keywords: Fuzzy methods · Markov diagram · Human error · Repair rate

1 Introduction

Industrial machine constitute systems composed of different components such as, in particular, the moving parts of the machine itself, the operator, the work-piece, and the command interface and circuit. Machines can furthermore contain forms of stored energies such as fluids under pressure which can be released. Accidents can occur which can injure the operator. In fact, an industrial machine can present many various hazards (sources of accidents) such as sharp edges, hazardous zones where entanglement, crushing, cutting can occur, slipping, tripping hazards, chemicals, poor work layout and, in some cases, biological hazards. As a way of illustrating the theory, brake press operation will be considered. A press brake is a machine commonly found in the metal manufacturing industry. It is used to bend sheet metal in different shapes. A typical press brake is illustrated here (Fig. 1).

The machine has been described in some detail a previous paper, ref. [1]. But briefly, in the operation of the press, the operator holds the sheet metal part and actuates the closing motion of the press with a foot pedal, in most applications. A hazardous situation is created from the proximity of the workers hands to the press closing motion. A possible undesirable event (often called a hazardous event) in such a situation is then that the worker gets his hands caught between the closing dies (the

P. M. Arezes and R. L. Boring (Eds.): AHFE 2020, AISC 1204, pp. 185–190, 2020.
https://doi.org/10.1007/978-3-030-50946-0_26

Fig. 1. Press brake

hazardous zone of the machine). Safety regulations and standards require brake presses to be equipped with safeguarding protective devices such as light curtains or laser beam sensors which stop the hazardous motion of the machine should the operator come too close to it.

A dynamic fault tree related to this situation, see Ref. [2], appears like in Fig. 2.

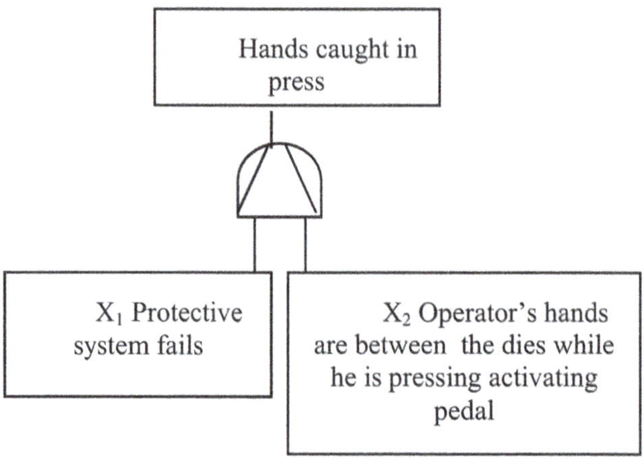

Fig. 2. Fault tree

The fault tree is dynamic in nature because event X_1, protective device failure, must occur before the worker puts his hands inadvertently between the bending dies. The proper operation of a brake press requires the operator to hold the part with one hand or both his hands in such a way as to properly position the piece according to the length of the bend to be created. Brake press operation can often involve large lots of pieces to be made. This creates a repetitive work pattern which can lead the worker to becoming distracted, bored and tired and hence to errors on his part such as placing his hands in wrong position. However, as the fault tree shows, if the protective device functions properly the hazardous closing motion of the press will cease and no accident will then occur.

2 Markov Analysis

The equivalent associated Markov diagram is then (Fig. 3):

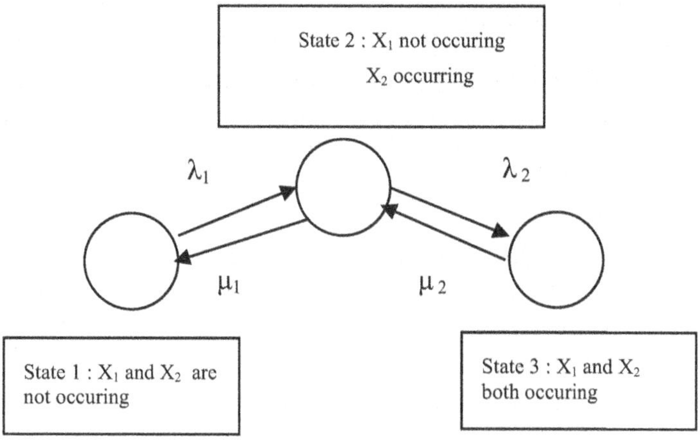

Fig. 3. Markov diagram with repair

In this diagram, State 1 represents the condition where the protective device functions properly and the operator keeps his hands out of the hazardous zone. State 2 describes the situation where the protective device fails to detect the hands of the worker and fails to stop the closing motion. In State 3, the worker commits a human error by putting his hands in the hazardous zone due to various contributing factors such as fatigue, repetitive task, distraction, etc.

The equations of state are then

$$\frac{d}{dt}P = \begin{bmatrix} -\lambda_1 & \mu_1 & 0 \\ \lambda_1 & -\lambda_2 - \mu_1 & \mu_2 \\ 0 & \lambda_2 & -\mu_2 \end{bmatrix} \begin{bmatrix} P_1 \\ P_2 \\ P_3 \end{bmatrix}$$

or

$$\frac{d}{dt}P = Q^T P \tag{1}$$

The initial conditions for the problem are: $P_1 = 1$ because state 1 represents the system in working conditions and no failures are occurring.

The basic assumptions used in the model are as follows:

- component failures and repair rates are statistically independent, constant, very small and obey exponential distribution functions;
- the product of the failure rate and repair time is small (less than 0.1);
- there are no simultaneous failures among the subsystems;
- maintenance and repair will be done by outside service outlet;
- the repair process begins soon after a unit fails but includes call to service firm and commuting to site;
- after repairs, the repaired component is considered as good as new;
- system structure is precisely known so that diagnostic and repair is efficiently performed;
- importantly, the safety protective device encompasses the entire control circuit including the device itself, the wiring, the piping, the control module (e.g. electrical relay, electronic controller) and all the downstream components such as valves or contactors which transfer power to the actuators which make the hazardous motions possible.

In this equation, in the right-hand side, the matrix Q terms have units hr^{-1}. Therefore, on the right hand side, the units should also be hr^{-1}. Hence, P has no units and therefore represents a probability, not a rate.

The meaning of the terms are:

λ_1 = the failure rate of the protective device in units of h^{-1} (a fuzzy number);
λ_2 = the human failure rate in units of h^{-1} (a fuzzy number);

These data come from an expert elicitation exercise performed by the authors. This process consisted in asking plant personnel familiar with the operation and maintenance of brake presses their qualitative opinions on the probability of occurrence of the specific failure modes being considered in this research. Fuzzy methodology was used to amalgamate and quantify these expert judgements. The manufacturing plants where the expert elicitations were conducted, data on failures, repairs were incomplete and scarce. As for the number of events where operators placed their hands in the hazardous zone between the dies of the press, no record were kept. Hence, the need to resort to other form of data gathering such as expert elicitation. Further details are provided in [3], wherein literature references can be found.

μ_1 = the repair rate of the protective device (a "fuzzified" number);
μ_2 = the repair rate following a human failure (a "fuzzified" number).

The repair rates merit further explanations. The repair rate is defined as in units of h^{-1} as it should

$$\frac{1}{MTTR(Mean\ Time\ to\ Repair)}$$

This definition is well suited for technical repairs a in the model. Again, the manufacturing plants visited, MTTR data was not recorded or the personnel involved with health and safety issues were not aware of this type of information. This illustrates again the point that, often, uncertainty surrounds attempts to analyze and improve safety issues at work. A brief search on internet led to a value of 2.5 h average repair time for a safety light curtain as given by various brake press service suppliers. Considering the uncertainty shrouding this value, this value was "fuzzified".

What about now human error? The failure due to human action has been discussed. But can one talk of "human error" repair? The concept does not seem to appear in the literature. In this research, the "repair" is defined as the actions that are deemed necessary to correct the human error observed. Since one is dealing with human performance, training appears to be the most adequate course of action to take. Hence repair in the situation considered will refer to re-training of the worker following the event "Putting his hands between dies". The chosen value is based on actual practical experience. It is common practice after incidents to give initially trained workers a re-training session focusing on the factors that affected the event. A typical length for this kind of re-training (a "refresher session") is 2 h. Hence, 0.5 h^{-1} (the inverse of the time to repair) is the adopted value in this research. Again, this value was "fuzzified" due to the uncertainty involved.

3 Results

The results are then, with the following data, $\lambda_1 = <2.85, 3, 3.15> \times 10^{-5}$; $\lambda_2 = <0.95, 1, 1.05> \times 10^{-5}$ (which comes from expert elicitation, as explained before, see [3]; $\mu_1 = <0.119, 0.125, 0.131>$; $\mu_1 = <0.475, 0.5, 0.525>$, which come from a (non-exhaustive, brief) literature search: $P_3 = <4.7878, 4.7979, 4.8070> \times 10^{-9}$.

This demonstrate that the objective of this paper which was to estimate quantitatively the probability of a work accident occurring can be achieved, even in the presence of uncertainty on data and incomplete knowledge. However, validation of the result is difficult in view of the scarcity of specific data available. It can be noted though that the result satisfies machine safety standards (such as [4]) which call for probabilities of occurrence of hazardous events of less than 10^{-6}.

4 Conclusion

In this paper we presented a method for calculating the top event probability of occurrence of a fuzzy dynamic fault tree and its associated Markov diagram which represented the operation of a brake press. The model included not only failure rates but repair rates as well. A novel feature of the model was the inclusion of a human-related repair rate. The Markov diagram was solved, and the desired probability was calculated.

Acknowledgments. The authors acknowledge financial support from ÉREST.

References

1. Venditti, T., Tran, N.P.D., Ngô, A.D.: Dynamic fuzzy safety analysis of an industrial system. In: 2017 8th International Conference on Applied Human Factors and Ergonomics (AHFE), Los Angeles, California, USA, 19th to 21st July 2017 (2017)
2. Venditti, T., Tran, N.P.D., Ngô, A.D.: System safety analysis of an industrial process using fuzzy methodology. In: 4th International Conference on Fuzzy Systems and Data Mining, FSDM 2018, Bangkok, Thailand, 16–19 November 2018 (2018)
3. Venditti, T., Tran, N.P.D., Ngô, A.D.: Expert elicitation methodology in the risk analysis of an industrial machine. In: 9th International Conference on Applied Human Factors and Ergonomics (AHFE), Orlando, Florida, USA, 21st to 25th July 2018 (2018)
4. Department of Defense USA: MIL-STD-882D Standard Practice for System Safety (1993)

Risk Level Assessment to Develop a Hand Disorder in a Bag Sealing Process

Luis Cuautle-Gutiérrez(✉), Luis Alberto Uribe-Pacheco, and Jesús Juárez-Peñuela

Facultad de Ingeniería Industrial y Automotriz, Universidad Popular Autónoma del Estado de Puebla, Puebla, Mexico
{Luis.Cuautle,Luisalberto.uribe, jesusjuarez01}@upaep.mx

Abstract. A cleaning and personal hygiene products company performs its bag sealing process manually with a production standard of 420 boxes in one shift. This situation has caused 33% of workers to present injuries to the wrists, which has led to a constant turnover, as well as significant economic losses. The objective of this work is the decrease in, at least, 5% of this type of injuries from the ergonomic evaluation and the implementation of improvements in the work area. First, the process flow is studied, as well as its cycle time. In the evaluation of the current working method, ergonomic analysis is used. The findings show that the number of repetitions is not adequate, which causes physiotherapeutic problems. Three proposals are raised. Once implemented, a survey of discomfort is applied and the level of discomfort of the worker is contrasted with the new method which validates the improvement.

Keywords: Hand disorder · Ergonomics · Industrial process

1 Introduction

1.1 Hand (Work Accidents and Injuries)

Traditionally, companies handle both manual and automatic processes to perform their daily operations. This condition has increased the frequency of occupational accidents mainly in the hand with the partial or total loss of the grip and grip functions (Traumatic hand injuries. Epidemiological study). Traumatic hand injuries refer to any damage caused by an external agent, acutely to the hand, in the anatomical space comprised of the metacarpals to the distal phalanges that include nail bed injuries [1]. According to labor accident statistics, the hand is the most affected anatomical region, since it is involved in 26% of the total cases. Among the first ten causes of occupational disease, two of them are in the hands [2].

The Secretary of Labor and Social Welfare of Mexico in its national report on accidents and occupational diseases in the period 2005-2016 (most recent report) points the wrist and hand as the main anatomical region of occupational accidents, shown in Table 1, with the occurrence in males almost double that presented in the female gender. In terms of the type of injury, the national report considers superficial injuries

P. M. Arezes and R. L. Boring (Eds.): AHFE 2020, AISC 1204, pp. 191–196, 2020.
https://doi.org/10.1007/978-3-030-50946-0_27

and dislocations, sprains and tears as the main injuries in the same period of time nationwide [3].

Table 1. Work accidents according to anatomical region and sex, 2015–2016, National.

Anatomical region	2015		2016	
	Men	Women	Men	Women
Wrist and hand	84,402	30,517	79,719	30,408
Ankle and foot	39,768	21,621	36,873	19,978
Head and neck	34,536	18,472	31,207	15,958
Lower limb (excludes ankle and foot)	33,140	15,027	30,886	14,050
Upper limb (excludes wrist and hand)	30,979	12,167	28,693	11,641

1.2 Distal Upper Extremity Disorders

The distal upper extremities (DUE) refer to the elbow, wrist, hand and fingers [4]. For musculoskeletal disorders in the distal upper extremities, the index that incorporates precise or repetitive movements and the muscular strength developed with the arms and hands is the most important physical demand variable discarding the psychosocial demands present at work [5].

The more common DUE musculoskeletal disorders are: Dupuytren disease, carpal tunnel syndrome, ulnar nerve entrapment, de Quervain's disease, tenosynovitis, and tendinitis [6]. Upper extremity musculoskeletal disorders are among the most prevalent and costly compensated disorders in worker's compensation systems [7].

De Dupuytren's disease is a disorder that is manifested by a progressive contracture of the fingers that impairs the function of the hand. Among the risk factors are genetic predispositions and environmental factors, their probability of occurrence increases with age and is higher among men. The basic treatment consists of surgical intervention, although it is also treated with pharmacology, ultrasound, physiotherapy and radiotherapy [8].

Carpal tunnel syndrome consists of compression of the median nerve as it passes through the carpal tunnel, manifests itself in numbness of the first three fingers of the hand, pain and weakness. The high pressure in the carpal tunnel can compress the ulnar nerve fibers and this can get trapped in the wrist together in the median nerve [9]. Kinesiotaping consists of an elastic bandage that lifts the skin and relieves compression to increase blood and lymphatic flow that reduces pain in muscle and joint injuries; In severe cases surgery is recommended [10].

De Quervain's disease is a tenosynovitis of the first extensor compartment of the wrist, its main treatment consists of injections of ultrasound-guided corticosteroids [11].

1.3 Ergonomic Assessment

One method to determine the risk of developing distal upper limb disorders in single task jobs is the calculation of the strain index (SI). This consists of the product of six variables present in the task or work: Intensity of exertion (IE), duration of exertion (DE), efforts per minute (EM), hand/wrist posture, speed of work (SW), and duration of task. Each of these elements is qualified on a five-level qualitative/quantitative scale. SI scores between 3 and 7 mean that the work could cause problems in the distal system while scores greater than 7 reveal that the work is associated with a distal disorder of the upper limb, similarly an SI score less than 3 represents safe work [12].

Generally, workers perform multiple tasks in their work centers at the same time, so the application of the SI is limited, which has led to the consideration of other components such as studies of time, frequency and repetitive work cycle in activities muscle, to quantify risk levels in different hazardous work situations [13]. Another method to assess the exposure of the finger-hand-arm area when working with physical objects is the key indicator method for manual handling operations (KIM-MHO) [14].

In this one, a score less than or equal to 20 is considered as a low level that does not imply any health risk. However, scores greater than 50 result in pain and even dysfunctions for normally resistant people.

2 Research

The research was carried out in a company that produces personal and hygiene products specifically in the process of sealing 50 g of cotton bags. This production area has a standard of 420 boxes per shift and 3 women work, who say they finish the 8-h day with wrist pain and excessive tiredness. Recently, one of them presented an injury to his hands due to the current development of the work. This situation generates staff turnover and significant losses to the company in insurance and additional expenses to workers' injuries. The objective is to reduce injuries in this work area by at least 5% through ergonomic evaluation and the development of process improvements.

First, a survey of discomfort at the work station was used. From the same, it was observed that 12 repetitions per minute are performed in each operator and that there is a probable risk due to frequency and static postures as indicated in Fig. 1.

Task information				
Risk factors selection		Description	Picture	
Repetitions/Frequency	X	12 repetitions per minute	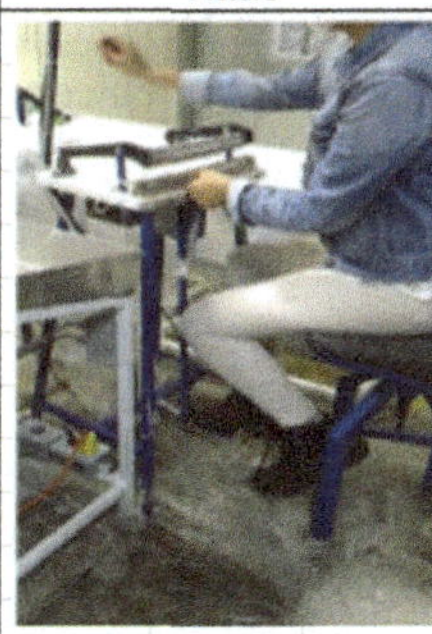	Discomfort manifested
Strength/Weight		50 grams		Severe pain in the wrists
Vibrations				
Temperature	X			
Ilumination	X			
Noise	X			
Contact Stress				
Risk Postures	X			
Static Postures	X			
Others: Work Stress (through efficiency)				

Fig. 1. Sealing process ergonomic risk identification.

Subsequently, a time study was carried out for which 200 cycle times were taken from each of the operators and a cycle time of sealing a cotton bag of 5.33 s on average was identified with the current method. Then, to validate the possible ergonomic risks, the work deformation index for each operator was calculated as shown in Table 2.

Table 2. Strain index calculation.

Operators (3)	IE	DE	EM	Hand/Wrist posture	SW	Task duration	SI score
1	1	3	1.5	2	1.5	1	13.5

Then, a KIM-MHO study was conducted in which a score of 126 was estimated, which implies that the task causes pronounced disorders or dysfunctions as well as significant structural damage. See Fig. 2.

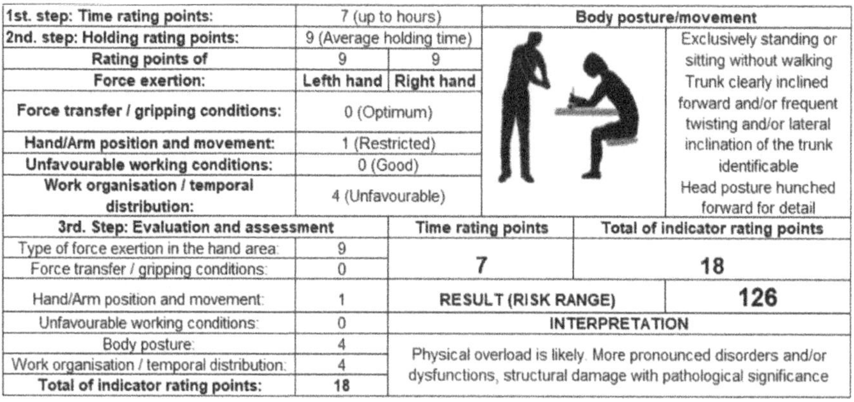

Fig. 2. KIM-MHO assessment.

Thus, it was determined that the task in question can produce the following cumulative trauma disorders: De Dupuytren disease, Carpio tunnel syndrome, Quervain tenosynovitis and probably epicondylitis and Guyon's canal syndrome.

3 Results

From the above, a scheme of rotation of tasks was proposed every hour and a half, as well as a break of five to ten minutes. In addition, the use of wristbands is considered to allow immobilization of the hand joint and provide heat and compression to the bones of the workers. Table 3 shows the reduction of the job strain index by more than 80%, which implies that the potential risk of injury is non-existent in the task.

Table 3. Job Strain Index method comparison.

Current method		Proposed method	
Job Strain Index	Qualification	Job Strain Index	Qualification
13.5	Task associated with a distal disorder	2.25	Task probably safe

A survey of discomfort was applied to the three operators after the implementation of the proposals and the tingling and numbness of the hand is eliminated and pain in the wrist and lower fingers is considerable reduced as indicated in Fig. 3.

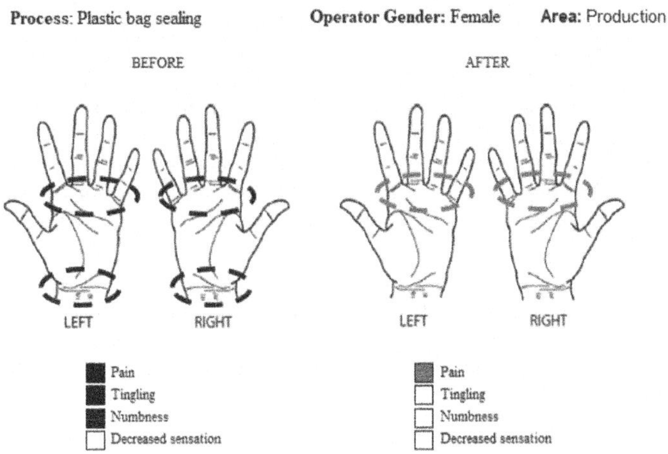

Fig. 3. Discomfort survey results.

With respect to the KIM-MHO study score, the implementations made estimate a value of 55 points which implies an improvement of the current process situation. But, a redesign of the workplace and other prevention measures should be considered. The authors propose the automation of the bag sealing process, however, the last decision rests in management when exercising the necessary economic resources.

References

1. González, C.E., Rodríguez, R.: Lesiones traumáticas de la mano. Estudio epidemiológico. Revista Mexicana de Ortopedia y Traumatología **15**, 206–209 (2001)
2. Instituto Mexicano del Seguro Social. http://www.imss.gob.mx/sites/all/statics/guiasclinicas/065GER.pdf
3. Secretaría de Salud, Gobierno de la Ciudad de México. https://www.salud.cdmx.gob.mx/comunicacion/nota/26-por-ciento-de-accidentes-laborales-son-en-munecas-y-manos-armando-ahued

4. Drinkaus, P., Bloswick, D.S., Sesek, R.: Job level risk assessment using task level strain index scores: a pilot study. Int. J. Occup. Saf. Ergon. **11**, 141–152 (2005)
5. Fonseca, N.: Factors related to musculoskeletal disorders in nursing workers. Revista Latino-Americana de Enfermagem **18**, 1076–1083 (2010)
6. Biron, C.: How not to overwork your hands. RDH 64–69 (2014)
7. Thiese, M.S., Hegmann, K.T., Kapellusch, J., Merryweather, A., Bao, S., Silverstein, B., Garg, A.: Associations between distal upper extremity job physical factors and psychosocial measures in a pooled study. Biomed. Res. Int. **2015**, 1–9 (2015)
8. Latusek, T., Miszczyk, L., Gierlach, G., Zajac, P.: An effectiveness evaluation of the palmar fascia irradiation of patients suffering from Dupuytren's disease. Nowotwory J. Oncol. **67**, 162–167 (2017)
9. Kaplan, C.B., Inan, R., Boru, U.T.: Investigation of the coincidence of idiopathic carpal tunnel syndrome and ulnar nerve entrapment neuropathy: role of the cutaneous silent period. SCIE **30**, 217–221 (2019)
10. Koca, T.T.: Kinesiotaping in the management of carpal tunnel syndrome. Ortadogu Med. J. **12**(1), 34–39 (2020)
11. Bing, J.H., Choi, S.J., Jung, S.M., Ryu, D.S., Ahn, J.H., Kang, C.H., Shin, D.R.: Ultrasound-guided steroid injection for the treatment of de Quervain's disease: an anatomy-based approach. Skelet. Radiol. **47**(11), 1483–1490 (2018)
12. Moore, J.S., Garg, A.: The strain index: a proposed method to analyze jobs for risk of distal upper extremity disorders. Am. Ind. Hyg. Assoc. J. **56**, 443–458 (1995)
13. Bao, S., Spielholz, P., Howard, N., Silverstein, B.: Application of the Strain Index in multiple task jobs. Appl. Ergon. **40**, 56–68 (2009)
14. Klussmann, A., Liebers, F., Gebhardt, H., Rieger, M.A., Latza, U., Steinberg, U.: Risk assessment of manual handling operations at work with the key indicator method (KIM-MHO) – determination of criterior validity regarding the prevalence of musculoskeletal symptoms and clinical conditions within a cross-sectional study. BMC Musculoskelet. Disord. **18**, 184 (2017)

Investigations of Human Psychology and Behavior in the Emergency of Subway

Ping Zhang[1,2], Lizhong Yang[1(✉)], Siuming Lo[2], Yuxing Gao[1], Fangshu Dong[1], Fei Peng[1(✉)], Danyan Huang[1,2], Han Cheng[1], Maoyu Li[1], and Jiajia Jiang[1]

[1] State Key Laboratory of Fire Science, University of Science and Technology of China, Hefei 230026, People's Republic of China
{zpinger, yanglz, yuxing95, dongfs, pfei, huangdy, chenghan, maoyuli, jomo13}@mail.ustc.edu.cn

[2] Department of Architecture and Civil Engineering, City University of Hong Kong, Hong Kong Special Administrative Region, China
bcsmli@cityu.edu.hk

Abstract. The rapid development of urban subways has greatly improved the travel efficiency. While accidents like fires, explosions and terrorist attacks in subway system always cause severe consequences due to the limited space and high population densities. The psychological and behavioral parameters in simulating subway evacuation are derived from observations of pedestrian movement under normal conditions, but not necessarily apply to emergency situations. Therefore, more statistical data on the psychological and behavioral characteristics of pedestrians in subway emergencies should be collected. In this study, we conducted a questionnaire survey in two different regions of China, namely Hefei and Shenyang. Correlation analysis was performed on the results of the questionnaire using the chi-square test and cross tabulation tables in SPSS. The relationship among people's basic characteristics, psychology and behavior were analyzed. The in-depth analysis of correlative factors influencing the evacuation psychology and behavior were done as well.

Keywords: Subway · Safety evacuation · Crowd panic · Human behavior · Questionnaire

1 Introduction

In recent years, with the emphasis on public safety, the issue of emergency evacuation and how to effectively evacuate a large number of people has caused more and more attention, especially in public places with limited space and high population density such as airport, subway station and so on. However, in actual disasters and terrorist attacks, secondary disasters such as trampling caused by crowd panic are the main cause of casualties.

In the past few decades, scholars in different fields have devoted themselves to study the pedestrian evacuation dynamics and improve crowd management methods [1–3]. There are many subway evacuation researches on both theoretical and simulating

P. M. Arezes and R. L. Boring (Eds.): AHFE 2020, AISC 1204, pp. 197–209, 2020.
https://doi.org/10.1007/978-3-030-50946-0_28

work [4, 5]. The psychological and behavioral parameters in simulating subway evacuation are derived from observations of pedestrian movement under normal conditions, but not necessarily apply to emergency situations. In order to make up for the lack of theoretical and simulation research, many scholars have made a lot of explorations on psychology and behavior of pedestrians in subway evacuation. Wang [6] designed a set of questionnaire to investigate the psychology and behavior of pedestrians in subway emergencies in Nanjing, China. It is found that the age and the safety knowledge do not have so strong impact on evacuation psychology and behavior. Mu [7] studied the psychological behavior of people in subway evacuation considering the peak and low peak traffic of the subway. She found that most passengers have good psychological qualities and women are not as sensible as men, however, people with low degree of education level and carrying the baggage are more likely to be panic. Ge [8] collected and analyze the psychological responses and behavioral characteristic data after evacuation drill hold in subway in Beijing, China. It is found that slight panic is the main psychological response during the pre-evacuation and evacuation phase. In addition, the behavior of pedestrians is often influenced by many factors like different regional cultures, work characteristics and living habits. Luo [9] conducted questionnaire survey on human's behavior and reaction in building fire with consideration of different multi-ethnics and work characteristics in Xinjiang, China. According to analyze human's basic characters, cognition degree of fire, social behavior and daily behavior statistically, he found that men are more willing to participate in the rescue process than women, and respondents with stronger toxic tolerance tend to follow other people's reaction rather than make their own decisions. He [10] investigated the safety awareness and evacuation behavior of subway passengers in three different regions of Beijing, Nanjing and Guangzhou through questionnaire survey and statistical analysis. The result shows that passengers' safety awareness is rising as the urban economy grows and people's quality improves. However, there are still some problems in the existing research. On the one hand, there are many different conclusions from the existing research results. On the other hand, differences in different regions are not reflected as the result of subway evacuation in emergency situations.

As of 2018, there are 35 cities in China that have opened subways. The questionnaire surveys were conducted in two cities, Hefei and Shenyang, China. The Shenyang Metro was opened in 2010 with 4 operating lines, a mileage of 89 km and the number of the average daily passenger traffic is 873,000. The Hefei Metro was opened in 2016 with 2 operating lines, a mileage of 52 km and the number of the average daily passenger traffic is 507,000. We have obtained 616 questionnaires in Hefei and 559 questionnaires in Shenyang, and the effective rates were 80.03% and 91.06%, respectively. Firstly, the statistical data of individual information such as the gender, age, education background, frequency of taking the subway, subway safety knowledge, and previous emergency experience were studied in SPSS. In addition, statistics of the evacuation psychology and behavior be analyzed as well. Secondly, the relationship among people's basic characteristics, psychology and behavior were analyzed by using the chi-square test. When p-values smaller than 0.05, it indicates that the two variables are related. Finally, we use cross tabulation tables to conduct in-depth analysis of correlative factors that affect evacuation psychology and behavior, so as to provide some suggestions on subway operation management and the design of subway evacuation model.

2 Method

2.1 Design of Questionnaire

Based on psychological and behavioral characteristics of pedestrians in subway emergencies, survey questionnaire included two parts, which is individual information and the evacuation psychology and behavior. The design of questionnaire is shown in Fig. 1.

Individual information

- Gender
- Age
- Education background
- Frequency of taking the subway
- Knowledge of subway safety
- Previous emergency experience

Evacuation psychology and behavior

- ➢ First psychology and behavioral reaction
  - the first psychology reaction
  - the first behavioral reaction
- ➢ Impact of panic
  - affected by the panic of the surrounding people
  - following which one can reduce the panic the most
- ➢ Selection of escaping routes
  - in a familiar subway station
  - in an unfamiliar subway station
- ➢ In case of getting lost
  - the first psychological reaction
  - the first behavioral reaction
- ➢ In case of caught in congestion
  - the first psychological reaction
  - the first behavioral reaction

Fig. 1. The design of questionnaire

2.2 Data Sources and Analysis Method

In this study, we conducted a questionnaire survey in two different regions of China, namely Hefei and Shenyang. 1175 questionnaires were obtained from 616 passengers in Hefei and 559 passengers in Shenyang, and the effective rates were 80.03% and 91.06%, respectively.

Correlation analysis was performed on the results of the questionnaire using the chi-square test and cross tabulation tables in SPSS. The relationship among people's basic characteristics, psychology and behavior were analyzed. The in-depth analysis of correlative factors influencing the evacuation psychology and behavior were done as well.

3 Results and Discussion

3.1 Individual Information

- Gender: Whether in Hefei or Shenyang, the proportion of women in the survey is higher than that of men, see in Table 1.

Table 1. Individual information about object of investigation in Hefei and Shenyang

Individual information		HF (%)	SY (%)
Gender	Male	37.3	48.1
	Female	62.7	51.9
Age	<18	6.9	16.7
	19–40	77.5	73.5
	41–65	15.2	8.4
	>66	0.4	1.4
Education background	Middle school or below	10.5	9.4
	High school or vocational high school	12.6	18.3
	College	18.7	22.4
	Postgraduate	41.2	43.6
	Master or higher	17.0	6.3
Frequency of taking the subway	More than 21 days	3.2	31.8
	11–20 days	3.0	19.8
	6–10 days	16.7	21.8
	Less than 5 days	77.1	26.6
Subway safety knowledge	Never	25.4	42.0
	Acquired in daily life	69.2	50.9
	Received drills or special training	5.4	7.1
Previous emergency experience	Yes	2.8	5.7
	No	97.2	94.3

- Age: The survey results show that the majority of people who take the subway are between 19 and 40 years old, and this result is more in line with the characteristics of China's population distribution.
- Education background: Most of the respondents in Hefei and Shenyang have bachelor degree or above.
- Frequency of taking the subway: The survey found that people take subways less frequently in Hefei, while Shenyang rides more frequently in subways, which is related to the opening time of subway lines in the two cities.
- Subway safety knowledge: Most people lack knowledge of subway safety.
- Previous emergency experience: The enormous majority of people don't have subway emergency experience.

3.2 Statistics of the Evacuation Psychology and Behavior

The survey found that passengers in Hefei and Shenyang have some similar psychological and behavioral characteristics in emergencies, and the specific findings are shown in Table 2.

Whether in Hefei or Shenyang, nearly half of the respondents can keep calm when facing emergencies in subway, which indicates that the respondents have better psychological qualities, however, there are still one-third of people are anxious and panic. The results show that most respondents thought they would be affected by the panic around them. In terms of first behavioral reaction, "Freeze", "Flight", and "Fight" are the common behavioral actions of human response to threats, and in our survey people attempt to escape rather than fight. When people are selecting escape routes in emergencies, people have a strong sense of dependence because they believe that relying on others is an effective way to alleviate their panic: authoritative people like subway staff is their first choice and then followed by their companions, people are not confident enough so that making decisions independently is the last choice.

The survey also found some differences of passengers' psychological and behavioral characteristics between Hefei and Shenyang under emergency condition. As for choosing a route in a familiar or unfamiliar subway, respondents in Hefei are more likely to go to places with fewer people, while respondents in Shenyang are prone to follow more people. When people are lost, following more people to escape is the main behavioral reaction of respondents in Hefei, while respondents in Shenyang are looking for companions first.

Therefore, it is important to take actions such as broadcasting or guiding to make the information accurate and clearly when facing emergencies. Better preparations of evacuation guidance can not only keep people calm down and avoid causing crowd panic, but also improve evacuation efficiency.

3.3 Correlation Analysis

Based on the statistics analysis of questionnaire, the relationship among people's basic characteristics, psychology and behavior were analyzed by using the chi-square test. When p-values smaller than 0.05, it indicates that the two variables are related. We found that the gender, age, educational level, frequency of taking the subway, and

Table 2. Evacuation psychology and behavior response in Hefei and Shenyang

Evacuation psychology and behavior		HF (%)	SY (%)
The first psychological reaction	Calm	54.0	50.1
	Anxious	11.6	15.1
	Panic	22.9	15.3
	Fluke mind	1.4	1.6
	Herding or conformity	10.1	17.9
The first behavioral reaction	Be at a loss	3.0	5.7
	Give an alarm phone	36.2	38.1
	Escape immediately	37.7	34.2
	Ask about the accident	6.1	12.0
	Participate in rescue	17.0	10.0
Affected by the panic of the surrounding people	Strongly agree	15.6	9.4
	Agree	53.1	54.3
	Neither agree nor disagree	6.5	9.2
	Disagree	14.0	16.1
	Strongly disagree	10.8	11.0
Following which one can reduce the panic the most	Subway staff	/	68.4
	Companion	/	18.5
	Most people	/	7.1
	Own feelings	/	6.0
Selection of escaping routes in a familiar subway station	Following the entrance route	5.5	5.3
	Following the more people	4.9	11.0
	Follow fewer people	22.9	6.3
	Following the exit signs	31.0	31.6
	Following the command of subway staff	35.7	45.8
Selection of escaping routes in an unfamiliar subway station	Following the entrance route	3.2	5.1
	Following the more people	8.7	10.6
	Follow fewer people	11.2	3.1
	Following the exit signs	25.6	20.9
	Following the command of subway staff	51.3	60.3
The first psychological reaction if getting lost	Calm	50.5	46.6
	Anxious	32.5	39.3
	Querulous	0.8	1.9
	Panic	16.2	12.2
The first behavioral reaction if getting lost	Be at a loss	0.6	2.8
	Looking for companions	17.6	45.0
	Find the exit on my own	18.3	19.8
	Following the more people	37.1	14.5
	Ask someone for help	26.4	17.9

(*continued*)

Table 2. (*continued*)

Evacuation psychology and behavior		HF (%)	SY (%)
The first psychological reaction if caught in congestion	Calm	57.2	47.9
	Anxious	30.6	41.7
	Querulous	2.0	4.7
	Panic	10.2	5.7
The first behavioral reaction if caught in congestion	Wait in line	53.8	50.5
	Push forward through the crowd	2.2	4.1
	Find a new route	39.4	41.3
	Others	4.6	4.1

safety knowledge have a significant impact on passengers' psychology and behavior in emergencies. Then we can get in-depth analysis of correlative factors that affect evacuation psychology and behavior by using cross tabulation tables.

- The influence of gender
  According to Tables 3, 4 and 5, it is found that: When facing emergencies, the first psychological reaction of women is more likely to be panic than men, since the

Table 3. Cross tabulation table of "Gender* The first psychological reaction" in Hefei and Shenyang

				The first psychological reaction					Total
				Calm	Anxious	Panic	Fluke mind	Herding or conformity	
HF	Gender	Male	Count	115	22	32	3	12	184
			Percentage (%)	62.5	12.0	17.4	1.6	6.5	100.0
		Female	Count	151	35	81	4	38	309
			Percentage (%)	48.9	11.3	26.2	1.3	12.3	100.0
	Total		Count	266	57	113	7	50	493
			Percentage (%)	54.0	11.6	22.9	1.4	10.1	100.0
SY	Gender	Male	Count	143	33	25	4	40	245
			Percentage (%)	58.4	13.5	10.2	1.6	16.3	100.0
		Female	Count	112	44	53	4	51	264
			Percentage (%)	42.4	16.7	20.1	1.5	19.3	100.0
	Total		Count	255	77	78	8	91	509
			Percentage (%)	50.1	15.1	15.3	1.6	17.9	100.0

Table 4. Cross tabulation table of "Gender* The first psychological reaction if getting lost" in Hefei and Shenyang

				The first psychological reaction if getting lost				Total
				Calm	Anxious	Querulous	Panic	
HF	Gender	Male	Count	108	61	3	12	184
			Percentage (%)	58.7	33.2	1.6	6.5	100.0
		Female	Count	141	99	1	68	309
			Percentage (%)	45.7	32.0	0.3	22.0	100.0
	Total		Count	249	160	4	80	493
			Percentage (%)	50.5	32.5	0.8	16.2	100.0
SY	Gender	Male	Count	134	90	6	15	245
			Percentage (%)	54.8	36.7	2.4	6.1	100.0
		Female	Count	103	110	4	47	264
			Percentage (%)	39.0	41.7	1.5	17.8	100.0
	Total		Count	237	200	10	62	509
			Percentage (%)	46.5	39.3	2.0	12.2	100.0

Table 5. Cross tabulation table of "Gender* Affected by the panic of the surrounding people" in Hefei and Shenyang

				Affected by the panic of the surrounding people					Total
				Strongly agree	Agree	Neither agree nor disagree	Dis-agree	Strongly disagree	
HF	Gender	Male	Count	25	78	16	37	28	184
			Percentage (%)	13.6	42.4	8.7	20.1	15.2	100
		Female	Count	52	184	16	32	25	309
			Percentage (%)	16.8	59.5	5.2	10.4	8.1	100
	Total		Count	77	262	32	69	53	493
			Percentage (%)	15.6	53.1	6.5	14.0	10.8	100
SY	Gender	Male	Count	20	109	25	50	41	245
			Percentage (%)	8.2	44.5	10.2	20.4	16.7	100
		Female	Count	28	167	22	32	15	264
			Percentage (%)	10.6	63.3	8.3	12.1	5.7	100
	Total		Count	48	276	47	82	56	509
			Percentage (%)	9.5	54.2	9.2	16.1	11.0	100

Table 6. Cross tabulation table of "Age* Affected by the panic of the surrounding people" in Hefei and Shenyang

				Affected by the panic of the surrounding people					Total
				Strongly agree	Agree	Neither agree nor disagree	Dis-agree	Strongly disagree	
HF	Age	<18	Count	4	14	2	6	8	34
			Percentage (%)	11.8	41.2	5.9	17.6	23.5	100.0
		19–40	Count	62	215	23	44	38	382
			Percentage (%)	16.2	56.3	6.0	11.5	10.0	100.0
		41–65	Count	11	33	7	18	6	75
			Percentage (%)	14.7	44.0	9.3	24.0	8.0	100.0
		>66	Count	0	0	0	1	1	2
			Percentage (%)	0.0	0.0	0.0	50.0	50.0	100.0
	Total		Count	77	262	32	69	53	493
			Percentage (%)	15.6	53.1	6.5	14.0	10.8	100.0
SY	Age	<18	Count	3	53	8	15	6	85
			Percentage (%)	3.5	62.4	9.4	17.6	7.1	100.0
		19–40	Count	34	211	34	61	34	374
			Percentage (%)	9.1	56.4	9.1	16.3	9.1	100.0
		41–65	Count	11	12	5	5	10	43
			Percentage (%)	25.6	27.9	11.6	11.6	23.3	100.0
		>66	Count	0	0	0	1	6	7
			Percentage (%)	0.0	0.0	0.0	14.3	85.7	100.0
	Total		Count	48	276	47	82	56	509
			Percentage (%)	9.4	54.3	9.2	16.1	11.0	100.0

proportion of panic is 26.2% for women and 17.4% for men in Hefei, and 20.1% for women and 10.2% for men in Shenyang. Women are more easily to be panic than man under certain scenes such as separating from peers during emergency evacuation, since the proportion of panic is 22.0% for women and 6.5% for men in Hefei, and 17.8% for women and 6.1% for men in Shenyang. In addition, women are more prone to be influenced by people around them with panic, since there are 76.3% of

Table 7. Cross tabulation table of "Education background* The first psychological reaction" in Hefei

				The first psychological reaction					Total
				Calm	Anxious	Panic	Fluke mind	Herding or conformity	
HF	Education background	Middle school or below	Count	34	3	8	3	4	52
			Percentage (%)	65.3	5.8	15.4	5.8	7.7	100.0
		High school or vocational high school	Count	43	4	8	0	7	62
			Percentage (%)	69.3	6.5	12.9	0.0	11.3	100.0
		College	Count	48	8	24	0	12	92
			Percentage (%)	52.2	8.7	26.1	0.0	13.0	100.0
		Postgraduate	Count	104	30	49	0	20	203
			Percentage (%)	51.2	14.8	24.1	0.0	9.9	100.0
		Master or higher	Count	37	12	24	4	7	84
			Percentage (%)	44.0	14.3	28.6	4.8	8.3	100.0
	Total		Count	266	57	113	7	50	493
			Percentage (%)	54.0	11.6	22.9	1.4	10.1	100.0

Table 8. Cross tabulation table of "The frequency of taking the subway* Affected by the panic of the surrounding people" in Hefei

				Affected by the panic of the surrounding people					Total
				Strongly agree	Agree	Neither agree nor disagree	Dis-agree	Strongly disagree	
HF	The frequency of taking the subway	>21 days	Count	2	7	0	1	6	16
			Percentage (%)	12.5	43.7	0.0	6.3	37.5	100.0
		11–20 days	Count	0	9	1	5	0	15
			Percentage (%)	0.0	60.0	6.7	33.3	0.0	100.0
		6–10 days	Count	11	39	5	14	13	82
			Percentage (%)	13.4	47.5	6.1	17.1	15.9	100.0
		<5 days	Count	64	207	26	49	34	380
			Percentage (%)	16.8	54.6	6.8	12.9	8.9	100.0
	Total		Count	77	262	32	69	53	493
			Percentage (%)	15.6	53.1	6.5	14.0	10.8	100.0

Table 9. Cross tabulation table of "Subway safety knowledge* The first psychological reaction" in Hefei

				The first psychological reaction					Total
				Calm	Anxious	Panic	Fluke mind	Herding or conformity	
HF	Subway safety knowledge	Never	Count	59	8	38	5	15	125
			Percentage (%)	47.2	6.4	30.4	4.0	12.0	100.0
		Acquired in daily life	Count	185	49	72	2	33	341
			Percentage (%)	54.2	14.4	21.1	0.6	9.7	100.0
		Received drills or special training	Count	22	0	3	0	2	27
			Percentage (%)	81.5	0.0	11.1	0.0	7.4	100.0
	Total		Count	266	57	113	7	50	493
			Percentage (%)	54.0	11.6	22.9	1.4	10.1	100.0

Table 10. Cross tabulation table of "Subway safety knowledge* The first psychological reaction if getting lost" in Shenyang

				The first psychological reaction if getting lost				Total
				Calm	Anxious	Querulous	Panic	
SY	Subway safety knowledge	Never	Count	104	73	5	32	214
			Percentage (%)	48.6	34.1	2.3	15.0	100.0
		Acquired in daily life	Count	109	118	5	27	259
			Percentage (%)	42.1	45.6	1.9	10.4	100.0
		Received drills or special training	Count	24	9	0	3	36
			Percentage (%)	66.7	25.0	0.0	8.3	100.0
	Total		Count	237	200	10	62	509
			Percentage (%)	46.5	39.3	2.0	12.2	100.0

the total 309 women think they will be affected by others' panic while the proportion of the total 184 men is 56% in Hefei, and 73.9% of the total 264 women think they will be affected by others' panic while the proportion of the total 245 men is 52.7% in Shenyang. It shows that women are not as calm as men when facing emergencies, so more attention should be paid to women.

- The influence of age
  As shown in Table 6, it is found that people whose age between 19 and 40 in Hefei will be more easily affected by the surrounding people's panic, since the proportion

of people who think they will be affected by panic is 53%, 72.5%, 58.7% and 0.0% for "<18", "19–40", "41–65" and ">66", respectively. Different from Hefei, younger people are more susceptible to the panic emotions of those around them in Shenyang, since the proportion of people who think they will be affected by panic is 65.9%, 65.5%, 53.5% and 0.0% for "<18", "19–40", "41–65" and ">66", respectively.

- The influence of education background
  According to Table 7, subway passengers in Hefei with higher education are more likely to be panic in the first psychological response to an emergency, since the proportion of panic is 15.4%, 12.9%, 26.1%, 24.1% and 28.6% for "Middle school or below", "High school or vocational high school", "College", "Postgraduate" and "Master or higher", respectively.
- The influence of the frequency of taking the subway
  As shown in Table 8, subway passengers in Hefei who take the subway less frequently in Hefei are more likely to be influenced by people around them with panic, since the proportion of people who think they will be affected by panic is 56.2%, 60.0%, 60.9% and 71.4% for ">21 days", "11–20 days", "6–10 days" and "<5 days", respectively.
- The influence of subway safety knowledge
  According to Tables 9 to 10, in the face of an emergency, people in Hefei who lack the knowledge of safety will be more likely to be panic, since the proportion of panic is 30.4%, 21.1% and 11.1% for "Never", "Acquired in daily life" and "Received drills or special training", respectively. When getting lost with their companions, people in Shenyang who lack the knowledge of safety will be more prone to be panic, since the proportion of panic is 15.0%, 10.4% and 8.3% for "Never", "Acquired in daily life" and "Received drills or special training", respectively.

4 Conclusion

In order to collect statistical data on the psychological and behavioral characteristics of pedestrians in subway emergencies. We conduct questionnaire survey in Hefei and Shenyang, China. Based on the statistical analysis, some interesting results were concluded as follows:

It is found that the gender, age, educational level, and subway safety knowledge have a significant impact on passengers' psychology and behavior in emergencies. Women and those who lack safety knowledge are more likely to be panic and influenced by the people around them. In particular, we found that people with higher educational levels and those whose age between 19 and 40 in Hefei would be easy to be panic. When people are selecting escape routes in emergencies, people have a strong sense of dependence because they believe that relying on others is an effective way to alleviate their panic: authoritative people like subway staff is their first choice and then followed by their companions, people are not confident enough so that making decisions independently is the last choice.

Therefore, more careful psychological and behavioral parameters of pedestrians should be considered in emergency simulation. The research in this article could not only provide a reference for safety management in metro operations, but also give supports to develop the evacuation strategies effectively.

Acknowledgments. This research was supported by National Key R&D Program of China.

References

1. Helbing, D., Farkas, I., Vicsek, T.: Simulating dynamical features of escape panic. J. Nat. **407**(6803), 487–490 (2000)
2. Jiang, C.S., Deng, Y.F., Hu, C., et al.: Crowding in platform staircases of a subway station in China during rush hours. J. Saf. Sci. **47**(7), 931–938 (2009)
3. Lei, W., Li, A., Gao, R.: Effect of varying two key parameters in simulating evacuation for a dormitory in China. J. Phys. A Stat. Mech. Appl. **392**(1), 79–88 (2013)
4. Kallianiotis, A., Papakonstantinou, D., Arvelaki, V., et al.: Evaluation of evacuation methods in underground metro stations. J. Int. J. Disaster Risk Reduct. **31**, 526–534 (2018)
5. Ding, Y., Yang, L., Weng, F., et al.: Investigation of combined stairs elevators evacuation strategies for high rise buildings based on simulation. J. Simul. Model. Pract. Theory **53**, 60–73 (2015)
6. Wang, J.H., Yan, W.Y., Zhi, Y.R., et al.: Investigation of the panic psychology and behaviors of evacuation crowds in subway emergencies. J. Proc. Eng. **135**, 128–137 (2016)
7. Mu, N.N., Xiao, G.Q., He, L., et al.: Investigation and correlation study on human evacuation psychological behavior in subway. J. Saf. Sci. Technol. **9**(6), 85–90 (2013)
8. Ge, X.X., Dong, W., Jin, H.Y.: Study on the social psychology and behaviors in a subway evacuation drill in China. J. Proc. Eng. **11**, 112–119 (2011)
9. Luo, L., Yang, L.Z., Fu, Z.J.: Investigation of human's evacuation behavior in multi-ethnics jumping-off of China. In: International Conference on Applied Human Factors and Ergonomics, pp. 278–287. Springer, Cham (2017)
10. Li, H., Maohua, Z., Congling, S., et al.: Experimental research on investigation of metro passenger evacuation behaviors in case of emergency. In: Pedestrian and Evacuation Dynamics, pp. 173–184. Springer, Boston (2011)

Risk Assessment in Operations of Static Large Format Out of Home (OOH) Billboards for Advertising

Oca Malagueno[1(✉)], Isachar Bernaldez[2], and Mariam Idica[3]

[1] Pixelboards Incorporated, San Juan City, Philippines
oca.malagueno@gmail.com
[2] Far Eastern University, Metro Manila, Philippines
ibernaldez@feu.edu.ph
[3] EcoSci Food Incorporated, Antipolo, Philippines
yam@ecoscifood.com

Abstract. This study used a risk assessment tool to evaluate the key hazards and degree of risks in the production unit for digital printing, and on-site installation and dismantling. It determines the nature of existing risk controls and their effectiveness. With this, the hazards with the highest probability of occurring and with the greater severity were observed and recorded to minimize workers' exposure to potential risk and ensure the safety of their workplace. Findings from the assessment showed that workers do not comply with the safety work measures for hazards that they perceived to be a low risk while they comply with safety work measures for hazards with medium to high risk.

Keywords: Risk Assessment · Billboard installation · Static large format billboards · Safety management

1 Introduction

Major thoroughfares of metro cities in the Philippines have rapidly changed for the last two decades in terms of the visual landscape. This resulted in the growing number of large-format Out-of-home (OOH) advertisements along the major roads. OOH is any visual advertising media found outside of homes. It could be classified into a small format (flyers, posters, etc.) and large format (static and digital billboards, bus ads, etc.). According to the Outdoor Advertising Association of the Philippines (OAAP), there are overwhelmingly 8,000 billboards in Metro Manila alone for the year 2004, and the number is rising fast [1]. Shown in Fig. 1 is an example of billboards advertisements with workers installing the tarpaulin.

The setting up a billboard advertisement is divided into two operations: digital printing of tarpaulin and on-site installation. The process of digital printing for tarpaulin, and actual photo is shown in Fig. 2 and Fig. 3, respectively.

Printing process starts with creating a file for advertisement, then loaded to ripping software to convert the pixels into dots of inks. Next is to check the ink of the printer. If ink is not enough, then a worker will load a solvent ink. Ink spills usually happen in

P. M. Arezes and R. L. Boring (Eds.): AHFE 2020, AISC 1204, pp. 210–215, 2020.
https://doi.org/10.1007/978-3-030-50946-0_29

Fig. 1. Actual installation of Static billboard advertisement in Metro Manila, Philippines

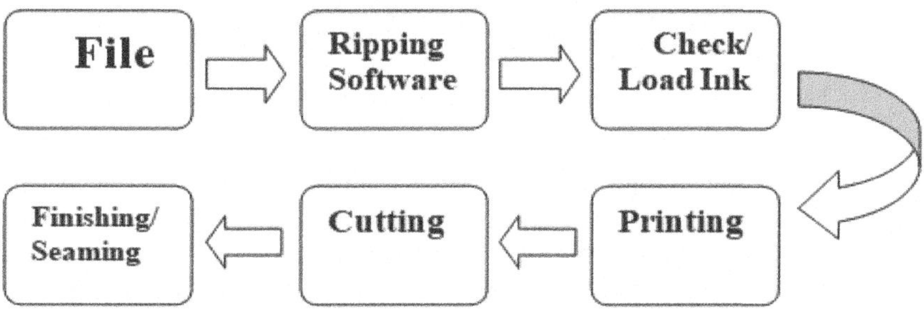

Fig. 2. Process flowchart of tarpaulin printing

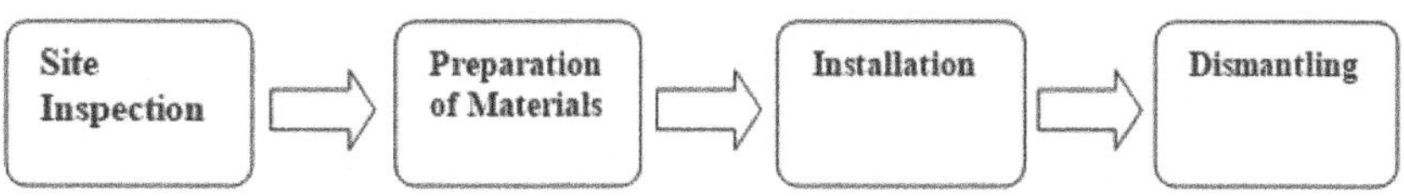

Fig. 3. Flow chart for the process of onsite installation

this process. Fourth stage is printing and then cutting it to the desired size. Lastly, the process of finishing or seaming. Here the sides were folded and glued.

According to Health and Safety Executive, UK, the most common types of hazards in printing are manual handling (27%), slips and trips (22%), and machinery (22%) while the long term effect of some inks cured can cause cancer and harm to the unborn child [2]. The common ink used for tarpaulin printing is solvent. Raw materials in solvents are hydrocarbon solvents, petroleum distillates, esters, ketones, alcohols, glycols, and others [3]. A study conducted in Denmark among the printing workers in 1970 and followed up over a 17-year period showed that lung, bladder, renal pelvis, and primary liver cancer were in excess [4] (Fig. 4).

Fig. 4. Installation of billboards advertisement with workers on action

Installation of static billboards starts with the site inspection. The team checks the strength of the structure, its size, and the possible material needed to install. Second is the preparation of materials. Third is the actual installation. This is the part where installers experience high hazards as they climb billboards with height around 70 ft from the ground or higher. The last part is dismantling when the advertising campaign ends.

The common hazards for installation and dismantling of billboards are falling due to human error, wind, lightning, structure failure, and electrocution.

Pixelboards Incorporated is an advertising company that is categorized as a media supplier in the Philippines. This company owns and maintains advertising sites such as static billboards in the country. Production unit is located in Quezon City where digital printing occurs while the installation and dismantling were done at different sites.

This study aimed at using the risk assessment tool to evaluate the key hazards and degree of risks in production unit for digital printing, and onsite installation and dismantling. It also determined the nature of existing risk controls and their effectiveness. With this, the hazards with the highest probability of occurring and with the greater severity must be addressed to minimize the worker's exposure to potential risk and ensure their workplace safety.

2 Methodology

The six hazards were observed if present in the site, namely, safety hazards, biological hazards, physical hazards, chemical hazards, ergonomic hazards, and work organization hazards.

The study was conducted in the production and installation department of Pixelboards, Inc in the Philippines. Production department prints large format tarpaulins to be installed by installation department in different billboard sites. Printer ink used solvent. Risk assessment on work hazards were conducted on the production and installation workers. The methodology involves data collection of existing hazards and risks in production and installation department. The data were collected from interviews with the workers involved. The workers were asked to rate the likelihood and severity of the hazards based on the Hazards Identification, Risk Assessment, and Risk Control (HIRARC) process.

Then the risk level was computed by multiplying the severity and the likelihood of a hazard as shown in the equation below.

$$Severity \times Likelihood = Risk\,Level \quad (1)$$

The risk level that ranges between 1 to 25 were averaged and then ranked. The variance was then computed in case there is a big difference between the risk levels made by the assessors. According to International Principles of Occupational Health and Safety [5], high risks are defined with risk level of 25-15. Moderate risks are defined with risk level 14-5. Low Risks are defined with risk level 4-1. Figure 5 shows a summary of the risk assessment methodology.

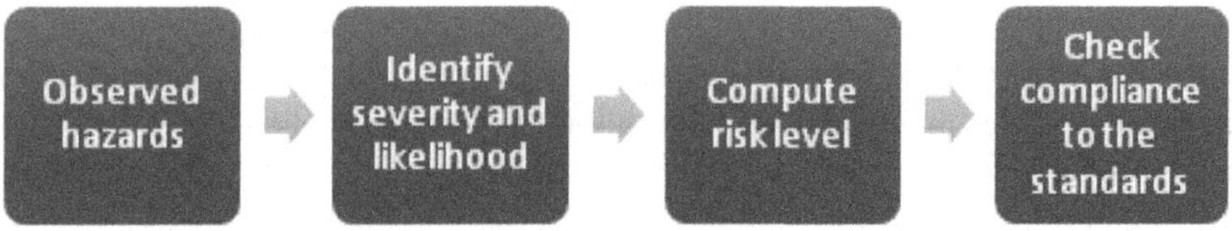

Fig. 5. Risk assessment methodology

3 Results and Discussion

3.1 Production Department Top Hazards

The results of the risk assessment showed that the top two hazards for printing are chemical hazards and ergonomic hazard.

The top one hazard is chemical exposure to ink. The workers inhale and touch printing chemicals every day and the workers see the hazards as low risk. They reported some symptoms of effect of chemical exposure, such as dizziness, difficulty breathing, skin irritation, and eye irritation. Furthermore, they are aware of the long-term effects of chemical exposure such as, cancer, lung problems, and infertility. Hence, this is a high-risk hazard in long term perspective.

The second is an ergonomic hazard that results in physical injuries. There are reported cases of muscle pain due to improper handling of heavy printed materials. Furthermore, there are also reported of ran over cases while handling heavy tarpaulin. This is also reported as a low-risk hazard but in the long run this could cause permanent injuries to the workers. The Table below shows the exposure and risk level of hazards (Table 1).

Table 1. Top two hazards of production department

Hazard type	Hazard	Exposure	Risk level
Chemical hazard	Chemical exposure	inhale ink, inhale vapor of solvent, splash ink to the eyes	**Low Risk Long term: High Risk**
Ergonomic hazard	Injuries	handling heavy materials, improper manual handling	**Low Risk Long term: High Risk**

The results are low risk but in the long-term perspective, it is high risk. No proper controls were used during printing operation.

3.2 Billboard Installation's Top Hazards

The results of the risk assessment showed that the top three hazards for billboard installation are safety hazard of falling, electrical hazard, and physical hazard.

The top one hazard is falling. The respondents reported that falling may occur due to the environment. Hence, falling might occur if the metals are slippery due to earlier rain, the sun's heat that causes dizziness, or falling due to the strong wind. Another factor is the physical maintenance of the billboard structure. Falling might occur if the metals are weak and suddenly break during installation. The third is the human factor. Falling might occur if the ropes and harness used were not put on properly. The second top hazard is electrocution. It was reported that there are chances of being electrocuted during billboards installation if open electrical wirings are present. Lastly, the third top hazard is having a lot of bruises due to heavy lifting, slipping, accidentally hit the metal, or broken bungee. The table below shows the hazard and its risk level of the installation department (Table 2).

The results are medium risk since the company has not experienced these hazards in the extreme level. They put on proper controls for these hazards.

- To avoid falling, harness and ropes with 9–10 mm diameter were used.
- To avoid falling, a proper maintenance of equipment and the billboard structure
- To avoid electrical hazard, insulating gloves, working vest, safety shoes, and fire-resistant clothing
- To avoid electrocution, the electric breakers are turned off before climbing

Table 2. Installation department top three hazards

Hazard type	Hazard	How will be exposed to this hazard	Risk level
Safety hazard	Falling	Slipping, weak structure, strong wind, electrocuted, no proper harness or ropes, dizziness	Medium risk
Electrical hazard	Electrocute	touch open electrical wirings	Medium risk
Physical hazard	Bruises	broken bungee, slipping, metal structure	Medium risk

- To avoid bruises, they wear hard hats, safety shoes, gloves, and proper clothing
- The company also makes sure that each operator has enough safety awareness

4 Conclusion and Recommendations

The study determined existing risk of production unit for digital printing, and on-site installation and dismantling. The hazards with the highest probability of occurring and with the greatest severity are documented to minimize worker's exposure to potential risk. Findings from the assessment showed that the printing operators perceived exposure to chemical and ergonomic related hazards as low risk and do not use personal protective equipment. On the other hand, the installation operators have medium risk exposure to safety, electrical, and physical hazards because of the controls used during installation. Therefore, workers do not comply with the safety work measures for hazards that they perceived to be a low risk while they comply with safety work measures for hazards with medium to high risk.

References

1. De Los Reyes, B.M., Santos, R.B.: Billboards: are they here to stay? A study of billboards in EDSA, Philippines. MUHON: J. Archit. Landscape Archit. Des. Environ. **1**(3), 38–43 (2009)
2. The Trust Insurance Group Information. https://www.thetrustgroup.co.uk/common-health-and-safety-hazards-for-printers (2018)
3. Ahmed, S.: Polymer Science, Coatings and Adhesives, 1st edn., pp. 11–12. Technology of Printing Inks, New Delhi (2007)
4. Lynge, E., Rix, B.A., Villadsen, E., Andersen, I., Hink, M., Olsen, E., Moller, U.L., Silfverberg, E.: Cancer in printing workers in Denmark. Occup. Environ. Med. **52**, 738–744 (1995)
5. Benjamin, O.A.: International Principles of Occupational and Safety. ILO Publications, Switzerland (2008)

Psychosocial Risk Factors at Work: The Legal Compliance Model in Mexico

Rodolfo Martinez-Gutierrez[1(✉)] and Concepción Cruz-Ibarra[2]

[1] Tecnológico Nacional de México/IT de Tijuana, Calzada Del Tecnológico S/N, Fraccionamiento Tomas Aquino, 22414 Tijuana, Baja California, Mexico
rodolfo.martinez@tectijuana.edu.mx

[2] Universidad de Sonora/Campus Nogales, Av Unisone no. 343, 84000 Nogales, Sonora, Mexico
concepcion.cruz@unison.mx

Abstract. In 2016, the draft Official Mexican Standard was generated that responded to the guidelines on psychosocial risk at work, through the National Advisory Committee for Standardization of Safety and Health at Work. The Federal Occupational Health and Safety Regulations, and 5, section III, and 24 of the Internal Regulations of the Ministry of Labor and Social Welfare, state that, for NOM035-STPS-2018 (Official Mexican Standard of the Secretary of Labor and Social Welfare of the Government of Mexico), on Psychosocial Risk Factors at Work, the objective should be to establish the elements to identify, analyze and prevent psychosocial risk factors, as well as to promote a favorable organizational environment in the workplace. Results of a design and proposal of intervention instrumentation to facilitate the organizational diagnoses that allow determining the roadmap of the implementation and compliance with the official Mexican norm. Conclusion: is a methodology with the appropriate instruments generates an efficient process of legal compliance.

Keywords: Organizational climate · Psychosocial risk · Work environment

1 Introduction

The research on the factors that determine psychosocial risks at work is based on the regulations established in the official Mexican standard, since 2018, in public sector and private sector organizations in Mexico. According to the approach of the NOM035-STPS standard [1], which indicates that it is an instrument that allows work centers to identify and analyze in a general way, psychosocial risk factors in work centers, as well as evaluate the organizational environment of workers in their work context. The limitations of the Official Mexican Standard is generated due to the fact that it does not contemplate a methodological instrumentalization of psychological evaluation of workers, generating a series of questions associated with the various reflections of information gaps for specialists, human resources professionals, safety and hygiene, psychologists, including labor lawyers, implying the need for design from consulting tools, design of an intervention methodology according to the characteristics of organizations, institutions or companies; the design and development of a battery of

P. M. Arezes and R. L. Boring (Eds.): AHFE 2020, AISC 1204, pp. 216–221, 2020.
https://doi.org/10.1007/978-3-030-50946-0_30

questionnaires to identify mental disorders, a method to know the internal psychological variables of individuals such as: attitudes, values, personality, and the level of exhaustion in the development of the functions of their responsibilities of a position. The foregoing necessitates a review of authors on the topic of Psychosocial Risk Factors, to determine intervention roadmaps, evaluation typologies and evaluation axes to determine the effective plans and programs for compliance with aspects become a requirement of mandatory legal compliance, involving new internal policies in organizations.

2 Framework

According to the International Labor Organization (ILO) [2], it groups psychosocial factors into two aspects; 1. Work content and 2. Work context. Being considered psychosocial risk factors (FRPS) in Table 1, the following groups are presented:

Table 1. Classification of psychosocial risk generators

Risk level	Types of psychosocial risks
1	Leadership and justice at work
2	The demands of the job
3	Work control
4	Social support
5	The physical environment
6	The environment between life and work
7	Recognition at work
8	Job security
9	Information and communication

The Ministry of Labor and Social Welfare establishes FRPS as a conceptual definition, which can cause non-organic anxiety disorders of the dream-wake cycle and severe adaptation stress, derived from the nature of job functions, the type of working hours and exposure to severe traumatic events or acts of labor violence developed. According to the STPS (Secretary of Labor and Social Security of the Government of Mexico) [3], FRPS are derived from deficiencies in the design, organization and management of work and the poor social context of work, see Table 2 on FRPS in content and context of work, which are associated with six actions within organizations, which are listed below.

García [4] points out that, the organizational climate is a key factor in business development, the study, diagnosis and improvement, affects the so-called spirit of the organization; to clarify the causes of satisfaction and dissatisfaction at work, and its impact on organizational changes.

Table 2. Psychosocial risks according to the work context

Risk level	Types of psychosocial risks
1	Excessive workloads
2	Lack of clarity of job functions
3	Lack of participation in decision making
4	Job insecurity
5	Inefficient communication
6	Psychological and sexual harassment

In the case of Méndez [5], he proposes the organizational climate as the analysis management and interpretation of particular methodologies carried out by consultants in the area of human management or organizational development of the company.

Derived from the limitations of the official Mexican norm on Psychosocial Risk Factors (NOM035-STPS), it generates questions see Table 3 on the implications, responsibilities and evidence of legal requirement to be met by organizations.

Table 3. Questions as challenges to identify psychosocial risks

Challenges	Questions to implement NOM 035-STPS
1	Do I need a psychologist?
2	Do I have to hire a certified external consultant?
3	Will there be fines for workers' stress?
4	Am I required to become certified?
5	Leadership courses?

2.1 Model and Methodology

For the investigation, they will be the product of the methodology developed with a systemic approach FHS (Fifth Systemic Helix Methodology) [6, 7], with the purpose of considering the business link and the experience of the specialists of the Occupational Health services sector. The ILO recommends for the development of Organizational Climate Surveys, topics that have been defined in International Technical Committees as elementary for all sectors, regardless of the functions performed by a worker, should be considered the following topics for the analysis of FRPS: a) Work environment, b) Working conditions, c) Relations between workers, d) The organization, e) Characteristics of the worker f) Their culture, g) Their needs, h) Personal situation outside of work.

Cardona [8] points out that there are 8 dimensions that are repeated in most of the organizational climate assessment instruments: decision making, organizational clarity, leadership, social interaction, institutional motivation, rewards and incentives system, organizational openness and supervision. Brito [9], establishes that the Quality Management Systems approach, the ISO 9001 standard, establishes: "the organization must determine and manage the work environment necessary to achieve compliance with the requirements of the product or service provided". Models of management excellence such as the North American Baldrige Prize [10] and the Deming Prize [11] are based on principles that have to do with people and their interaction with the organizational environment. Standard NOM035-STPS-2018, states that, for proper compression and the way to implement it in an organization or institution, you should consult the official Mexican standards and the Mexican standard listed below:

1. NOM-019-STPS-2011, Constitution, integration, organization and operation of health and safety commissions. NOM-030-STPS-2009, Servicios preventivos de seguridad y salud en el trabajo-Funciones y actividades.
   NMX-R-025-SCFI-2015, On Labor Equality and Non-Discrimination.

The Secretary of Labor and Social Welfare (2018), through NOM035-STPS-2018, defines a Psychosocial Risk Factor to what can cause non-organic anxiety disorders, sleep-wake cycle and severe and adaptive stress, derived from the nature of job functions, type of workday and exposure to severe traumatic events, or acts of labor violence to the worker, for the work performed. Additionally, establishes the following obligations of the employer through the following points: 1. Establish in writing, implement, maintain and disseminate in the workplace a policy of prevention of psychosocial risks that includes: a) The prevention of factors of psychosocial risk; b) The prevention of workplace violence, and c) The promotion of a favorable organizational environment. 2. Identify and analyze psychosocial risk factors, in accordance with the provisions of numerals 7.1, a), and 7.2, of this Standard, in the case of work centers that have between 16 and 50 workers. 3. Identify and analyze the psychosocial risk factors and evaluate the organizational environment, in accordance with the provisions of numerals 7.1, b), 7.2 and 7.3, respectively, of this Standard, in the case of work centers with more than 50 workers. 4. Adopt measures to prevent and control psychosocial risk factors, promote the favorable organizational environment, as well as to address the practices opposite to the favorable organizational environment and acts of labor violence, based on the provisions of Chapter 8 of the This Standard. 5. Identify the workers who were subject to severe traumatic events during or on the occasion of work and channel them for their attention to the social or private security institution, or the doctor of the workplace or the company. 6. Practice medical examinations and psychological evaluations of workers exposed to workplace violence and/or psychosocial risk factors, when there are signs or symptoms that indicate any alteration to their health and the result of the identification and analysis of psychosocial risk factors, to which the numeral refers. 7. According to the needs of this Standard, or so it is suggested and/or there are complaints of labor violence, practice Medical examinations and psychological evaluations may be carried out through the social or private security institution, doctor, psychiatrist or psychologist of the workplace, 8. Disseminate and provide information to workers about: a) The psychosocial risk prevention policy; b) The measures adopted to combat practices opposed to the favorable organizational environment and acts of labor violence; c) Prevention measures and actions and, where appropriate, control actions of psychosocial risk factors; d) The mechanisms to present complaints about practices opposed to the favorable organizational environment and to denounce acts of labor violence; e) The results of the identification and analysis of psychosocial risk factors for workplaces that have between 16 and 50 workers, and of the identification and analysis of psychosocial risk factors and the evaluation of the organizational environment in the case of centers of work of more than 50 workers, and f) The possible alterations to health due to exposure to psychosocial risk factors. Records should be kept on: 1. The results of the identification and analysis of psychosocial risk factors and, in addition, in the case of work centers of more than 50 workers, of the evaluations of the organizational environment; 2. The control measures adopted when the result of the identification and analysis of psychosocial risk factors and evaluation of the organizational environment indicates it, and 3. The names of the workers who underwent clinical examinations or evaluations and who exposure to psychosocial risk factors, acts of workplace violence or severe traumatic events were verified. Obligations of the Worker: observe the prevention measures, collaborate to have a favorable organizational environment and prevent acts of labor violence. And specifically, the following points must be followed: 1. Refrain from practices contrary to the favorable organizational environment and acts of workplace violence. 2. Participate in the identification of psychosocial risk factors and, where appropriate, in the evaluation of the organizational environment. 3. Report on practices opposed to the favorable organizational environment and report acts of workplace violence, using the mechanisms established by the employer for this purpose and/or through the health and safety commission, referred to in NOM-019-STPS- 2011, or those that replace it. 4. Inform the employer in writing directly, through preventive health and safety services at work or the safety and hygiene commission; have witnessed or suffered a severe traumatic event. The document must contain at least: the date of preparation; the name of the worker who writes, where appropriate, the name of the workers involved; the date of occurrence, and the description of the event.

5. Participate in information events provided by the employer. 6. Undergo the medical examinations and psychological evaluations that determine this Standard and/or the official Mexican standards issued in this regard by the Ministry of Health and/or the Ministry of Labor and Social Welfare, and in the absence thereof, those indicated by the social or private security institution, or the doctor or psychologist or psychiatrist of the workplace or company.

3 Conclusions

This document presents different references of international research and reference frameworks, as well as the substantive aspects of NOM035-STPS-2018 to consider the core aspects, the design and proposal of a model and methodological instruments for evaluation, diagnostics, as well as programs and systems of compliance with the regulations that companies must comply with as a mandatory requirement, and must show evidence of possible audits and inspections by the Ministry of Labor and Social Welfare in Mexico. The evolution of Occupational Safety and Health (SST, SSO or H&S of Hygiene and Safety), evidences that, from simple training programs, audits, or risk analysis, to sophisticated techniques such as HAZOP and FMEA, Systems of Management, multidisciplinary work is necessary.

References

1. Norma Oficial Mexicana NOM035-STPS-2018. https://www.dof.gob.mx/nota_detalle.php?codigo=5541828fecha=23/10/2018
2. OIT: Trabajo decente - Trabajo seguro. In: Congreso Mundial sobre Salud y Seguridad en el Trabajo, Orlando, Estados Unidos (2005). http://www.ilo.org/global/about-the-ilo/newsroom/news/WCMS_006117/lang–es/index.htm. visitado el 6 de Enero 2013
3. Norma NOM035-STPS. http://www.stps.gob.mx/bp/secciones/dgsst/normatividad/normas/Nom-025.pdf. fecha de Consulta 3 de octubre de 2019
4. García Solarte, M.: Clima Organizacional y su Diagnóstico: Una aproximación Conceptual. Cuadernos de Administración (42) (2009). https://www.redalyc.org/articulo.oa?id=2250/225014900004. fecha de Consulta 3 de octubre de 2019. ISSN 0120-4645
5. Méndez, C.: Clima organizacional en Colombia. El IMCOC: Un método de análisis para su intervención. Colección de lecciones de administración. Universidad del Rosario, Bogotá (2006)
6. Martinez-Gutierrez, R.: Methodology of the fifth systemic helix for the development of public sector policies. J. Compet. Stud. **22**(3–4), 147+ (2014). https://go.gale.com/ps/anonymous?id=GALE%7CA427666157&sid=googleScholar&v=2.1&it=r&linkaccess=abs&issn=23304103&p=AONE&sw=w
7. Martínez Gutiérrez, R.: Quinta Hélice Sistémica (QHS), un método para evaluar la competitividad internacional del sector electrónico en Baja California, México. Investigación Administrativa, Julio–Diciembre 2012. http://www.redalyc.org/articulo.oa?id=456045338003. Fecha de consulta: 20 de agosto de 2019. ISSN 1870-6614
8. Cardona Echeverri, D.R., Zambrano Cruz, R.Z.: Revisión de instrumentos de evaluación de clima organizacional. Estudios Gerenciales **30**(131) (2014). https://www.redalyc.org/articulo.oa?id=212/21231108010. fecha de Consulta 3 de noviembre de 2019. ISSN 0123-5923
9. Brito, Y., Jiménez, J.C.: Evaluación del Clima Organizacional Universitario. Caso: Facultad de Ingeniería - Universidad de Carabobo. Ingeniería Industrial. Actualidad y Nuevas Tendencias **I** (2) (2009). https://www.redalyc.org/articulo.oa?id=2150/215016874006. fecha de Consulta 3 de octubre de 2019. ISSN 1856-8327
10. Premio Malcolm-Baldrige. http://calidad.overblog.com/el-premio-malcolm-baldrige. fecha de Consulta 3 de noviembre de 2019
11. Premio Deming para la ergonomia y salud ocupacional. http://ergonomiaysaludocupacionaliin.blogspot.com/2013/10/william-edwards-deming-guru-de-la.html. fecha de Consulta 3

Emergent Issues in Safety Management

Safety Requirements for the Design of Collaborative Robotic Workstations in Europe – A Review

Carlos Faria[1(✉)], Ana Colim[1], João Cunha[1], João Oliveira[1], Nelson Costa[2], Paula Carneiro[2], Sérgio Monteiro[2], Estela Bicho[2], Luís A. Rocha[1], and Pedro Arezes[2]

[1] DTx-Colab, Guimarães, Portugal
{carlos.faria, ana.colim, joao.cunha, joao.oliveira, luis.rocha}@dtx-colab.pt
[2] Centre Algoritmi - Universidade do Minho, Guimarães, Portugal
{ncosta, pcarneiro, parezes}@dps.uminho.pt, {sergio, estela.bicho}@dei.uminho.pt

Abstract. Industrial manufacturing is moving towards flexible and intelligent processes. Human-Robot Collaboration (HRC) has a pivotal role in smart factories due to a more versatile resource allocation that ultimately drives higher productivity and efficiency. The physical barriers that separate robots' and humans' workspaces are removed to facilitate HRC, which raises new safety concerns. To cope with this new robotics paradigm, regulatory legislation and international safety standards have been issued and are enforced for any machinery placed in factories. In this paper, we aim to shorten the gap between research projects and industry-ready robotic systems, by providing the guidelines and general requirements for collaborative robotic applications. We review the current international safety standards, certification procedures under the scope of European jurisdiction, and elaborate a literature review of papers related to safety for collaborative workstations.

Keywords: Industrial safety · Collaborative robotics · International standards

1 Introduction

Industrial robots are an integral part of modern manufacturing processes. Robots are programmable agents capable of repetitively and consistently performing tasks with a high degree of precision. They are also applied to tasks deemed unhealthy or too dangerous for humans [1]. At the same time, robots are powerful machines, capable of generating large forces and torques with unanticipated or unpredictable movements for the operator; a prospect that poses a serious and immediate risk to the operator. The safety of the human operator is generally guaranteed by the physical separation of the operator's and the robot's workspaces [2–4].

Even though industrial robots are a staple in today's industry, there are some tasks that cannot be easily robotized due to their complexity or variability. To overcome this limitation, roboticists explored the idea of closing the distance between robot and

P. M. Arezes and R. L. Boring (Eds.): AHFE 2020, AISC 1204, pp. 225–232, 2020.
https://doi.org/10.1007/978-3-030-50946-0_31

operator in order to create a collaborative space, where the robot's consistency and precision are complemented by the operator's flexibility and fast decision making. This idea crystallized into a concept known as *collaborative robotics* a branch of robotics where the human-robot interaction is permissible and often desirable.

This paradigm shift means that barriers which once separated the workspaces of operators and robots are now removed to facilitate human-robot collaboration (HRC). The human/robot proximity and hybrid task assignment raise questions about the safety of the operators and of their surroundings. Moreover, the shared, unstructured and dynamic environments only add to the complexity of the risk assessment process.

Regulatory agencies such as the International Organization for Standardization (ISO), the Canadian Standards Association (CSA) or the American National Standards Institute (ANSI), have produced standards directed to the design and development of robot systems, their integration, and more recently, standards that specifically target collaborative robots - cobots. These serve as basic guidelines to identify possible hazards, evaluate them and reduce the risk of accidents.

In this document, we sought to address the key documentation and findings regarding safety for collaborative robotic workstations under the scope of European jurisdiction. The legal documents and International Standards applicable in this domain were also revised.

2 Legal Base

The first step into the certification process passes through the regulatory terms incited by the European Commission (EC). In Europe, machinery is required to comply with EC directives in order to be eligible for commercialization or usage within European borders. Robot systems available in Europe should be ratified against the Machinery Directive (2006/42/EC) [5] and the Use of Work Equipment Directive (2009/104/EC). These directives are upheld in each member state by their own set of law decrees. The directives define a set of legally binding obligations for manufacturers and sellers that favors the integration of safety parameters during the project phase.

The manufacturer or integrator of the robot system shall conduct a risk assessment procedure and meet the relevant essential health and safety requirements from the Annex I of the Machine Directive. Moreover, the technical files (indicated in Annex VII, part A), as well as any other necessary information, shall be provided. Once appropriate procedures to assess conformity are carried out, the EC declaration of conformity can be drawn up and the CE marking consequently affixed.

If the robot application is not listed in Annex IV – categories of machines capable of inducing severe injuries – the system integrator shall apply the procedure for assessment of conformity with internal checks on the manufacture of machinery (Annex VIII). Instead, if the robot system is listed in Annex IV, the integrator shall conduct one of the following procedures:

1. Internal checks on manufacture (Annex VIII);
2. EC type-examination (Annex IX) combined with the internal checks on manufacture (Annex VIII, point 3);
3. Full quality assurance procedure (Annex X).

3 Safety Standards

International Standards play a key role in setting the baseline to what constitutes best practices in safe system behavior, and in design methodology. Although compliance with standards is not mandatory, they are considered as valid regulatory guidance when recognized by the European Committee for Standardization (ECS). Standards capture the international consensus and set general guidelines to certify the "mission-worthiness" of new products/systems, which also accelerates their commissioning process.

The ECS cooperates with international bodies such as the International Organization for Standardization (ISO) and the International Electrotechnical Commission (IEC) to produce harmonized European norms. In the scope of Collaborative Robotics, the relevant international standards are summarized in Fig. 1.

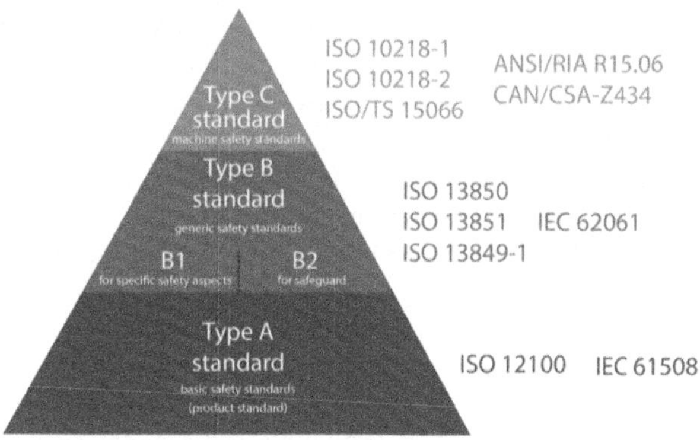

Fig. 1. Categories of Safety Standards according to International Standards Organizations. Other associations such as the American National Standards Institute (ANSI) and the CSA Group (formerly the Canadian Standards Association) exist and have published norms related to Collaborative Robotics. Due to regulatory differences between the EU/EEA and the U.S., the EC tends to work with organizations that produce globally relevant standards, for example, ISO and IEC.

IEC/ISO machine safety standards are organized in a three-level hierarchy from the most general and relatable to generic machinery – *type A* – to machine-specific – *type C*. *Type C* norms should be used as templates and take precedence over standards that are more general. In any case, different standard types are not dissociated, as *type C* norms often refer to standards of types A and B to address, for example, risk assessment and other safety-related details.

Type A norms establish basic concepts, conception principles and general requirements applicable to machinery. These define the strategy for risk assessment and risk reduction. The risk assessment involves the i) determination of the robotic system limits, ii) identification of risks and hazards that may occur during standard operation, commissioning, or maintenance, iii) risk estimation, and iv) risk evaluation. In the risk assessment process, the evaluator may combine different tools to determine the risk level based on damage severity, frequency or time of exposure, and the possibility to limit or avoid damage.

Type B norms specify safety aspects or safeguards applicable to machinery. These norms are split into two categories, the *B1* norms related to general safety aspects (e.g., safety distances, surface temperature, noise levels, etc.), and the *B2* norms about safeguard (e.g., bimanual commands, interlock devices, pressure-sensitive devices, protective equipment, etc.). *Type B* norms describe mechanisms to identify the safety functions, to determine the category[1] and required performance levels (PLr) for safety-related parts of the control systems (SRP/CS). The required performance level is determined after the risk level previously determined during risk assessment (ISO 12100). It is the robotic system integrator's responsibility to guarantee that the robot and any other parts of the system are accompanied by a certificate of conformity that attests to the part's performance level (PL) and category, and that it meets the expected PLr.

Type C norms identify specific safety requirements applicable to a machine category. The norm ISO 10218 divides into part 1 – provides guidance for the design and construction of robots, and part 2 – that focuses on the integration process of robotic systems. Integrators of cobot workstations that operate with commercial robots should focus on part 2, which defines guidelines for safeguarding personnel, commissioning, functional testing, programming, operation, and maintenance.

The norm ISO 10218-2 narrows down the scope of the risk assessment process, by focusing on topics relatable to robotic systems. First, a list of robot limits is proposed and organized by topic: usage, space, time, and others. Then, ISO 10218-2, Annex A, compiles a list of the most significant hazards of different sources (mechanical, electrical, thermal, noise, etc.) relatable to robotic applications. The norm proceeds to address the safety requirements and protective measures. It extends and particularizes requirements for SRP/CS in the aforementioned *type B* norms to the address robot applications. The layout design contemplated in this norm rests primarily in the use of physical or electro-sensitive equipment to delimit the robot and operator's workspace. Finally, the Annex G of the ISO 10218-2 includes a comprehensive table to verify the compliance of the safety requirements and measures for robotic systems.

The topic of collaborative robotics is referred to in the ISO 10218-2 (section 5.11 and Annex G), but given the complexity of the subject, it is supplemented by the technical specification ISO/TS 15066. In collaborative applications, the interaction between humans and robots is permitted and often desired. To facilitate this interaction, the physical limits that safeguard the robot's workspace are often removed. With a shared workspace, the safety of the operator is guaranteed by inherently safe design measures or safety-rated limiting functions.

The foreseeable human-robot interaction is categorized by the ISO/TS 15066 in four collaborative operation modes as depicted in Fig. 2.

[1] The control architecture typologies (number and relationship between input, logic, test, and output elements) are labeled in different categories (B, 1, 2, 3, 4).

The SMS is a black and white approach to collaborative robotics applications. The operator and robot may co-exist in the same space, although they shall not operate within the collaborative workspace at the same time. If the safeguard perimeter is breached, a safety-rated monitored stop ceases the robot motion.

The HG mode is an extension to the SMS collaborative mode, it functions based on the same premise, only this time, the system also comprehends a hand-guidance device to directly maneuver the robot when the operator is within the safeguard perimeter.

The SSM mode permits the robot and the operator to move concurrently within the collaborative workspace. The risk reduction is achieved by maintaining the protective separation distance at all times. A tracking device continuously monitors the distances between the hazardous robotic system parts and any person within the collaborative space. The robot operating speed is proportional to the shortest calculated distance.

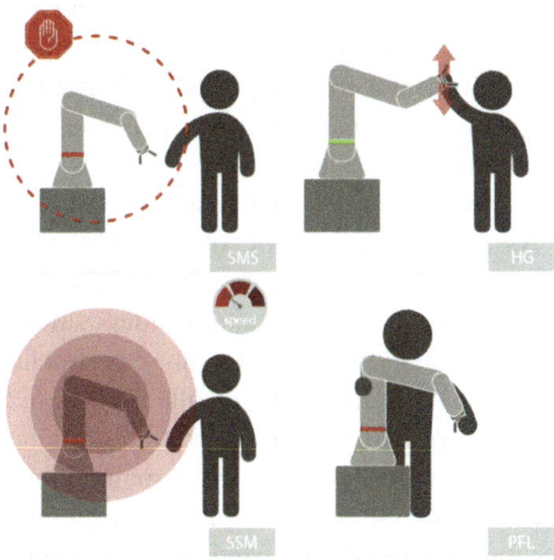

Fig. 2. Collaborative operation modes, Safety-rated Monitored Stop (SMS), Hand-Guiding (HG), Speed and Separation Monitoring (SSM), and Power and Force Limiting (PFL).

The PFL mode is the only collaborative mode that permits contact between the robot and the operator. Instead of separation, the risk reduction is achieved through inherently safe means or safety-rated control systems that keep hazards below a threshold. These include *passive measures* in the robot design (larger contact surface area through edge smoothing, padding or deformable parts, and the limitation of moving masses); and *active measures* in the robot control system (torque/force limiting, use of safety-rated soft axis/space limits, safety-rated monitored stop functions).

4 Literature Review

To extend the revision of the legal and normative requirements, we conducted a literature review focused on safety requirements for cobot workstations using the keywords *safety, collaborative robotics, HRC, and international standards*; in four databases: Scopus, PubMed, IEEEXplore, and ScienceDirect. Papers from the last 5 years were revised and grouped into state-of-the-art reviews, innovative safety solutions, risk assessment and analysis methods.

Extensive state-of-the-art reviews on safety standards and measures can be found in [1, 6] and [7]. Robla-Gomez et al. analyzed current legal regulations and the main safety systems applied in industrial robotics that contribute to safe HRC [1]. Finkemeyer discussed the potential benefits and possible use cases of HRC, safety functions, standards, and technical issues [6]. Murashov et al. summarized safety standards and recommended measures that guarantee the safety of human workers at all times [7].

Different innovative safety solutions were proposed, and experimentally validated, in [8–11], and [12]. Navarro et al. presented an adaptive damping controller that fulfills ISO 10218 and enables the operator to manipulate the robot safely without the risk of exceeding the velocity, power and force constraints [8]. The controller was validated in an industrial screwing application. Rojas et al. developed a trajectory planning method based on safe minimum-jerk trajectories that ensure physical safety and reduces psychological strain [9]. As a proof-of-concept, the authors presented an experimental setup and provide insights on the implementation of the proposed method. Vysocky et al. proposed a motion planning strategy that considers safety precautions stipulated in ISO/TS 15066 [10]. The strategy was tested in an experimental workstation with the assistant co-working manipulator PaDY. The results show a reduction of velocity and of the risk of harming workers during robot operation. Zanchettin et al. developed a kinematic control strategy that enforces safety while maintaining productivity [12]. The methodology was experimentally validated on a dual-arm concept robot performing a manipulation task. Lasota and Shah conducted an experiment in which participants performed a collaborative task with an adaptive robot that incorporates human-aware motion planning and with a baseline robot using shortest-path trajectories [11]. Results show that working with the adaptive robot allowed participants to complete the task faster, with more concurrent motion, with less human and robot idle time, and with a larger human-robot separation distance. Participants also indicated they were more satisfied with the adaptive robot as a teammate and felt safer and more comfortable working with it.

Methodologies and guidelines for risk assessment are provided in [4, 13], and [14]. Vicentini et al. proposed a risk analysis methodology for HRC applications compatible with ISO standards that relied on a formal verification technique to automate traditional risk analysis methods [13]. Rosenstrauch and Kruger introduced safety standards and guidelines for risk assessment in general and industrial robots specifically, followed by a detailed insight of ISO/TS 15066 [4]. Poot, Johansen, and Gopinath presented a thorough risk assessment for an automated warehouse, where mobile robots and humans collaborate in a shared workspace [14]. Risk assessment is performed using a Hazard and Operability Study (HAZOP) to identify hazards, coupled with Unified

Modeling Language (UML) and risk estimation. In this work, one can find a systemic description of collaborative scenarios and interactions, human roles, and unsafe scenarios.

5 Conclusions

HRC is an appealing prospect to the industry in general: to small and medium-sized enterprises (SME) due to the high degree of adaptability and flexibility, as well as to mass production companies that are rapidly shifting into mass customization. Moreover, the interaction between humans and robots is growing wider – with an increasing range of products fabricated with the assistance of precise and repeatable robotic platforms - and deeper as the human-machine interaction moves beyond the physical cages. The quest for a safer human-robot workstation grew considerably more complex with the lack of physical separation between the human and the robot workspaces. Labeling a product or system as safe is a substantial claim, one that is necessarily supported by results and data, and confirmed by satisfactory compliance assessment. We reviewed the key documentation related to the design of collaborative robotic applications, the legislation enforced in Europe, the harmonized international standards, and the latest publications on the field.

Acknowledgments. This work was supported by NORTE-06-3559-FSE-000018, integrated into the invitation NORTE-59-2018-41, aiming to hire highly-qualified human resources, co-financed by the Regional Operational Programme of the North 2020, thematic area of Competitiveness and Employment, through the European Social Fund (ESF).

References

1. Robla-Gomez, S., Becerra, V.M., Llata, J.R., Gonzalez-Sarabia, E., Torre-Ferrero, C., Perez-Oria, J.: Working together: a review on safe human-robot collaboration in industrial environments. IEEE Access **5**, 26754–26773 (2017)
2. International Organization for Standardization: ISO 10218 Robots and robotic devices - Safety requirements for industrial robots - Part 2: Robot systems and integration (2011)
3. International Organization for Standardization: ISO/TS 15066 Robots and robotic devices - Collaborative robots (2016)
4. Rosenstrauch, M.J., Kruger, J.: Safe human-robot-collaboration-introduction and experiment using ISO/TS 15066. In: 2017 3rd International Conference on Control, Automation and Robotics, ICCAR 2017, pp. 740–744 (2017)
5. The European Parliament and the Council of the European Union: Machine Directive 2006/42/EC (2006)
6. Finkemeyer, B.: Towards safe human-robot collaboration. In: 2017 22nd International Conference on Methods and Models in Automation and Robotics, MMAR 2017, pp. 883–888 (2017)
7. Murashov, V., Hearl, F., Howard, J.: Working safely with robot workers: recommendations for the new workplace. J. Occup. Environ. Hyg. **13**, D61–D71 (2016)

8. Navarro, B., Cherubini, A., Fonte, A., Passama, R., Poisson, G., Fraisse, P.: An ISO10218-compliant adaptive damping controller for safe physical human-robot interaction. In: Proceedings - IEEE International Conference on Robotics and Automation, 2016-June, pp. 3043–3048 (2016)
9. Rojas, R.A., Garcia, M.A.R., Wehrle, E., Vidoni, R.: A variational approach to minimum-jerk trajectories for psychological safety in collaborative assembly stations. IEEE Robot. Autom. Lett. **4**, 823–829 (2019)
10. Vysocky, A., Wada, H., Kinugawa, J., Kosuge, K.: Motion planning analysis according to ISO/TS 15066 in human-robot collaboration environment. In: IEEE/ASME International Conference on Advanced Intelligent Mechatronics, AIM. 2019-July, pp. 151–156 (2019)
11. Lasota, P., Shah, J.: Analyzing the effects of human-aware motion planning on close-proximity human–robot collaboration. Hum. Factors **57**, 21–33 (2015)
12. Zanchettin, A.M., Ceriani, N.M., Rocco, P., Ding, H., Matthias, B.: Safety in human-robot collaborative manufacturing environments: metrics and control. IEEE Trans. Autom. Sci. Eng. **13**, 882–893 (2016)
13. Vicentini, F., Askarpour, M., Rossi, M.G., Mandrioli, D.: Safety assessment of collaborative robotics through automated formal verification. IEEE Trans. Robot. **36**(1), 42–61 (2020). https://doi.org/10.1109/TRO.2019.2937471
14. Poot, L., Johansen, K., Gopinath, V.: Supporting risk assessment of human-robot collaborative production layouts: a proposed design automation framework. Proc. Manuf. **25**, 543–548 (2018)

Reviewing Tools to Prevent Accidents by Investigation of Human Factor Dynamic Networks

Salvador Ávila[1(✉)], Lucas Pereira[1], Rita Ávila[1], Camila Pena[2], Pedro Arezes[3], and Elvis Renan Fagundes Lima[4]

[1] Federal University of Bahia, Polytechnic Institute, Rua Professor Aristides Novis 2, Salvador 40210-630, Brazil
avilasalva@gmail.com, lucasmpereira13@hotmail.com, ritaavila2@gmail.com

[2] Federal Institute of Bahia (IF Baiano), Rua Barão de Camaçari, 118, Catu 48110-000, Brazil
pontespena@hotmail.com

[3] Minho's University, Largo do Paço, 4704-553 Braga, Portugal
parezes@dps.uminho.pt

[4] Dass Group, Santo Estevão, BA, Brazil
rennan20@hotmail.com

Abstract. A network of serial, parallel and cross-related factors transfers the hazard energy until the accident occurs. This paper aims to discuss the failure of barriers initially built to prevent the accident, the factors that need to be included in the analysis to review human elements and their safeguards, and the need for critical judgment for resilient and accident-alert safe behavior. It is intended to study past accidents and establish relationships between these factors, antecedents and consequences. The availability of a competent team to interpret the chain of events that can lead to the accident is crucial in ending the hazardous energy migration. It is expected to measure the ability to increase or decrease the hazardous energy considering ship collision with oil leak. Analysis of human elements in the project before and after the accident layer factors may suggest changes in safeguards or recommendation of new barriers to the accident.

Keywords: Accident control · Human factors · Ship collision · Hazard energy · Oil leak

1 Introduction

Failing to comply with procedures and carry out work by informal practices—still frequent in high-risk processes—can generate failures that transfer hazard energy towards an accident or even a disaster. Human error is a determining factor and it is considered to be the main cause of 70–80% of accidents [1].

This work is based on the model of Human Elements and Factors Network, and the main purpose is to discuss the method to find the best safeguards to reduce the concentration of hazard energy in cases of ship collision.

P. M. Arezes and R. L. Boring (Eds.): AHFE 2020, AISC 1204, pp. 233–240, 2020.
https://doi.org/10.1007/978-3-030-50946-0_32

An accident is the sum of several reactions that require an analysis based on various fields of knowledge [2]. The culture of managing large industries arises from technological and administrative sciences but does not include human factors. When evaluating an accident, the (famous) "Swiss-cheese" model suggests that multiple factors from the most diverse socio-technical scopes, represented by holes in cheese slices, must be aligned so that adverse events occur and then provoking the accidents [3]. A qualified socio-technical view of an organization's activities can strongly detect inadequate conditions that encourage the transmission of hazard energy through the various sectors that compose the company. Incorrect operations are usually camouflaged under management decisions and are often not perceived until other factors are combined and then cause an accident, so that deep reflection is made about unsafe operating conditions that could be avoided [3].

According to the International Association of Oil & Gas Producers (IOGP), the causes related to fatalities and high-risk events from 2010 to 2014 were mainly linked to low competence for risk assessment, improper procedure and supervision, and inattention along with a lack of judgment in decision-making [4]. Due to technological growth, it is increasingly common to have systems tied and composed of interdependent smaller parts and with high potential for the spread of hazard energy among its components. In addition, in these complex systems failures usually occur simultaneously, which gives total conditions of them to turn into accidents and even disasters [5]. In view of the considerable frequency of informal rules adoption, it is recommended to review safety procedures as an attempt to adapt them to the organization's reality and make the rules sufficient to ensure plant safety [6].

The Marine Accident Investigation Branch (MAIB) states that a single factor still prevails in the vast majority of maritime accidents: human error [7]. Replacing the crew with the automation of tasks has caused an overload on the work team by the need to now monitor automatic operation panels in addition to taking care of their original manual tasks, thus causing fatigue (stress added to tiredness) and lack of concentration, which is more likely to cause accidents [8].

If the higher hierarchical levels define conditions and restrictions that are not adequate for the accomplishment of the tasks, the hazard energy is unnecessarily propagated with a high degree since then and may reach even higher levels depending on how operational levels work [1].

A historical review of the various types of accidents involving ships, mainly oil tankers—as they have a higher risk of pollution—was carried out [9]. The investigation of the root cause of an incident (that could have serious consequences because of an oil spill) is always important, even if there is neither human/cargo loss nor damage to the vessel, as the results serve for the review and adaptation of procedures, which create safeguards through the lessons learned from the event [10].

The proposed methodology involves layers of human factors in a network, which determines the probability of failure, accident and human error by comparing the saturation level with the accumulated hazard energy.

2 Dynamic Bayesian Network of Human Factors

The Bayesian network is a tool used for the representation and inference of knowledge under conditions of uncertainty, mainly in conceptual modeling. When using it, a division of a complex system into smaller parts is done in order to understand the relations among the parts as compounding a whole. It is important to identify the dynamic relations between human, managerial, social and organizational factors in the workplace for the elaboration of preventive actions. The causal situations in the model are clear and can be submitted to discussion and review, with great potential for use in human decision-making [11]. The analysis of human reliability is related to the identification of causes and consequences of human decisions, and actions and the usual methods of analysis are quite simplistic. There is a need for more sophisticated methods, which present human errors as a causal fact, so that they are able to capture factors that influence human performance and the relationships between themselves [12].

3 Methodology

3.1 Proposed Human Factors Network Adapted for a Ship Collision Event

A model analysis adapted through the use of human factors for ship collision was constructed using a Bayesian network. This tool allows interdependent sociotechnical factors to be analyzed together in order to consider them part of a chain event that results in an accident. They were represented in the flow of hazard energy (operating load) for human factors and their ability to resist the passage of this energy through human elements. The 9 human performance factors at work are related to culture, social phenomena, company, management, group/individual, operational control, failure control, accident control, and disaster control.

The application context of the model was based on reports and articles investigating ship-related accidents that resulted or could result in loss of vessel structural water tightness and consequent presence of oil at sea. The main risk involved is the presence of oil in marine waters that could have serious public health consequences such as contamination of people and environment by affecting the entire marine ecosystem.

3.2 Accumulated Hazard Energy Along the Human Factors Network

A usefulness of the human factors network is to locate the region where the hazard energy and its respective flows are concentrated. The consequences of each layer of factors can lead to social phenomena, in which "apparently" there are no visible economic losses. From the hazard energy in the layer of individuals and groups, unsafe behavior can be found, which motivates not only deviances in routine, but also equipment failures, accidents and even disasters.

4 Results

Based on the structure of the network related to collision presented in [13], an original method was used to create Fig. 1, in which the human factors on the analysis of a ship collision accident are represented. The factors were organized within circles which the blue color refers to human factors; yellow is used for technological factors and green for those directly related to the disaster. With regard to hazard energy, the red color (high hazard energy: 3, 4 or 5) means higher flows or low ability to resist the passage of that energy. The black arrows and the black-bordered circles (low hazard energy: 1, 2) represent the low potential of the hazard energy.

In order to the hazard energy become lower, it is necessary to insert safeguards in the Bayesian Network that allow risk control which will be discussed below.

With regard to communication (A in Fig. 1), maritime operations have official language codes for the tasks, and the proposed model by the research group is dynamic. The level of communication has a strategic position and the Accident Control is related to technological factors, directly linked to the ship command that acts in surveillance by the officers. Communication also receives information from Bridge Resource Management (BRM) (C in Fig. 1), which prevents communication barriers to elevate hazards [14]. Thus, in this safeguard it is important to clarify the responsibilities (and duties) and to perform them quickly and safely, in addition to getting to know the crew composition and using universal procedures, mainly in safety.

The legislation requirements (B in Fig. 1) related to maritime activity are located in the Global and Regional Culture layer (D in Fig. 1). On the proposed model, they are directly linked to the type and size of the ship (E in Fig. 1). The starting point is considered the alignment of the standards and equipment specifications to meet national and international requirements. The International Maritime Organization (IMO)—which is part of the United Nations (UN)—acts in the development and maintenance of a regulatory structure that is met globally. It comprises items such as maritime safety, environmental concerns, legal issues, technical cooperation, maritime accident investigation records and efficiency of maritime transport. [15]

Competence (F in Fig. 1) is a sum of qualities, known today as socio-technical competence and involves not only the performance observed while accomplishing the designated tasks, but also the commitment to provide motivation, clarification and help to workmates. Several organizational and personal factors influence individual performance and make installed competence not to match effective competence. It is also essential to act on competence in technical decisions, such as the use of good quality equipment, as well as making spare parts available. Training should be available with simulations of standard procedures so that there will not be signs of crisis or panic when fault and accident control is necessary.

Intrinsic human factors can interfere with human performance during their work routine and are pertinent to internal worker issues such as personality, competence, communication, circadian cycle, age group, health status (G in Fig. 1), in addition to alcohol consumption. In order to avoid problems of worker's competence and communication, a healthier, more cooperative and safer working climate should be proposed, with the use of active listening among all members and the management team,

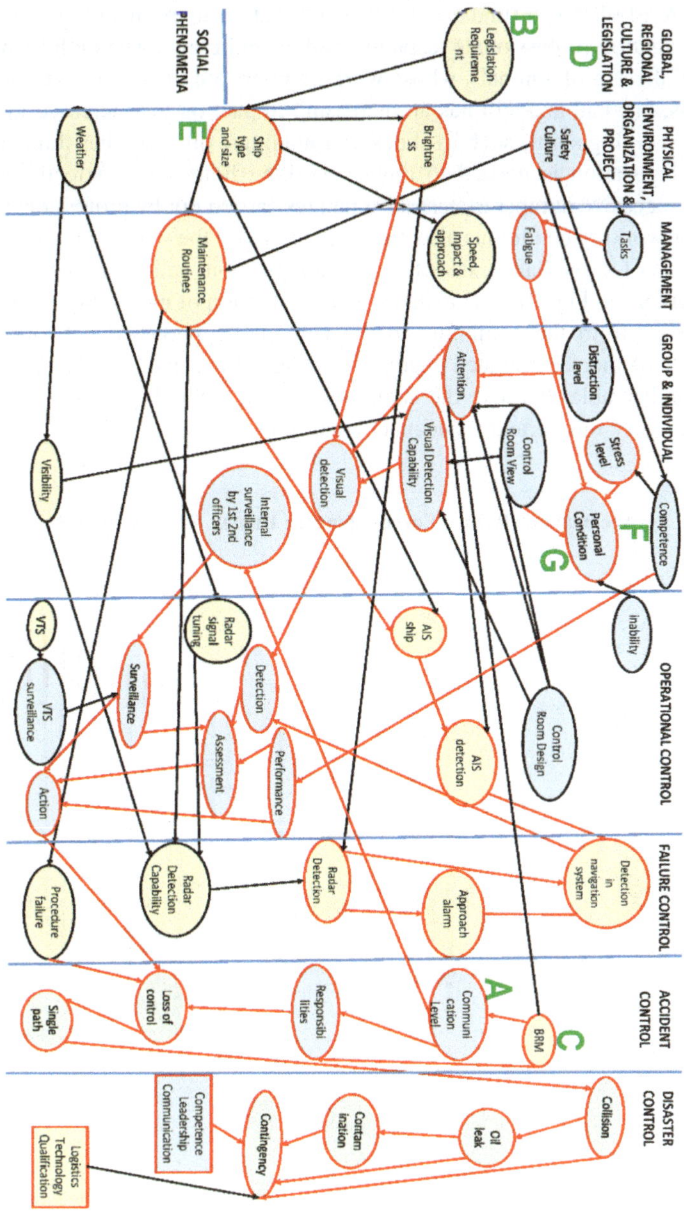

Fig. 1. Proposed human factors network, adapted for a ship collision event

which should be closer to the work team in order to strengthen communication and understanding of critical issues for the performance on the tasks and also the attempt to achieve maximum performance through the more efficient use of skills.

The state of alert is a fundamental demand that requires surveillance from people such as navigation professionals (captains, sailors and crew). The high tiredness faced by many of these professionals, whose working hours are the second largest among all occupations, has had and still has an increasing impact on the increased likelihood of accidents involving ships, such as shocks, groundings, oil and other fluid spills. More protective legislation is needed to reduce working hours, also extend vacation and reduce the excessive number of vessel exchanges carried out by professionals in a short time. Better working conditions also reduce pressure on the work team, thus improving the efficiency of surveillance, detection, communication, action and control.

A bar graph (in Fig. 2) was created with the intention of measuring the accumulated hazard energy from the beginning to the end of the proposed network. From the visualization of the accumulated energy, it is noticed that there is always an increase, which is undesirable, because the saturation region (in red) appears, which reveals operation in very unsafe and inadequate conditions.

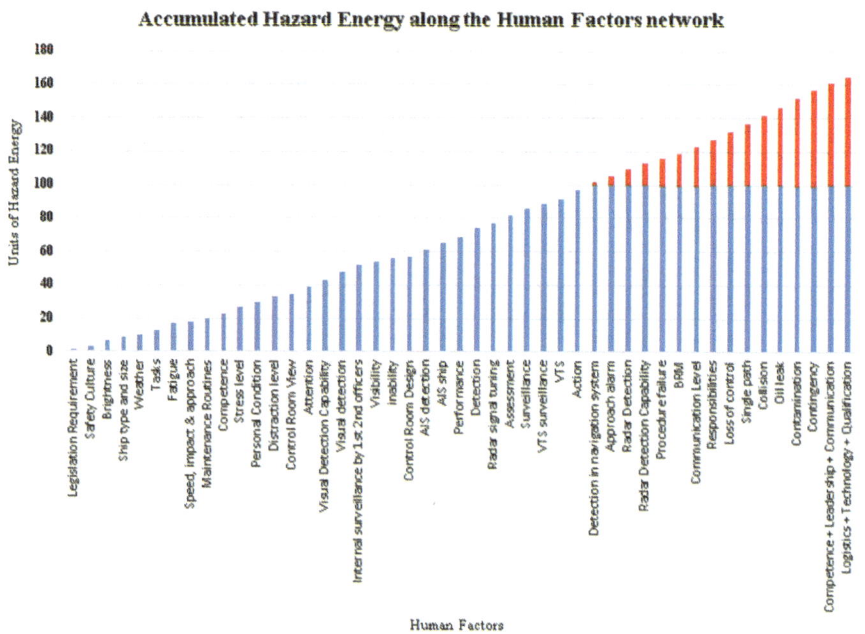

Fig. 2. Accumulated hazard energy along the human factors network.

5 Conclusion

An accident involving an oil industry ship has the potential to produce an environmental catastrophe as the result of an oil leakage at sea. This case leads to the need to comply with preventive standards by regulators and industry. Thus, the community and the environment are preserved. Several accidents have occurred in recent decades and this has shown the severity related to safety in the operation and handling of ships to avoid collisions or any other type of event that may result in oil at sea.

This article has built a network of serial, parallel and cross-powered factors that demonstrate the hazard energy transfers until a ship collision occurs. In the model it was possible to analyze and correlate the factors in order to evaluate the existing safety and operation barriers, as well as the possible paths for the flow of hazard energy. The gaps found throughout the analysis were translated into safeguards in order to create new limitations for the flow of this energy. The most important safeguards in reducing hazard energy should be related to Accident Control and Disaster Control, since both have higher influence on the amount of oil leaked.

Ship collision can be avoided when there is a commitment to safety, through immediate actions in operational abnormalities and in the effectiveness of internal communication tools used in high-risk situations. Technological tools can reduce human error, but overconfidence in the efficiency of these equipment and the apparent comfort may result in lack of awareness and then in the accident. For maintaining alertness, it is necessary that the rules governing maritime work are adjusted.

The standardization of procedures and the understanding of the maritime legislation should also be prioritized, as well as preventive maintenance of the vessel and the operational control of loading/unloading must be carried out rigorously to ensure the structural stability of the ship. It is necessary to motivate the work team and perform periodic training so that the crew members are prepared to deal with an adverse situation. The healthy work environment ensures safe and efficient conditions of performing tasks. Thus, when implementing the aforementioned safeguards, a reduction in the risk of ship collision is expected, which makes this activity safer.

References

1. Rasmussen, J.: Risk management in a dynamic society: a modelling problem. Saf. Sci. **27**, 183–213 (1997)
2. Llory, M.: Acidentes industriais: O custo do silêncio. Multimais Editorial Produções Ltda., Rio de Janeiro (1999)
3. Reason, J., Hollnagel, E., Paries, J.: Revisiting the Swiss Cheese Model of Accidents. Eurocontrol Experimental Centre, EEC Note No. 12/06, Brétigny-sur-Orge (2006)
4. IOGP: International Association of Oil & Gas Producers, Safety Performance Indicators – 2014 Data (2015). www.iogp.org
5. Perrow, C.: Normal Accidents: Living with High Risk Technologies. Princeton University Press, Basic Books, New York, New Jersey (1984)
6. CCPS: Guidelines for Management of Change for Process Safety, Wiley Editors, EUA (2008)
7. MAIB: Annual report 1999. Department of the Environment Transport and Regions, London (2000)
8. Hetherington, C., Flin, R., Mearns, K.: Safety in shipping: the human element. J. Saf. Res. **37**, 401–411 (2006)
9. Primorac, B., Parunov, J.: Review of statistical data on ship accidents. Marit. Technol. Eng. **3**, 809–814 (2016)

10. Ozen, M., Arslan, O., Kececi, T.: Root cause evaluation, the case analysis of ship contact accident to canal lock. In: Seahorse Conference Maritime Safety and Human Factors, Glasgow (2016)
11. Jackson, L., Alt, M.: Bayesian Networks for the Human Element. US Army Training and Doctrine Command (2013)
12. Groth, K., Mosleh, A.: Deriving causal Bayesian networks from human reliability analysis data: a methodology and example model. J. Risk Reliab. **226**(4), 361–379 (2012)
13. Hanninen, M., Kujala, P.: The effects of causation probability on the ship collision statistics in the Gulf of Finland. Int. J. Marine Navig. Saf. Sea Transp. **4**(1), 79–84 (2010)
14. Ayala, J., Reis, M.: Gerenciamento de recursos do passadiço. Curso de Formação de Oficiais de Náutica, Marinha do Brasil, Rio de Janeiro (2017)
15. af Geijerstam, K., Svensson, H.: Ship Collision Risk - An identification and evaluation of important factors in collisions with offshore installations. Department of Fire and Safety Engineering and Systems Safety. Lund University, Sweden, Report 5275 (2008)

How Accurate Is It to Measure Noise with Smart Mobile Devices?

Rui B. Melo[1,2(✉)], Filipa Carvalho[1,2], and Rafael Assunção[2]

[1] CIAUD – Centro de Investigação em Arquitetura, Urbanismo e Design, Faculdade de Arquitetura, Universidade de Lisboa, 1349-063 Lisboa, Portugal
{rmelo, fcarvalho}@fmh.ulisboa.pt
[2] Laboratório de Ergonomia, Faculdade de Motricidade Humana, Universidade de Lisboa, 1499-002 Cruz Quebrada, Portugal
rafael.apfa@gmail.com

Abstract. Traditionally, sound meters and noise dosimeters have been worldwide recognized as the most adequate devices to measure sound pressure level. In recent years, smart mobile devices have been incorporated with many different sensors and multiple useful applications have been developed, turning them into multi-purpose measuring units. Compared to sound meters, smart devices can be less expensive, are not task specific and allow sharing collected data in real time. Fourteen smartphones and five tablets based on iOS and Android platforms were used to conduct sound level measurements with three and two apps respectively. Seven different sound signals of pink noise were utilized at different levels (60, 65, 70, 75, 80, 85 and 90 dBA) along with an industrial noise signal. This study concluded that some mobile device/app combinations may be considered adequate as a screening tool but not as accurate as sound level meters.

Keywords: Noise · Measurement · Smartphone · Tablet · Accuracy

1 Introduction

Despite the actual knowledge and technological advances concerning noise emission, transmission and reduction, it still is a common occupational hazard in many workplaces and industries across the world [1]. Managing risks arising from occupational exposure to noise demands accurate and precise measurements, in first place. Traditionally, sound level meters (SLM) and personal sound exposure meters (dosimeters) have been worldwide recognized as the adequate devices to measure sound pressure levels. For occupational health and safety purposes, class 2 SLM are recommended in Europe, but class 1 should be preferred whenever possible namely because they are more accurate. Legally, they must be integrating averaging meters and measure at least the equivalent continuous A-weighted sound pressure level ($L_{A,eq}$) and the maximum C-weighted peak sound pressure level ($L_{C,peak}$). SLM and dosimeters must comply with International Electrotechnical Commission's IEC 61672-1: 2013 [2] and IEC 61252: 2017 standards [3] respectively.

P. M. Arezes and R. L. Boring (Eds.): AHFE 2020, AISC 1204, pp. 241–248, 2020.
https://doi.org/10.1007/978-3-030-50946-0_33

In recent years, smart mobile devices such as smartphones, tablets and smartwatches have become ubiquitous, because both hardware and software have improved. They have been incorporated with many different sensors and multiple useful applications have been developed. These innovations turned these devices into multipurpose units.

Compared to SLM and dosimeters, smart mobile devices can be less expensive as they are not task specific and allow sharing collected data in real time which may be an advantage for small and medium-sized enterprises in which financial resources may be limited and human resources rarely are trained to operate SLM.

Smartphones have been tested to measure illuminance, but the results reveal extremely variable and sometimes large deviations from the luxmeter readings and suggest these devices are not appropriate for use in occupational lighting assessments [4].

The possibilities and limitations of using smartphones for acoustical measurements have been discussed [5] and recent studies have dealt with the accuracy of smartphones used to measure either environmental or occupational noise [6–8]. The obtained results are very encouraging. Nevertheless, ways to improve smartphone performance when measuring noise were analyzed [9–11] and it was found that relying on external microphones rather than on the ones mounted on the mobile device delivers more accurate results.

People have started to use mobile devices to measure noise professionally, sometimes not selecting a proper device or application and risking obtaining less accurate and less precise measures, and ultimately putting workers at risk [12].

Since applications are often improved and replaced in App Store and Google Play Store, while new ones are also developed, and mobile devices are in continuous development, this study aims to provide some insights on the accuracy of smartphone and tablet models that have not been tested before.

2 Materials and Methods

2.1 Samples

All applications relied on the internal microphone of the mobile device and were downloaded for free either from the App Store or the Google Play Store. Their main characteristics are summarized in Table 1. All of them were set with Fast time

Table 1. Identification and main characteristics of the selected applications.

Application	Manufacturer	Operating system	Main characteristics
Noise meter	Jinasys	Android	Calibration/L_{Aeq}
OpeNoise	Arpa Piemonte	Android	Calibration/L_{Aeq}
NIOSH	EA LAB	iOS	Calibration/L_{Aeq}/L_{Cpeak}/TW
dB Meter	Vlad Polyanskiy	iOS	Calibration/L_{Aeq}/TW
SLA Lite	TOON, LLC	iOS	Calibration/L_{Aeq}/TW

weighting (TW) and provided L_{Aeq} values whereas only one of them provided L_{Cpeak} values.

Nineteen mobile devices (14 smartphones and 5 tablets) from 7 different manufacturers, using iOS and Android operating systems only, were gathered from university students and professors who volunteered their equipment to be tested (Table 2).

Table 2. List of the mobile devices used in the experiments.

Type of device	Manufacturer	Model	Operating system
Tablet	Acer	Iconia B1-830	Android
Smartphone	Alcatel	Star Pop	Android
Smartphone	Apple	iPhone SE	iOS
Smartphone	Apple	iPhone 5S	iOS
Smartphone	Apple	iPhone 6S	iOS
Smartphone	Apple	iPhone 7	iOS
Smartphone	Apple	iPhone 11	iOS
Tablet	Apple	iPad Pro 9.7	iOS
Tablet	Asus	K010	Android
Tablet	Asus	ZenPad C7.0	Android
Smartphone	Huawei	GR3	Android
Tablet	Huawei	Media Pad M3 Lite 10	Android
Smartphone	Huawei	P9 Lite	Android
Smartphone	Huawei	P10 Lite	Android
Smartphone	LG	E620	Android
Smartphone	Samsung	Galaxy J3	Android
Smartphone	Samsung	GT 19082	Android
Smartphone	Samsung	GT 19301L	Android
Smartphone	Samsung	SM T560	Android

2.2 Experimental Procedures

The objectives of the study and the procedures to be accomplished were passed to the mobile device owner so that informed written consent could be freely obtained. The second step was to download the apps into the device.

Two different sound signals (pink noise and industrial noise) were previously downloaded from a sound library to be reproduced from a laptop through a loudspeaker as noise sources.

All experiments were done in the Ergonomics Laboratory where background noise was measured before running trials to ensure it was low enough and would not be a confounding variable. This was accomplished with a hand-held programmable sound analyzer from Brüel & Kjær, type 2260 Investigator, recently calibrated by a certified laboratory, mounted on a tripod 0.5 m from the loudspeaker and 1.0 m from the ground. Prior to each testing session the SLM was checked using a Brüel & Kjær Sound Calibrator Type 4231.

Pink noise record set at 60 dB(A) was used for calibrating the apps before starting the trials. L_{Aeq} values were always checked against the measurement provided by the SLM. Each device was then tested with a set of applications, in accordance with its operating system, against 7 different L_{Aeq} values - 60, 65, 70, 75, 80, 85, 90 dB(A) – of pink noise and against the industrial noise record set between 80 and 85 dB(A). The devices' microphones were oriented toward the speaker and each trial lasted 1 min or was interrupted earlier when readings would not oscillate more than ±0.5 dB(A). Single measurements were recorded for each app, on each smartphone and tablet model, at each sound level, for each of all 8 sound signals.

2.3 Data Analysis

IBM SPSS Statistics (version 26) was used for data processing and analysis. Instead of analyzing L_{Aeq} and L_{Cpeak} raw values, difference between values obtained with the SLM and the mobile device were computed according to Eqs. (1) and (2).

$$\mathrm{DifL}_{Aeq} = L_{AeqSLM} - L_{AeqDev}. \tag{1}$$

Where L_{AeqSLM} is the value of L_{Aeq} provided by the Sound Level Meter and L_{AeqDev} the mobile device reading.

$$\mathrm{DifL}_{CpeaK} = L_{CpeakSLM} - L_{CpeakDev}. \tag{2}$$

Where $L_{CpeakSLM}$ is the value of L_{Cpeak} provided by the Sound Level Meter and $L_{CpeakDev}$ the mobile device reading. Positive/negative values of these two variables would mean that the mobile device readings are lower/higher, than those provided by the SLM and the larger the absolute value of the variables the worst the accuracy revealed by the mobile device.

Descriptive statistics based on median, maximum and minimum values allowed characterizing applications and mobile devices performance. Standard boxplot analysis was used to assess the variability in the relative differences across applications and mobile devices.

Additionally, Kruskal-Wallis and Mann-Whitney tests to assess the difference in deviation values associated to each app, type of device and device models were performed. A significance level of 0.05 was adopted as a criterion to reject the null hypothesis.

3 Results and Discussion

In this section, DifL_{Aeq} and DifL_{Cpeak} values are always presented as follows: median, minimum and maximum. The effect of noise level and noise type is not analysed in this manuscript.

Globally, considering all testing conditions, DifL_{Aeq} values were −0.30, −23.90 and 12.30 dBA. As for DifL_{Cpeak} the obtained results were −0.65, −5.20 and 9.20 dBC, revealing that mobile devices presented less dispersion when measuring $L_{C,peak}$.

Unfortunately, from the selected apps NIOSH was the only one able to measure L_{Cpeak}. Nevertheless, results obtained when measuring noise are much better than those we have obtained in a previous study where smartphones were tested to measure illuminance [4]. Still, some are unacceptable for occupational health and safety purposes and apps and mobile devices should be carefully chosen.

There were no statistically differences between Smartphones and Tablets (Table 3) in terms of $DifL_{Aeq}$ ($p = 0.631$) and $DifL_{Cpeak}$ ($p = 0.654$).

Table 3. Comparison of median, minimum and maximum values of $DifL_{Aeq}$ (dBA) and $DifL_{Cpeak}$ (dBC), obtained with smartphones and tablets.

	Smartphones		Tablets	
	$DifL_{Aeq}$	$DifL_{Cpeak}$	$DifL_{Aeq}$	$DifL_{Cpeak}$
Median	−0.30	−0.60	−0.40	−1.00
Minimum	−23.90	−5.20	−7.80	−1.50
Maximum	4.30	13.50	12.30	9.20

According to Fig. 1, which compares performance of all tested tablets, Zen Pad C7.0 from Asus was the single device reading L_{Aeq} values above the SLM ($p < 0.001$). Best and worst accuracy levels were obtained with Apple's iPad Pro 9.7 (−0.90, −3.40, 0.40 dBA) and Asus' Zen Pad C7.0 (1.70, −3.40, 12.30 dBA), respectively.

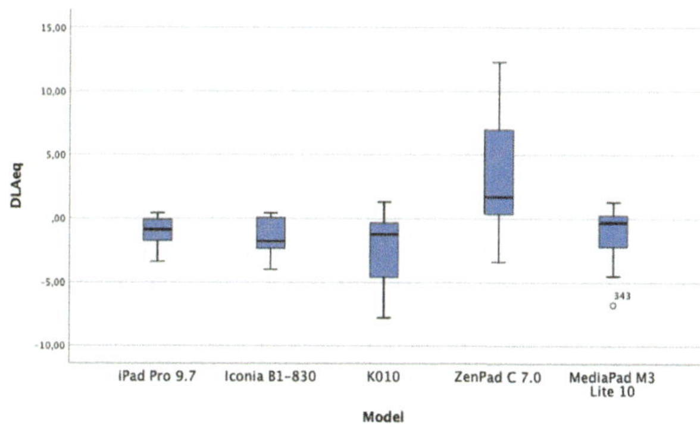

Fig. 1. Box plot of $DifL_{Aeq}$ (dBA) obtained with different tablets.

When comparing the performance of iOS smartphones concerning L_{Aeq} measurements (Table 4), it is possible to conclude that the older models - iPhone SE, 5S and 6S - do not present statistically different accuracy levels ($p > 0.05$). iPhone 11 was the best performing smartphone (0.45, −1.60, 1.80 dBA) while iPhone SE showed worst accuracy levels (-0.60, −8.50, 1.30 dBA). On the other hand, iPhone 11 revealed the worst performance regarding L_{Cpeak} measurements (0.20, −0.60, 8.30 dBC), whereas iPhone 6S was closest to the SLM results (−0.85, −1.80, 0.80 dBC).

Table 4. Accuracy comparison of iOS smartphones when measuring L_{Aeq}.

	iPhone 5S	iPhone 6S	iPhone 7	iPhone 11
iPhone SE	p = 0.844	p = 0.363	**p < 0.05**	**p < 0.05**
iPhone 5S		p = 0.521	**p < 0.001**	**p < 0.001**
iPhone 6S			**p < 0.001**	**p < 0.001**
iPhone 7				**p < 0.001**

Android smartphones' accuracy differences are more complex to describe but are summarized in Table 5. Nevertheless, Huawei's P10 Lite revealed less differences from the SLM (-0.05, −0.70, 1.50 dBA) and Samsung's Galaxy J3 delivered less accurate values (-2.00, −23.90, 0.00 dBA).

Table 5. Accuracy comparison of Android smartphones when measuring L_{Aeq}.

	GR3	P9 Lite	P10 Lite	E610	Galaxy J3	GT 19082	GT 19301L	SM T560
Star Pop	p = 0.678	**p < 0.05**	p = 0.925	p = 0.558	**p < 0.05**	p = 0.136	p = 0.121	**p < 0.05**
GR3		p = 0.127	p = 0.678	p = 0.450	**p < 0.05**	p = 0.266	p = 0.141	p = 0.080
P9 Lite			**p < 0.001**	p = 0.062	p = 0.141	**p < 0.05**	p = 0.405	p = 0.777
P10 Lite				p = 0.806	**p < 0.001**	p = 0.193	**p < 0.05**	**p < 0.05**
E610					**p < 0.05**	p = 0.079	p = 0.131	**p < 0.05**
Galaxy J3						**p < 0.001**	p = 0.083	p = 0.086
GT 19082							**p < 0.05**	**p < 0.05**
GT 19301L								p = 0.835

In Fig. 2 it is possible to note that most apps provided L_{Aeq} values below those obtained with the SLM (exception for the SLA Lite app). Both iOS and Android apps performed in a statistically different manner regarding L_{Aeq} measurement ($p < 0.05$).

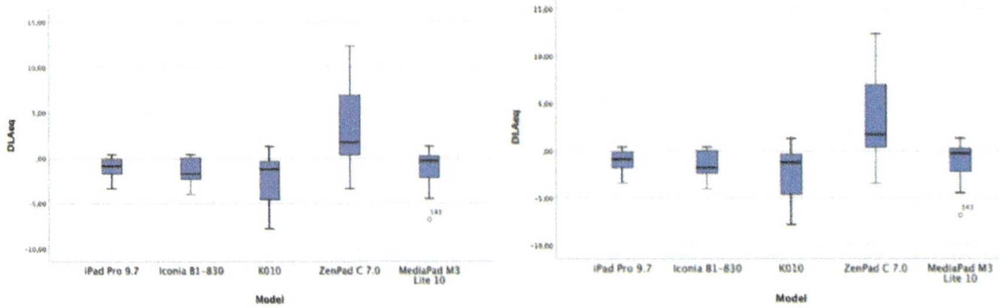

Fig. 2. Boxplot of $DifL_{Aeq}$ (dBA) obtained with different applications, according to the operative system of the mobile device (Android on the left side and iOS on the right side).

Best accuracy levels were achieved with iOS SLA Lite app (−0.05, −1.10, 2.10 dBA), while Android's Openoise app presented the worst results (0.10, −23.90, 2.90 dBA).

The obtained results with some apps and some mobile devices are far from those achieved by Kardous and Shaw [9] that were able to reduce deviation values from ±2 dB to ±1 dB when relied on external microphones to conduct sound measurements with smartphones. Some apps did not allow to choose time weighting and calibration was possible only in the paid version that was not included in this study.

4 Conclusions

From the results obtained in this study, it can be concluded that many of the mobile devices and apps tested do not comply with sound level meters' standards. In fact, in most cases sound level was underestimated, and therefore risk associated to workers exposed to noise may be neglected or considered tolerable. On the other hand, the results present some inconsistency which prevents the use of some sort of a correction factor.

As a limitation of this study we can highlight the reduced number of Apps that provide L_{Cpeak}, limiting the assessment of this variable to iOS operating systems devices, which can explain some of the results in particular the non-statistically significant difference found among devices (smartphones and tablets).

References

1. Eurofound: Sixth European Working Conditions Survey – Overview report (2017 update). Publications Office of the European Union, Luxembourg (2017)
2. IEC 61672-1: 2013 - Electroacoustics - Sound level meters - Part 1: Specifications, International Electrotechnical Commission, Geneva (2013)
3. IEC 61252: 2017 - Electroacoustics - Specifications for Personal Sound Exposure Meters, International Electrotechnical Commission, Geneva (2017)
4. Cerqueira, D., Carvalho, F., Melo, R.B.: Is it smart to use smartphones to measure illuminance for occupational health and safety purposes? In: Arezes, P. (ed.) Advances in Safety Management and Human Factors. Advances in Intelligent Systems and Computing, vol. 604, pp. 258–268. Springer, Cham (2017)
5. Faber, B.M.: Acoustical measurements with smartphones: possibilities and limitations. Acoust. Today **13**, 10–17 (2017)
6. Keene, K., Merovitz, A., Irvine, E., Manji, N., Everett, M., Chung, I., Moodie, S., Scollie, S., Gamble, A., Zimmo, S., Morrison, C., Chan, B.: Accuracy of smartphone sound level meter applications. Can. Hear. Rep. **8**, 24–28 (2013)
7. Murphy, E., King, E.A.: Testing the accuracy of smartphones and sound level meter applications for measuring environmental noise. Appl. Acoust. **106**, 16–22 (2016)
8. McLennon, T., Patel, S., Behar, A., Abdoli-Eramaki, M.: Evaluation of smartphone sound level meter applications as a reliable tool for noise monitoring. J. Occup. Environ. Hyg. **19**, 620–627 (2018)
9. Kardous, C.A., Shaw, P.B.: Evaluation of smartphone sound measurement applications (apps) using external microphones—a follow-up study. J. Acoust. Soc. Am. **140**, 327–333 (2016)

10. Roberts, B., Kardous, C., Neitzel, R.: Improving the accuracy of smart devices to measure noise exposure. Journal of Occupational and Environmental Hygiene **13**, 840–846 (2016)
11. Celestina, M., Hrovat, J., Kardous, C.A.: Smartphone-based sound level measurement apps: Evaluation of compliance with international sound level meter standards. Appl. Acoust. **139**, 119–128 (2018)
12. Iulietto, M.F., Sechi, P., Gaudenzi, C.M., Grispoldi, L., Ceccarelli, M., Barbera, S., Cenci-Goga, B.T.: Noise assessment in slaughterhouses by means of a smartphone app. Ital. J. Food Saf. **7**, 79–82 (2018)

Safety in Drilling Offshore Operations: A Narrative Literature Review

Carolina Maria do Carmo Alonso[1(✉)], Luciano do Valle Garotti[2], Eliel Prueza de Oliveira[1], Janaína Silva Rodrigues da Costa[3], William Silva Santana de Almeida[3], and Francisco José de Castro Moura Duarte[3]

[1] Department of Occupational Therapy, Medical School, Federal University of Rio de Janeiro, Rio de Janeiro, Brazil
carolmarial@gmail.com, elielprueza@gmail.com
[2] Petrobras, Rio de Janeiro, Brazil
luciano.garotti@petrobras.com.br
[3] Production Engineering Program, COPPE, Federal University of Rio de Janeiro, Rio de Janeiro, Brazil
janaina.s.r.costa@gmail.com, wilalmeida@gmail.com, fjcmduarte@gmail.com

Abstract. This paper sought to collect and analyze evidence reported in studies that address the safety dimension in drilling offshore operations. A bibliographic search was made at different databases and following the exclusion of the duplicates and applying the inclusion/exclusion criteria 19 papers were reviewed by the authors. This literature review showed that the documents still address the issue of security in drilling offshore operations from a behavioral perspective with little regard to the human and organizational factors of industrial security. Even studies that used a qualitative or mixed approach based on interviews and focus groups as data collection procedures, which are interesting strategies, do not reach the complexity of real work situations. To overcome this gap is therefore suggested to develop and disseminate researches in drilling offshore operations that adopt frameworks that access actual work situations and allow to address the security beyond the focus on individual behavior.

Keywords: Well safety · Oil and gas industry · Ergonomics

1 Introduction

The offshore oil and gas well drilling process encompasses complex and diverse activities that require special attention to the safety of these operations. The criticality of safety in well construction is due to uncertainties and unknowns in geological modeling and rock and fluid properties [1], exposure of facilities and workers to physical, chemical and ergonomic risks [2–7].

However, despite the importance of this theme, only one literature review was identified, this being the study by Assad and colleagues, which analyzed life-threatening disasters and the potential risks associated with onshore and offshore oil

P. M. Arezes and R. L. Boring (Eds.): AHFE 2020, AISC 1204, pp. 249–256, 2020.
https://doi.org/10.1007/978-3-030-50946-0_34

and gas drilling operations [7]. However, the specifics linked to the safety dimension in offshore drilling operations have not yet been analyzed by literature reviews.

Thus, aiming to outline an overview to identify gaps and point out ways for future research in this field, this study sought to collect and analyze evidence reported in literature that address the safety dimension in drilling offshore operations.

2 Methods

This article presents a narrative review of the literature about safety in drilling offshore operations carried out from the following databases Science Direct, Emerald, Web of Science and One Petro. The choice of the narrative review has been made because this type of review allows the acquisition and updating of knowledge about a specific theme [8].

The question that guides this research was: how the safety dimension has been addressed in documents that explores drilling offshore operations? The search strategy was carried out in January 2020 in the databases mentioned above using the terms "Safety" or "Safety Culture" or "Well safety" and "Drilling" or "Drilling workforce", "Completion", "Workover", "Plug and abandonment Offshore well" with this strategy was retrieved 1.213 documents and 2 other papers were included by hand search. All of the articles were saved and forwarded to an EndNote Web account to start the screening process which is demonstrated in Fig. 1.

Were included documents written in English without date limitation. Were excluded articles that didn't relate to the safety in drilling offshore operations. Following the exclusion of the duplicates and applying the inclusion/exclusion criteria 19 papers were reviewed by the authors, which allowed to assess mainly common themes to summarize the main conclusions.

3 Results and Discussion

Table 1 summarizes the 19 articles selected and their goals. The literature analysis results going to be shown by two axes. The first one analyzes the characteristics of the documents included in this paper. The second axis presents the analysis of the content of those documents that explores how the drilling offshore operations have addressed the safety dimension.

It is important to highlight that amidst the 19 articles included in this paper, 7 were multicenter studies [4, 9–11, 14, 15, 23], 7 focused on only one country [2, 12, 17–19, 21, 22] and 5 did not specify the location of the research [1, 13, 16, 20, 24]. We verified the following distribution of these studies among the world regions 7 studies were developed in Asia [4, 9–12, 15, 19], 4 in Europe [2, 9, 14, 17], 2 in North America [9, 18], 1 in South America [21] and 1 in Africa [22]. Related to this aspect, we call attention to a discrepancy in the number of papers carried out in Asia compared to the other world regions especially South America and Africa which are two substantial producers of the oil and gas.

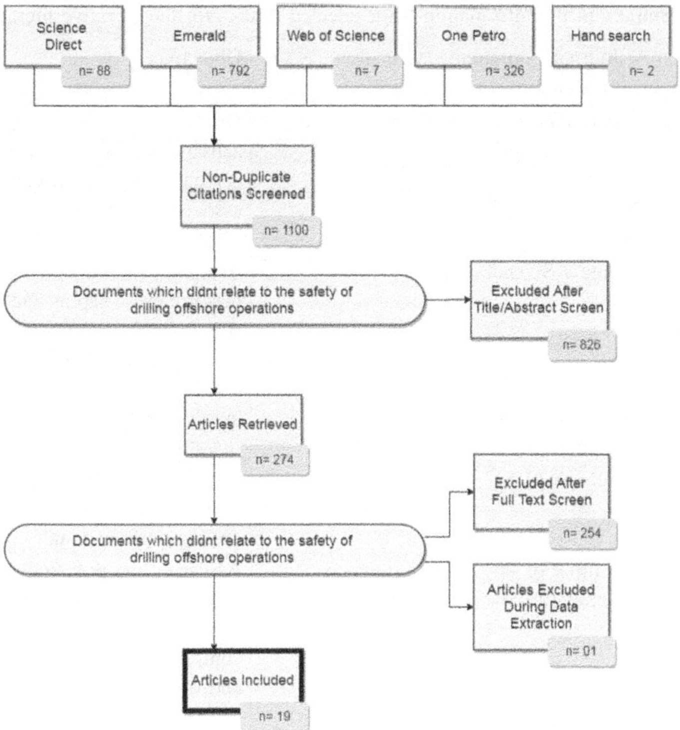

Fig. 1. Flowchart of study search and inclusion. Prepared by the authors.

When it comes to the sources of publications of the papers analyzed in this literature review, 10 are conference articles, 9 have been published in peer-review journals and one is a book chapter. Concerning the design of the documents, 11 were not based on empirical research, 9 of which were experience reports and the other two essays. The quantitative studies (n = 5) predominated, followed by mixed design studies (n = 2) and a study with a qualitative approach (n = 1).

This frame shows that although the safety in drilling offshore operations is a critical issue to the oil and gas companies, they still do little empirical researches about this subject. Hence, it is also important to remark that experience reports are important to disseminate successful initiatives or exemplary situations but, they are not evidence-based and consequently, they can be biased.

Regarding the empirical studies, those which use quantitative methods intended to assess: stress and fatigue, risks, training satisfaction rates and likewise identify how specific group membership characteristics influence perceptions of trust and self-reported safety behavior on exposure to involvement in incidents and major hazardous activities during the casing and cementing operation.

With respect to studies that used mixed methods, one sought out to identify effective safety risk-mitigating factors for well control drilling operations [10] and the other one demonstrated the development of one tool of knowledge-based decision

Table 1. Summary of the information from selected studies for the narrative literature analysis

Author (s)	Journal/Conference (year of publication)	Objective
Hassan R. B. et al. [4]	Pertanika Journal of Social Sciences & Humanities (2017)	Discusses the major hazardous activities during the casing and cementing operation with potential associated hazards in the on and offshore oil and gas industries
Sneddon A. et al. [2]	Safety Science (2012)	Examines the influence of the performance shaping factors of stress and fatigue upon WSA and the relationship between WSA, unsafe behaviour and accident involvement
Assad M. M. et al. [9]	Journal of Physics: Conference Series (2018)	Identify the satisfaction level for occupational safety and health training activities among oil and gas drilling crew and staff
Assad M. M. et al. [10]	Journal of Physics: Conference Series (2019)	Assessing the potentially hazardous activities associated with well control along with their appropriate controls and risk reduction factors and mitigating measures
Lievre T. [11]	SPE/IADC Drilling Conference (1995)	Describes a system which enables an early detection of failures of the safety management system, before such failures lead to injuries
Boucif M. N. et al. [12]	SPE/IADC Middle East Drilling Technology Conference and Exhibition (2016)	Demonstrate ADCO's duty of care and to show a strong and visible commitment toward labour safety from both ADCO and contractor Management
Karish J. et al. [13]	SPE International Conference on Health, Safety, and Environment in Oil and Gas Exploration and Production. (2004)	Report BP's experience regarding the Enterprise Development Network to improve the HSE leadership competencies of its wellsite and other operational leaders
Tharaldsen J. E. et al. [14]	Safety Science (2010)	Examines how specific group membership characteristics influence perceptions of trust and self reported safety behaviour on exposure to involvement in incidents across two shelves

(*continued*)

Table 1. (*continued*)

Author (s)	Journal/Conference (year of publication)	Objective
Assad M. M. et al. [15]	Journal of Engineering, Design and Technology (2019b)	Aims to focus on the design and development of knowledge base decision support system for the prevention of hazardous activities onshore and offshore oil and gas industries during drilling operations
Cohen M. A. et al. [16]	Vand. L. Rev. (2011)	Define and characterize the economic and policy forces that affect safety culture and identifies reasons why those forces may or may not be adequate or effective from the public's perspective
Krahn D. R. et al. [17]	SPE Internat. Conf. on Health, Safety and Environ. in Oil and Gas Expl. and Production (2000)	Documents the launch and evolution of the UK Cross Industry Safety Leadership Forum (also know as the Step Change Initiative)
Grayson M. B. [18]	SPE Americas E&P Environmental and Safety Conference (2009)	Describe the processes and variations, and the technologies used with MPD. It will further detail how MPD techniques are applied to a variety of operations to increase both drilling efficiencies and the overall safety of the operation
Levy J. et al. [19]	International Petroleum Tech. Conf. (2005)	Demonstrate that safety can be improved with direct participation of the workforce in a behavior-based observation program
Macpherson J. D. et al. [20]	SPE Drilling & Completion (2013)	Examines and defines drilling-systems automation, its drivers, enablers and barriers, and its current state and goals
Alteren B et al. [1]	Probabilistic Safety Assessment and Management (2004)	Presents a process used in the drilling operation of an oil production platform for safety and efficiency improvement
Cabral J. M. et al. [21]	SPE Internat. Conf. on Health, Safety and Environ. in Oil and Gas Expl. and Production (2010)	Documents the development of the "Red Alert" program
Okoli E. C. et al. [22]	IADC/SPE Asia Pacific Drilling Tech. Conf. (2014)	This paper presents learning/experience gained in the sustained safety performance for the ramped-up drilling operations in SPDC Wells starting from 2011

(*continued*)

Table 1. (*continued*)

Author (s)	Journal/Conference (year of publication)	Objective
Blank B. et al. [23]	SPE/IADC Drilling Conference (2003)	Presents management system created and implemented by one offshore drilling contractor to manage safety. An overview of the Safety Management System (SMS) and its most important components are provided
Miura K. et al. [24]	Journal of Petroleum Science and Engineering (2006)	This paper proposes the use of Quantitative and Dynamic Risk Assessment (QDRA) to assess the degree of safety of each planned job

support system for industrial safety management at the drilling process [15]. The paper that used a qualitative approach collected data (interviews, documents, written materials) about the platform, performed analysis and initiated a dialogue about future work [1]. However other papers include the worker's point of view, this is the only study that discusses the bottom-up participatory perspective in depth. In this sense, even studies that used a qualitative or mixed approach used interviews and focus groups as data collection procedures, which are interesting strategies but do not reach the complexity of real work situations.

So far, this section has presented the analyses related to the characteristics of the documents included in this research. Next will be offered the results of the thematic analysis which sought to reveal how the safety dimension has been addressed in documents that explores drilling offshore operations.

Documents included in this review identify and characterize risk factors related to the offshore drilling operation [24], highlighting the cementation phase [4] and the well control operations because of the oil-based mud, confined space at the site, pinch points and drop while working on eon blow out preventers [10]. Other documents addresses strategies for mitigating risks involved in drilling such as the Managed Pressure Drilling (MPD) methodology [18], the Monitoring Performance for Drilling system [11], or the stand up for safety program [12], the knowledge-based decision support system for industrial safety management in the drilling process called HAZFO Expert 1.0 [16], the management system for drilling contractors [23].

It was observed also that the documents still address the issue of security from a behavioral perspective [2, 9], with little regard to the human and organizational factors of industrial security. In this sense, the solutions for improving safety are linked with training [9, 12, 13, 19, 22].

4 Conclusions

This paper sought to collect and analyze evidence reported in studies that address the safety dimension in drilling offshore operations. Thus, it was possible to identify the scarcity of empirical studies that address the issue of safety in this field, as well as the concentration of publications developed in the Middle East in contrast to other regions of the world. In addition, although the publications identify issues related to the human and organizational aspects of safety, actions aimed at improving this dimension are still focused on individuals.

In short, to overcome this gaps, therefore, it is suggested the development and dissemination of empirical research that adopts frameworks that access actual work situations and allow a safety approach that go beyond the focus on individual behavior.

References

1. Alteren, B., et al.: "Smarter Together" in offshore drilling-a successful action research project? In: Probabilistic Safety Assessment and Management (2004)
2. Sneddon, A., Mearns, K., Flin, R.: Stress, fatigue, situation awareness and safety in offshore drilling crews. Saf. Sci. **56**, 80–88 (2013)
3. Blackley, D.J., et al.: Injury rates on new and old technology oil and gas rigs operated by the largest United States onshore drilling contractor. Am. J. Ind. Med. **57**, 1188–1192 (2014)
4. Hassan, R.B., et al.: Severity of the casing and cementing operation with associated potential hazards in the drilling process in the on and offshore oil and gas industry: a cross-sectional investigation into safety management. Pertanika J. Soc. Sci. Humanit. **25**, 129–138 (2017)
5. Adedigba, S.A., et al.: Data-driven dynamic risk analysis of offshore drilling operations. J. Petrol. Sci. Eng. **165**, 444–452 (2018)
6. Orimolade, A.P., Larsen, S., Gudmestad, O.T.: Vessel stability in polar low situations: case study for semi-submersible drilling rigs. Ships Offshore Struct. **13**, 303–309 (2018)
7. Asad, M.M., et al.: Oil and gas disasters and industrial hazards associated with drilling operation: an extensive literature review. In: 2nd IEEE International Conference on Computing, Mathematics and Engineering Technologies (2019)
8. Rother, E.T.: Revisão sistemática X revisão narrativa. Acta paulista de enfermagem. **20**, 5–6 (2007)
9. Asad, M.M., et al.: Level of satisfaction for occupational safety and health training activities: a broad spectrum industrial survey. J. Phys.: Conf. Ser. **1049**, 012021 (2018)
10. Asad, M.M., et al.: Identification of effective safety risk mitigating factors for well control drilling operation. J. Eng. Design Technol. (2019)
11. Lievre, T.: Safety performance monitoring for drilling: a more comprehensive system. In: SPE/IADC Drilling Conference (1995)
12. Boucif, M.N., et al.: Stand down for safety; powerful tool in enhancing drilling contractors safety. In: SPE/IADC Middle East Drilling Technology Conference and Exhibition (2016)
13. Karish, J., Siokos, G.: Improving safety leadership in drilling and completion operations. In: SPE International Conference on Health, Safety, and Environment in Oil and Gas Exploration and Production (2004)
14. Tharaldsen, J., Kathryn, J.M., Knudsen, K.: Perspectives on safety: the impact of group membership, work factors and trust on safety performance in UK and Norwegian drilling company employees. Saf. Sci. **48**, 1062–1072 (2010)

15. Asad, M.M., et al.: Design and development of a novel knowledge-based decision support system for industrial safety management at drilling process. J. Eng. Design Technol. (2019)
16. Cohen, M.A., et al.: Deepwater drilling: law, policy, and economics of firm organization and safety. Vand. L. Rev. **64**, 1851 (2011)
17. Krahn, D.R., Mearms, C.S., Richards, S.C.H.: The UK cross industry safety leadership forum-a drilling sector perspective. In: SPE International Conference on Health, Safety and Environment in Oil and Gas Exploration and Production (2000)
18. Grayson, M.B.: Increased operational safety and efficiency with managed-pressure drilling. In: SPE Americas E&P Environmental and Safety Conference (2009)
19. Levy, J., et al.: Evolution of the offshore drilling safety culture. In: International Petroleum Technology Conference (2005)
20. Macpherson, J.D., et al.: Drilling-systems automation: current state, initiatives, and potential impact. SPE Drill. Complet. **28**, 296–308 (2013)
21. Cabral, J.M., et al.: Red alert program in drilling rigs: a strong decision to show leadership and to involve work force towards zero serious or fatal incidents. In: SPE International Conference on Health, Safety and Environment in Oil and Gas Exploration and Production (2010)
22. Okoli, E.C., Amadi, A.: Sustainable safety performance in drilling operation. In: IADC/SPE Asia Pacific Drilling Technology Conference (2014)
23. Blank, B., Prewitt, A.: Drilling contractor safety management system. In: IADC/SPE Drilling Conference (2003)
24. Miura, K., et al.: Characterization of operational safety in offshore oil wells. J. Petrol. Sci. Eng. **51**, 111–126 (2006)

Resilience and Recovery

Implications of Narcissistic Personality Disorder on Organizational Resilience

Ronald Laurids Boring(✉)

Idaho National Laboratory, Idaho Falls, ID, USA
ronald.boring@inl.gov

Abstract. In the Fifth Diagnostic and Statistical Manual (DSM-5) of the American Psychiatric Association, narcissistic personality disorder is characterized by having feelings of self-importance, seeking for admiration, and lacking empathy. These traits map well into qualities of leadership such as having vision, having high achievement, and being able to make hard decisions. Unfortunately, narcissism often carries with it a number of negative traits such as manipulation or subversion of individuals, ethical lapses, and a need for constant change. On an organizational level, these traits may serve to undermine workers and the workplace, causing considerable damage in their wake and leaving personnel feeling helpless to intervene. This paper reviews an anonymized organizational case study of narcissistic personality disorder at a university. The dysfunction caused by a narcissistic leader directly led to the collapse of a program at the university, unusually high levels of attrition, and a diminished institutional reputation. This paper considers the implications of opportunistic leadership in eroding organizational resilience. Current research models focus on many of the factors that may erode organizational resilience, but they overlook the internal threat posed by narcissistic leaders. This paper reviews opportunities to consider narcissism as a causal factor in organizational resilience and human reliability analysis.

Keywords: Organizational resilience · Narcissistic personality disorder · Leadership · Human reliability analysis

1 Introduction

Organizational resilience is defined by [1] as "the ability of an organization to anticipate, prepare for, respond and adapt to incremental change and sudden disruptions in order to survive and prosper." Much of the research on organizational resilience focuses on case studies for external disruptions such as changes in the underlying business model or technologies associated with an organization's means of production. Such external disruptions include new competitors with more efficient or cost effective ways of doing things or making products; geopolitical changes that threaten production or the ability to sell products; changes in regulations that impact long-standing ways of doing business; natural disasters including pandemics that impact facilities, distribution channels, and personnel; or even malicious actions such as cyberattacks that undermine a company's infrastructure.

P. M. Arezes and R. L. Boring (Eds.): AHFE 2020, AISC 1204, pp. 259–266, 2020.
https://doi.org/10.1007/978-3-030-50946-0_35

In the safety-critical industries, a major external disruption can also be caused by an accident such as the wellhead oil leak associated with Deepwater Horizon [2] or the nuclear reactor meltdowns triggered by the tsunami at the Fukushima Daiichi plant [3]. Such events can have catastrophic effects well beyond the organization, including large-scale economic and environmental implications. Organizational resilience should prevent such incidents but also provide ready opportunity to recover from such disruptions.

One threat to organizational resilience that is not widely discussed is the impact of individual opportunism within the organization. In this paper, I will discuss how narcissism can manifest as an internal threat to an organization. Narcissism serves as a gradual erosion to organizational resilience that may leave the organization unable to recover from its effects.

2 Narcissistic Personality Disorder

The American Psychiatric Association treats narcissism as a personality disorder [4]. Personality disorders are characterized by significant, enduring dysfunction in self-definition and interpersonal behaviors. In [4], personality disorders fall within three clusters: (A) odd or eccentric disorders including paranoid, schizoid, and schizotypal; (B) dramatic, emotional, or erratic disorders including antisocial, borderline, histrionic, and narcissistic; and (C) anxious or fearful disorders including avoidant, dependent, and obsessive-compulsive. As a member of the dramatic, emotional, or erratic personality disorder cluster, narcissism features feelings of grandiosity, attention seeking, and a marked lack of empathy toward others. As an impairment, narcissism may manifest in exaggeration of self-accomplishments, excessive need for admiration and recognition, and severe disregard for others.

Narcissism is comorbid with antisocial personality disorders like psychopathy (and the often synonymous sociopathy), particularly in terms of deceitfulness, disregard for rules and consequences, and impulsiveness. The focus in this paper is specific to this co-occurrence. In the author's view, the personality disorder of narcissism is especially prone to escalation to psychopathy given the right workplace. In fact, earlier definitions of narcissism [5] featured many overlapping traits with psychopathy such as taking advantage of others and desire for elevated status. Psychopaths are narcissists, but not all narcissists are psychopaths.

Traits associated with narcissism may actually help drive productive workplace behavior in some cases—what [6] calls healthy narcissism. Charm and decisiveness combined with seeming high self-confidence and accomplishments are hallmarks of leadership. On the opposite end of the spectrum, narcissism combined with psychopathy can result in destructive patterns of undermining others and the organization for personal gain. Toxic narcissistic leadership may feature seeking of attention or affection from higher-ups, bullying, gaslighting, compulsive lying, manipulation or exploitation of others, and retaliation. As described in [7], climbing the corporate ladder by a narcissist may result in significant collateral damage to co-workers and the organization. Good workers who are threatening to the narcissistic psychopath are manipulated into compromising situations; subordinates are bullied and left in a diminutive role [8]; higher-ups are baited and flattered, and ultimately eliminated as competition. Moreover, organizational checks and balances may be slowly eroded as greater power is concentrated into the hands of the narcissist, a culture of following is instilled, and new organizational thinking is indoctrinated [9]. Furthermore, the immediate production capabilities of the organization suffer, since the good workers' efforts are subverted and their output stymied.

One common characteristic of dramatic, emotional, or erratic personality disorders like narcissism and psychopathy is a reduced sense of object constancy. Object constancy refers to relationships as objects, such as the relationship between two individuals. Diminished object constancy may result in a detachment to other individuals and a corresponding lack of empathy. Another form of diminished object constancy is in planning and decision making. While a narcissist may seem very confident and decisive, decisions may exert less permanence than with non-narcissists. There may be a pervasive sense of change for the sake of change, not necessarily based out of necessity. The whimsical nature of decision making can result in significant organizational change when a narcissist is put into a position of leadership. Personnel may be moved to new roles, without careful consideration of their skills and experience. Likewise, policies may be changed through executive orders, though the rationale for such changes may not always be apparent. Such changes are possible in part because of a cult of personality surrounding the narcissistic leader. Narcissism demands followers, and decisions are issued as a display of power. In some cases, these may be a form of conscious manipulation to sow discord, confuse staff through gaslighting, and maintain leadership. In other cases, it may not be conscious so much as a result of poor object constancy.

3 Factors in Organizational Resilience

Denyer [1] refers to the organizational resilience tension quadrant, marked by the junction between consistency and flexibility on one axis and defense and progress on the second axis. Progressive consistency, the first quadrant, may result in performance optimization; progressive flexibility may result in adaptive innovation; flexible defense may result in mindful action; consistent defense may result in preventative control. Too much conservatism in an organization centered on consistency and defense prevents evolution of the organization, while too much flexibility and progress may destabilize an organization. The latter realm of flexibility and progress are potential danger areas for narcissistic leadership if not kept in check by organization elements of consistency and defense. There is no clear guidance for when flexibility and progress move too far at the hands of narcissistic change, nor when consistency and defense are used to protect the hierarchy that may be desired through narcissistic leadership.

Denyer [1] proposes the 4Sight method. This approach proposes four approaches to overcoming organizational resilience tensions:

- *Foresight* to anticipate, predict, and prepare for the future
- *Insight* to interpret and respond to present conditions
- *Oversight* to monitor and review what has happened and assess changes
- *Hindsight* to learn the right lessons.

The 4Sight method does not specify who should be in charge of these roles, nor does it offer a clear prescription for how these roles can be enforced. The goal of the 4Sight method is to avoid erosion, which results from complacency (e.g., consistent defense) or short-term profit goals (e.g., flexible progress). Organizational resilience is depicted as drifting toward failure [10] through gradual erosion. This approach does not, however, explain the opportunity for a single individual to accelerate this drift.

4 Treatment of Organizational Factors in HRA

The field of HRA treats vulnerabilities of systems triggered by human error. Historically, HRA has centered on modeling the performance of reactor operators in control rooms at nuclear power facilities, but the scope of HRA has broadened considerably. Organizational factors became an area of research in HRA with the emergence of safety culture. HRA generally treats the error context through performance shaping factors (PSFs), which are the conditions that increase or decrease the human error probability. As noted in [11], organizational factors through PSFs remain underrepresented in HRA.

Groth [12] created a multifactorial model of PSF concepts found across HRA methods. Her research was especially useful in nuancing the facets of PSFs that might not necessarily be operationalized in each HRA method. In Groth's PSF taxonomy, one of the parent PSFs is Organization, which includes the child PSFs depicted in Table 1.

Table 1. Organizational performance shaping factors (from [12]).

• Training Program	• Resources	• Safety Culture
— Availability	— Procedures	• Management Activities
— Quality	⇒ Availability	— Staffing
• Corrective Action Program	⇒ Quality	⇒ Number
— Availability	— Tools	⇒ Qualifications
— Quality	⇒ Availability	⇒ Team composition
• Other Programs	⇒ Quality	— Scheduling
— Availability	— Necessary Information	⇒ Prioritization
— Quality	⇒ Availability	⇒ Frequency
	⇒ Quality	• Workplace adequacy

To understand this list of PSFs, consider a PSF like "Training Program." This PSF includes two child nodes—"Availability" and "Quality." This can be interpreted as two separate PSFs encompassing "Availability of the Training Program" and "Quality of the Training Program." Each of these PSFs might be nuanced with levels, e.g., "Good Quality of Training Program," "Average Quality of Training Program," or "Poor Quality of Training Program." These levels would typically correspond to multipliers on a nominal human error probability. Although HRA and resilience have not always made good bedfellows, [13] a common interpretation would be that a resilient organization would encounter fewer human errors and, thus, lower human error probabilities. Recent research on creating quantitative resilience models [14] highlights the relationships between these organizational PSFs and other factors like team robustness and adaptation.

While Groth's model remains arguably the most complete model of PSFs, the question remains if these organizational PSFs can account for the effects of narcissism. An initial observation suggests that the vulnerabilities captured by the organizational PSFs refer primarily to factors beyond the purview of individual management quality. Only at the point that a manager degrades the training program, for example, would the effects of narcissism become apparent. While narcissistic behavior is hardly covert, its effects are rarely as blatant as changing a program for the worse. For example, narcissistic behavior may erode factors influencing the training program, leading to its gradual decline, but the effects are not so incendiary to be captured immediately. Clearly, the effects of narcissism elude HRA until such time as they manifest as a problem. Current HRA practice fails as a useful tool to identify or prevent toxic narcissism in an organization.

5 Brief Case Study of Failed Organization Resilience

To illustrate the effects of narcissistic leadership, a factual but anonymized case study is presented. A midsized public university in the U.S. had experienced a downward turn in enrollment. This downward turn was precipitated by a high-profile scandal involving the university that damaged its reputation. Further factors were a booming regional economy that offered high-paying jobs without the need for higher education and a nationwide trend away from the liberal arts focus of the university. Since state funding was tied to enrollment numbers, the quality of programs was threatened as funding declined with enrollment.

The university leadership underestimated the ongoing severity of the enrollment decline, assuming that enrollment numbers would normalize given time. The general out-of-touch nature of the leadership later resulted in their dismissal, but not before another annual double-digit decline in enrollment occurred. The leadership team practiced a high degree of cronyism, and with the attrition of senior administrative staff with the decreases in enrollment, hand-picked replacements arrived.

Consolidation became necessary at the university. Two related but successful programs were combined under a newly appointed chair who was a favorite of the university president. Despite no experience with these programs, he was put in charge of the programs and their heads, each with over thirty years of successful experience in their respective programs.

A number of signs of toxic narcissism were observed by the staff. The lack of competence exhibited by the new chair was in stark contrast to his perceived competence by the university president. The new program chair fostered this relationship and, in fact, put in place practices to obfuscate any negative news that might paint him in bad light. Several staff were actually fired as part of a coverup for his own actions.

Because the new chair lacked competence in the chair role, he was threatened by the two highly experienced heads under him. He summarily assigned them to diminutive roles. Many of their roles were reassigned to junior staff, while one 60-year-old former head of the program was reassigned as a social media specialist despite not using social media. A secretary who specifically requested not to work on accounting tasks due to innumeracy was put in charge of accounting, to be personally assisted by the new chair. Student recruiting, one of the strong areas for these programs, was also reassigned, resulting in the departure of two highly successful recruiters.

After less than a year, several of the senior staff had departed, opting for early retirement or other, external jobs. Three mission-critical personnel were fired. Recruitment went from around 200 new students per year to less than ten. A junior staff member, who was revealed to be having a sexual relationship with the new chair, was promoted to a senior position. And, more than $400,000 were unaccounted for in expenditures.

By the time the new chair was fired less than two semesters after starting, the program was decimated. Long-standing staff and key institutional knowledge were gone. The very strength of the program—its strong recruitment of new students—had vanished. Necessary funds for running the program were not only missing but had to be paid by borrowing against future revenues.

This example highlights a classic and tragic case of a narcissistic leader. Schmoozing, bullying, gaslighting, sexual misconduct, and misappropriation of funds occurred in a longstanding program at a venerable university. While the university should have epitomized a resilient organization, the changing enrollment conditions eroded the ability of the university to withstand the onset of the narcissistic leader. The new program chair seized an emerging vulnerability at the university due to declining enrollment and manipulated the situation for personal gain. The sudden overall decline in enrollment forced the university to make changes such as consolidating programs and power. Rather than follow an altruistic course that would help the university weather this disruptive event, the narcissistic leader further eroded the stability of the university, resulting in a far less resilient organization. The university prevailed albeit in a weakened state but only due to the short-sightedness of the narcissistic leader. His entanglements precipitated a more rapid decline than he could escape and were eventually uncovered by the new administration. An investigation led to his dismissal and legal proceedings against him. His aftermath was a decimated program with little prospects of recovery.

This case study illustrates how a narcissistic leader may play off emerging vulnerabilities that are beginning to erode organizational resilience. Such times of change often merit further changes in an attempt to fix the underlying problem. In this illustration, the narcissistic leader is shown to exploit those cracks that emerged due to external forces on the organization. In such a case, the narcissistic leader presents a so-called double whammy: external forces erode the organization, but the internal threat of the opportunistic leader undermines the very forces that hold together the organization. Thus, a vulnerability opens the door to more destruction from within.

6 Discussion

Narcissism serves as an insider threat to resilient organizations. The factors associated with narcissism do not appear to be treated in current organizational research. Research on organizational resilience tends to focus on outside threats such as disruptions to the organization. Research on risk within HRA may not adequately address organizational factors, especially the strong impact of individuals within that organization.

This paper introduces the insider threat of narcissism to organizational vulnerability. It suggests areas where further research is needed. First, the current organizational resilience frameworks need to account for the threat vector associated with leaders who do not operate in the best interests of the organization. Models of organizational resilience should prescribe screening tools to prevent the emergence of narcissistic opportunism and ways to keep checks and balances in place in the face of disruptive changes to the organization. Second, current risk modeling does not adequately account for the relationship between leaders and the organization. Organizational weaknesses are not solely the product of the institution or its work processes. Rather, they are influenced by the level of control individual leaders within that organization exert and their ability to drift the organization beyond its resilient safety zone. Future human error models must account for personality as a potential pathological influence on the organization.

Disclaimer. This work of authorship was prepared as an account of work sponsored by Idaho National Laboratory, an agency of the United States Government. Neither the United States Government, nor any agency thereof, nor any of their employees makes any warranty, express or implied, or assumes any legal liability or responsibility for the accuracy, completeness, or usefulness of any information, apparatus, product, or process disclosed, or represents that its use would not infringe privately-owned rights. Idaho National Laboratory is a multi-program laboratory operated by Battelle Energy Alliance LLC, for the United States Department of Energy under Contract DE-AC07-05ID14517.

References

1. Denyer, D: Organizational Resilience, British Standards Institute (2017)
2. Bureau of Safety and Environmental Enforcement: Deepwater Horizon Joint Investigation Team Report (2011)
3. American Nuclear Society Special Committee on Fukushima, Fukushima Daiichi ANS Committee Report (2012)
4. American Psychiatric Association: Diagnostic and Statistical Manual of Mental Disorders, 5th ed. (2013)
5. Crego, C., Widiger, T.: Psychopathy and the DSM. J. Pers. **83**, 665–677 (2015)
6. Malim, C: Rethinking Narcissism, Harper Perennial (2015)
7. Babiak, P., Hare, R.D.: Snakes in Suits: When Psychopaths Go to Work. HarperCollins, New York (2006)
8. Boddy, C.R.: Corporate psychopaths, bullying and unfair supervision in the workplace. J. Bus. Ethics **100**, 367–379 (2011)

9. Maccoby, M.: Narcissistic leaders: the incredible pros, the inevitable cons. Harvard Bus. Rev. **82**, 92 (2004)
10. Dekker, S.: Drift into Failure: From Hunting Broken Components to Understanding Complex Systems. Ashgate, Farnham (2011)
11. Alvarenga, M.A.B., Melo, P.F.F.F., da Fonseca, R.A.: A review of the models for evaluating organizational factors in human reliability analysis. In: International Nuclear Atlantic Conference (2009)
12. Groth, K., Mosleh, A.: A data-informed model of performance shaping factors for use in human reliability analysis, University of Maryland Report (2009)
13. Boring, R.L.: Reconciling resilience with reliability: the complementary nature of resilience engineering and human reliability analysis. In: Proceedings of HFES, vol. 53, pp. 1589–1593 (2009)
14. Kim, J.T., Park, J., Kim, J., Seong, P.H.: Development of a quantitative resilience model for nuclear power plants. Ann. Nuclear Energy **122**, 175–184 (2018)

Forward and Backward Error Recovery Factors in Digital Human-System Interface Design

Torrey Mortenson(✉) and Ronald L. Boring

Idaho National Laboratory, Idaho Falls, ID, USA
{torrey.mortenson,ronald.boring}@inl.gov

Abstract. This paper examines the idea that just as there are specific factors that can affect the performance of a task in human reliability analysis (HRA) models, there are also specific characteristics that will impact the likelihood of a recovery from an error state. The first set of factors is commonplace in many HRA models and methods and fits well with the common understanding of task completion and human error. This paper seeks to set out a brief list of measurable factors that would influence a recovery in both negative and positive means and various design practices and tools that can assist in these processes. By focusing on recovery, and specifically only actions that come after the human error or post initiator, this paper seeks to highlight positive design changes, areas of improvement, and a more robust design framework that can capture the second half of an erroneous action and empower the operator to successfully guide a system to safety. Two ideas, forward and backward error recovery, are discussed in terms of what potential changes they may make to designing complex process control interfaces and displays. As the nuclear power industry seeks to implement digital control schemes in various systems, there could be clear design decisions that can be made to aid the future operator in recovering from an erroneous action in a way that prevents any deleterious effects on the plant.

Keywords: Human reliability analysis · Error recovery · Human-system interface design · Nuclear process control · Usability

1 Introduction

Human factors as a field outside of nuclear power was born as a means of increasing worker productivity, safety, and the easing the use of various tools and pieces of machinery. These interests have extended into modern digital systems in many cases with the aim of increasing productivity and the overall experience of performing a task. Nuclear power employs these concepts and methods as well, but the primary events that drove human factors as a discipline in the consideration nuclear power operations were the accidents at Three-Mile Island, Chernobyl, and others. The specific construct of human factors in nuclear power was born in accident investigation and future error prevention in the operation of nuclear power stations and that legacy still requires an active investment of time and research in how to help humans perform well in these

P. M. Arezes and R. L. Boring (Eds.): AHFE 2020, AISC 1204, pp. 267–273, 2020.
https://doi.org/10.1007/978-3-030-50946-0_36

scenarios. Whether that manifests in human error research, human-machine interface design, or human reliability analysis (HRA), we are still concerned with the risk of human operation of dangerous systems.

The reality of modern nuclear operations and future advanced reactors is digital in design and in operation. Legacy control rooms were and continue to be significantly analog. While some new control rooms may still have analog components for specific safety-related or redundant purposes, the majority of human-system interfaces (HSIs) in future reactor control rooms will be digital. This new paradigm presents key challenges in areas of cybersecurity, fidelity, and operability, but it will also require a different perspective on designing interfaces and displays for these high-risk, process control applications. A key way for designers of these systems to streamline usability and operating experience is to align operations with other digital systems that their users are likely to be familiar with and competent at operating. In nearly all modern digital interactions there are methods of error recovery, and this then requires designers to ask how to employ recovery methods in a system that is highly complex, interdependent, and in many ways impossible to undo.

A specific aspect of designing in this new space is the absence of true physical requirements for instrumentation and controls, whereas previous systems were bound by engineering practicalities such as the need to have a one-to-one correspondence between sensor and indicator or the need to place all indicators in parallel on a control panel. Moreover, physical, mechanical systems could be activated but not readily backtracked, making recovery mechanisms challenging for an analog instrumented control system. These limitations are absent when constructing within a digital environment. Information may be nested within a multifunction single physical screen, and digital control logic opens the door for fail-safe checks and recovery scripts. With such an expanded canvas to work on there is no reason to not design in recovery methods for operators.

The nuclear power industry has had to contend with characterizations of being an unsafe or risky practice for decades. Just as human factors was brought to the forefront after the accidents of the 1970s and 1980s, the public's negative perception of nuclear power has persisted, despite an overall high safety rate within the industry. Nuclear has taken direct actions to mitigate any errors or risky circumstances to a large degree of success, and plants are significantly more resilient and safer than decades ago. However, there are continual efforts through HRA, automation, and design to prevent human error with an apparent goal to make human error extinct in nuclear power. Unfortunately, the only way to truly remove any possibility of human error is to remove the humans. Even then, this does not obviate latent errors in design. Therefore, while error prevention, work design, and HRA are all critical angles of this problem, there needs to be an addition of understanding what impacts human performance after an error and how the industry can make recovering from the inevitable errors faster, simpler, and safer.

2 Errors, Faults, and Recovery

2.1 Errors

This paper will not rehash the notion and nuances of human error. For ease of application, human error is defined here as an action that is outside of the standard, correct performance of the task being undertaken. This definition separates the result of the erroneous action or the requirement of an unintended outcome, because an erroneous action in terms of its inherent risk is still a concern even if a non-negative or no-result outcome follows. There are ample business and technical reasons to maintain attention on errors, regardless of consequence, as even mundane errors can impact efficiency, usability, or operations experience in a negative fashion. However, these errors are not always completely negative, as they open the door to correcting errors and making the process better.

This paper identifies three categories of error contexts and seeks to capture the consequential and causal characteristics of erroneous actions (see Table 1). The second row is the focus area of this paper—those errors that don't immediately manifest but could later trigger a fault. These errors are elsewhere referred to as latent errors, and they represent the best opportunity for recovery. The time before the error triggers the fault is the recovery period. There is a need to adequately test, design, and iterate these scenarios to develop processes that are resilient to human error as well as develop recovery processes that make the best use of human abilities and create the highest likelihood of recovery for the inevitability of human error.

Table 1. Categories of errors and recovery opportunities.

#	Description	Example	Recovery?
1	Errors without impact	Dropping a tool or piece of equipment or assembling a component incorrectly prior to installation	Recovery assumed – no contribution to a fault state
2	Errors that could lead to a fault	Positioning a valve that operates on a time function, or measured rate, in an erroneous way	Recovery possible within the relevant time increment
3	Errors that trigger fault	Turning off a pump for the incorrect system and creating a system transient as a result	Recovery not possible – in mitigation and correction space

2.2 Faults

The term *fault* may be used interchangeably with errors, but for this paper a fault is the point where an error or errors impact the overall system state or a poor operational action is completed which results in a deleterious system state. Errors that don't have a direct effect on the system are still errors, but a fault is a particular instance when the errors have made an impact on the operations of the plant. The second category of errors are the focus of a recovery task, as those errors allow for a recovery period. HRA makes use of performance shaping factors (PSFs) to identify characteristics or context of a task that are in some ways incompatible with human performance or that can easily result in errors. These factors are not exclusive to HRA, however, and common usability problems are usually paralleled in specific ways with PSFs of HRA. Figure 1 shows how error precursors and PSFs can contribute to an error that will lead to a system fault, the inciting error, and the time to recover before a system fault.

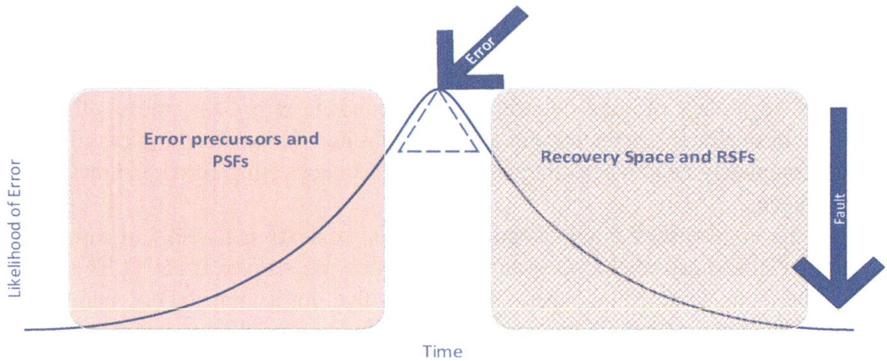

Fig. 1. Relationship between error and recovery.

The primary goal of human factors and usability in a nuclear power plant is to prevent faults that are caused or driven by human actions. A fault can occur absent an error due to a mechanical failure, or other event outside of the bounds of human action. However, these arguments will focus on the specific envelope surrounding human errors where faults and recoveries can occur. The question asked here is: in an instance where an error that will lead to a fault has been committed, how do human factors practitioners, designers, system engineers, and procedure writers include recovery tasking or guidance in order to steer operators to correct their mistake? Another aspect of these designs are decision support tools and other advanced, computational methods to assist the operator. These interactions and tools will need to be tested and validated so they comprise the best communications and interfaces possible.

2.3 Recovery

Figure 1, above, shows a key point in which errors have created the conditions for a fault from a causal perspective. However, it is expected that most human errors that we experience can be recovered from. For example, a dropped item can be caught, or an accidental deletion can be undone. The prevalence of computers and modern error prevention and recovery schemes has increased the potential for the mitigation of error implications and may have made some users slightly more careless in their digital interactions. A question that can make some system engineers or procedure writers blanch is, “What if the operator doesn’t do that step?” A core understanding and foundation of nuclear safety culture is that procedures *are* followed and executed correctly. This is understandable considering the public and environmental safety that nuclear operators are charged with managing on a day-to-day basis. However, it does beg the question as to why there are human errors in a nuclear facility at all? If procedures prescribe the correct actions, and operators take those actions, then operators could never take an incorrect action. Yet that isn't what seems to happen.

The reality is that humans are error prone and have, at times, predictable and expected failure points given certain task conditions. In other instances, humans can err in manners ranging from comical to confusing to dangerous. There is no way to ensure perfect human performance, because human action and ability is constantly variable. Many industries build this understanding into systems and controls, and most commercial product design teams have learned that not designing for error is a bad option, as is not designing for recovery. Oftentimes in nuclear power, human factors operates in a triage style where human factors practitioners are given an as-built or already designed system and constrained by those previously made engineering decisions. The human factors practitioner may be required to ensure human factors compliance to standards but not have been given opportunity to influence the specification they inherited. Similarly, in an HRA the analyst may find difficulty in crediting an operator's ability to recover from an error or to mitigate an error-likely scenario, simply because the available actions can be so constrained.

Two of Nielsen's usability heuristics [1] that every system should have are the focus here: *Error prevention*; and *Help users recognize, diagnose, and recover from errors*. These characteristics are to be viewed as system requirements when designing for human interactions, as humans will commit errors and are accustomed to being able to recover from said errors within a reasonable amount of time. The first is really a challenge for designers to engage with early and ensure that the system removes as many sources of error as possible prior to launch of the product or the system coming online. The second heuristic is more the focus of the design arguments in this paper. Two specific design concepts, forward and backward error recovery, are discussed and their design impacts are discussed.

Recovery shaping factors (RSFs) operate like PSFs to frame the context of the impending fault following a human error. RSFs can serve to enhance the likelihood that that the error will be detected and mitigated prior to its manifestation as a fault. Dependence is the propensity for another error to happen once a previous error has occurred. It is typically modeled in HRA by an increase in the human error probability subsequent to an initial error. One of the greatest contributors to error dependence is operator mindset—if nothing happens to shake the train of thought and make the operator aware of the error, then the error is propagated forward. RSFs work the opposite of dependence. They are factors that help make the operator aware of the error (e.g., additional indicators, a human second checker, or steps in the procedure requiring verification of states) and take actions (e.g., turn off an inadvertently activated system) to limit the fault.

3 Forward and Backward Error Recovery in Process Control

When designing a system such that users can recover from errors, the two methods, forward and backward recovery, are used to reference specific temporal states of the action [2]. Backward error recovery is focused on undoing or going back to the pre-error state without requiring new actions to be taken by the user in response to the error. If we examine user actions in a chain or event tree format, then recovery actions are additive to the chain. For example, if a user is following directions while driving to a new location and misses a key turn then the user must make several new turns and navigational actions, in addition to the existing direction set, to get back on the correct path. Backward error recovery would describe the user having the ability to just teleport back to a time prior to the error. In software this functionality is generally manifested in confirmation screens, prior to closing an unsaved document, and most famously the CTRL + Z or undo in most Microsoft Windows systems. One warns the user of an action with potentially negative implications and the second is a digital instance of our teleportation button. When designing digital interfaces, even in the process control industries, there are ample opportunities to work with system engineers and operators and identify key areas where these actions are possible.

If we return to the driver who made an incorrect turn, when a user is required to take new actions outside of the bounds of the defined task boundaries then the user is undertaking forward error recovery. Rather than attempting to go backward to a past time or state, the user is looking forward at the required actions to return to the correct state [3]. In the proposed navigation task humans are able to successfully negotiate these circumstances and take appropriate turns to return to the correct location, perform a U-turn, or other driving tasks. Our driver is able to undertake these tasks and innately know what is required because of the spatial reality involved. If you pass a location, you must go back. However, in process control, most operator-initiated control actions set specific systems and mechanical steps into motion and those are not steps that can be undone at the click of a button. Therefore, the interface and potentially the system must be designed for such recovery in mind. Further in digital systems there is an absence of these physical corrective actions, which is why understanding critical digital uses of these recovery modalities is necessary to ensure the operator can learn quickly and apply competently those recovery actions that align with their existing mental model. Given the complexity of these systems and processes there will be a need for operator flexibility in response and tasks to enable creative problem solving and not overly constrain a capable operator from navigating their plant. Many of these recovery methods will require a careful touch with communication and interactivity with the operators [4].

These novel interactions will require robust research methods and long-term commitments by industry stakeholders to understand how such a rigid industry designs for and accepts failure. In addition to qualitative information gathering and standard design iteration [5] there are interesting modeling and quantitative approaches that may yield additional insights into human error structures, causes, and sensitivities [6, 7] which can add to the designer's toolbox.

4 Discussion

As nuclear power begins to incorporate digital interfaces and components into aging analog plants or into new builds and advanced reactors, a new set of design concepts and challenges face the individuals who will design these future systems and interfaces.

Users will expect some fashion of error recovery and, due to the development of mental models supporting these tasks in other areas of digital interaction, human factors practitioners can lean on these methods in order to algin these future displays with user expectations. Additionally, with the strides made in computing and data analysis, there is very little reason why prognostic engines and operator support tools cannot assist in these recovery tasks.

Balancing the realities of the mechanical systems that operators control with computing resources and good digital design asks new questions of HSI designers. The positive potential for collaboration between these disciplines is immense if captured correctly, but new systems will have to be accepting of the realities of human errors, be willing to allow flexibility in operational procedures, and communicate well with operators to allow for maximum recovery and usability.

Acknowledgments. This work of authorship was prepared as an account of work sponsored by Idaho National Laboratory, an agency of the United States Government. Neither the United States Government, nor any agency thereof, nor any of their employees makes any warranty, express or implied, or assumes any legal liability or responsibility for the accuracy, completeness, or usefulness of any information, apparatus, product, or process disclosed, or represents that its use would not infringe privately-owned rights. Idaho National Laboratory is a multi-program laboratory operated by Battelle Energy Alliance LLC, for the United States Department of Energy under Contract DE-AC07-05ID14517.

References

1. Nielsen, J.: 10 heuristics for user interface design: Article by Jakob Nielsen, 24 April 1994. https://www.nngroup.com/articles/ten-usability-heuristics/. Accessed 10 Feb 2020
2. Campbell, R.H., Randell, B.: Error recovery in asynchronous systems. IEEE Trans. Softw. Eng. **SE-12**(8), 811–826 (1986)
3. Issarny, V., Tartanoglu, F., Romanovsky, A., Levy, N.: Coordinated forward error recovery for composite web services. In: 22nd International Symposium on Reliable Distributed Systems, Proceedings, Florence, pp. 167–176 (2003)
4. Suhm, B., Myers, B., Waibel, A.: Interactive recovery from speech recognition errors in speech user interfaces. In: Proceedings of Fourth International Conference on Spoken Language Processing, ICSLP 1996, Philadelphia, PA, USA, vol. 2, pp. 865–868 (1996)
5. Ulrich, T.A., Boring, R.l., Lew, R.: Qualitative or quantitative data for nuclear control room usability studies? A pragmatic approach to data collection and presentation. In: Proceedings of the Human Factors and Ergonomics Society Annual Meeting, vol. 62, no. 1, pp. 1674–1678 (2018). https://doi.org/10.1177/1541931218621380
6. Wang, C., Hwang, S.: A stochastic maintenance management model with recovery factor. J. Qual. Maint. Eng. **10**(2), 154–165 (2004). https://doi.org/10.1108/13552510410539222
7. Lüdtke, A., Osterloh, J.P., Mioch, T., Rister, F., Looije, R.: Cognitive modelling of pilot errors and error recovery in flight management tasks. In: Palanque, P., Vanderdonckt, J., Winckler, M. (eds.) Human Error, Safety and Systems Development. Lecture Notes in Computer Science, vol. 5962. Springer, Heidelberg (2010)

A Discussion of Quantitative Stress Analysis in Long-Term Embarked Work

Salvador Ávila[1(✉)] and Ronald Boring[2]

[1] Polytechnic Institute, Industrial Engineering Program, Federal University of Bahia, Salvador, BA, Brazil
avilasalva@gmail.com

[2] 2525 Fremont Avenue, Idaho Falls, ID 83402, USA
Ronald.Boring@inl.gov

Abstract. A network of factors in a serial, parallel and cross way transfers the hazard energy until it causes the accident. Ávila adjusted the Swiss cheese model to 8 layers of human elements and 9 layers of dynamic-fallible HF. This discussion will test tools to measure the level of stress and relate causal factors. The relationships between factors in critical activities are studied to prevent loss of perception of new emerging problems. The level of stress is impacted by activities confined progressively. The mind map is altered by the level of stress and affects the outcome tasks. The quantitative relationships try to signal a max stress level. Tasks, skills, technologies, risks are based on maintaining motivation around the psychological contract. Cognitive degradation caused by chronic stress, and, failure caused by acute stress are present in critical tasks and are discussed in divers JOB, oil production operators and astronauts to Mars.

Keywords: Progressive stress · Hazard energy · Task · Health · Operation mode

1 Stress and New Concepts

Stress is discussed by [1] as a factor that affects human performance and that can cause error and failure by altering human reliability in technical systems. Several authors [1–3] discuss technical factors, organizational, managerial and intrinsic-human as human performance factors in the task causing human error. The premise that environmental risk is dynamic indicates the need for new tools to study the net-works between these factors.

According to [4], pressure in the workplace generates symptoms associated with emotional illness. In addition, it confirms that stress has a high incidence in companies due to human and work-related characteristics. [5] shows that the state of high stress, the lack of physical activities and being overweight are the main factors of disease in the developed countries. [3] discussed that psychological stress can be caused by unexpected start of activities, high-speed tasks, excessive workload, high risk of exposure to risks, monotonous work, and others. [4] states that the stressed individual performed a connection between the stressor agent and the individual's suitability

P. M. Arezes and R. L. Boring (Eds.): AHFE 2020, AISC 1204, pp. 274–281, 2020.
https://doi.org/10.1007/978-3-030-50946-0_37

feedback in the face of the stimulus. If the individual does not adapt to the stressor agent, he will suffer.

The behavior experienced by each individual or group in the organization depends on specific psychological and physical factors to indicate the level of stressors' effect on the body or work. Thus, the stress level curve in relation to the effectiveness of the task depends on characterizing the level of influence of stress on labor for each peri-od, under the command of each leader, in each geographical location and with the respective cultural characteristics. It is difficult to say how many levels are made up of human factors (HF) under stress and how to lessen their impact. Thus, it is difficult to affirm when the group reaches the stress phases defined by [6].

According to [7], stress is a state produced by a change in the environment, which is perceived as challenging, threatening or harmful to the person's dynamic balance. The nature of stress is variable. An event or change that produces stress at one time may not generate the same reaction at another. A person enjoys challenging situations and deals with them. The desired goal is to adapt or adjust to the change, so that the person regains balance and has energy and the ability to meet new demands. This consists of the process of dealing with stress, a compensatory process with physiological and psychological components.

The research on work stress carried by the GRODIN [8–14] states that there is a high influence of this factor or property on action, diagnosis, planning of man in relation to the task. The developed tools will be tested in situations of critical activities in the areas of industry and services in long-term confinement. The tools present a protocol of the stress level, causes, consequences and the respective intensities. The premise adopted is that the stressor environment influences the quality of services through human factors, being thus considered as an environmental property that can threaten the stability of activity. The study on stress should be part of the installation project and in this work, it is named as a human element together with the leadership, quality of worker behavior and technological characteristics.

According to [15], stress is not a specific human performance factor for the calculation of risk in the Bayesian network as considered by [16], but it acts in a broad way and permeating these HPF [3]. This network is composed of subjective factors such as culture and social phenomena, or objective as equipment failures from poorly applied interfaces and skills. Stress is studied in the normal operation mode and in the contingency treatment besides the accident. Why is stress considered an environmental property and not a specific human factor in this study? Because, the stress agent or the "stressor vector" that promotes the stressful environment is characteristic by the group of task-technology that permeates all human factors, having a progressive reaction [11] in the loss of quality of the cognitive functions, which affect the decision in the routine, and in the stability of the human body, being able to generate the state of mental or physical illness.

Thus, stress is not exactly a human factor, but a state that alters the probability of human error based on the relationship with other human factors that preceded or resulted from the event. Here, we will study the relationships between these factors in the task, human types, management and technology. The network of human factors that involves different modes of operation is dynamic and has associations with serial, parallel and crossed formats. Stress influences this network that transfers hazard energy, in a continuous or intermittent flow, until it causes the accident and the chain reaction that generates the disaster. Ávila adjusted the Swiss cheese model to a more complex model that involves 8 layers of barriers or human elements; and 9 layers of operational factors of a human, social, organizational and technical nature. These factors are dynamic and consider the possibility of degradation of organizational culture project and the possibility that different failures can start the accident process. Stress can be present in all HF and is discussed in Project to avoid the accident.

1.1 Risked Activities and Stress Level [9, 11]

This investigation of risk and stress activity was carried out in a LPG plant where an exercise of BLEVE phenomenon: chronology; amount of hazard energy released by an event; and analysis of the behavior of the operation team. The interpretation of risk is based on an investigation that uses failure and consequences tree tools, valuing human errors and technical failures and applying progressive stress.

This progressive stress in the levels of deviation, failure in task equipment, accident and disaster brings different behaviors in routine, emergency and contingency teams. Routine leaders often do not detect that this event can grow "silently" to the level of disaster. Understanding the time for actions and the appropriate profile for each stress level brings up the subject of cognitive gaps in uncontrolled situations, where there are regions of great gaps that can facilitate the development of dangerous energy and the promotion of accidents. Can you measure the level of stress and check if the mode of operation has been shifted to deviance and to failure?

The target is identification of stressor environment level impacting on health and task. After data collection, we assign weights to each stressor environment given the complexity of the task, the psychological and organizational type to measure the impact of this stress installed on losses. Causality is established to enable the development of safeguards against stress. In institutional terms is prepared a protocol for further analysis of job stress. This exercise was based on a real case of purification in sulfuric acid plant where the result was 6,4 indicating not manageable situation.

1.2 Impacts of the Stressful Environment on Health and Oil Production [12]

LESHA [12] is a tool that intends to monitor the level of physical and psychological stress from environmental factors and monitor events related to the task, which can transform a pathogen resident in a symptom [17] and after in an occupational disease. The events that initiate failure and cause stressor environments can cause progressive stress state on individuals, activating, or causing the human body fault through the stages: resident pathogens; uncontrolled organic process; symptom, disease and mortality from replication, or after stopping work permanently. Each level of stress is analyzed for signs and/or indicators of failure, with impact on worker health, allowing taking preventive or corrective measures (biosecurity barriers). The result of this analysis allows to build programs to prevent diseases also in confined Work [14].

Accidents in oil fields, at sea or on land, continue to happen despite new technologies for oil exploration with the increase in remote control, greater care in process and operations controls, and equipment more robust.

Some questions are asked in relation to current references about accidents in on-shore production, because there are four times more accidents in relation to incidents? Probably daily incidents and deviations are not recorded, but only the most serious accidents. The rules of social coexistence within the production platforms/regions must be based on an explicit and evident sense of justice for all its residents. These rules are naturally applied in the routine, without which written standards that are not practiced imply the formation of "social castes" and, in a worse situation, "feuds", separating contractors from their own personnel.

In some real cases, the discourse presented is that the contractors are lucky when they meet a just platform manager improving team social life during the boarding period. The unjust situation allows formation of isolated groups, where the negative effect increase in the critical periods: first day on board and the final two days on board. In these cases, communication and natural physical contact in the work environment admit to the worker that he is isolated, including from his own existence in the labor society. It is necessary to humanize the platform in regions of isolation.

The monotony of performing the same activities over a long period of time and the lack of affective and more human coexistence can induce the operator to depression and psychotic breakdown, melancholic behavior of looking at the horizon.

2 Methodology

The methodology involves identifying the stress level as an initial and final diagnosis (NEI, NEF) for the analysis of compliance and the need for adjustment. Depending on the mode of operation (MOPEE) in the function to be performed, the consequence of the stress level may become more critical. The discussion of the mode of operation is related to the amount of time and hazard energy was released (AEPT), so, to assist in the design of safeguards, the analysis of the discourse of experienced operators is made and the requirements for reducing the impact are evaluated. The health, safety and production consequences are analyzed by LODA and LESHA techniques to prioritize safeguards that will decrease the concentration of hazard energy in regions of human factor networks (EPS). In this way, the safeguards are implemented in a smaller format so that they can later be disseminated throughout the organization. A new stress level measurement (NEF) is made for system compliance. Follow the quantitative methodology by Fig. 1.

3 Application of Tools in Cases of Services and Industry

The subjectivity involved suggests a cultural study to structure a cognitive map that is resilient for different types of tasks and that manages to keep the motivation in a continuous and controlled flow. This work treat a stressful environment will require a multidisciplinary discussion involving psychology, engineering, and medicine. Stress

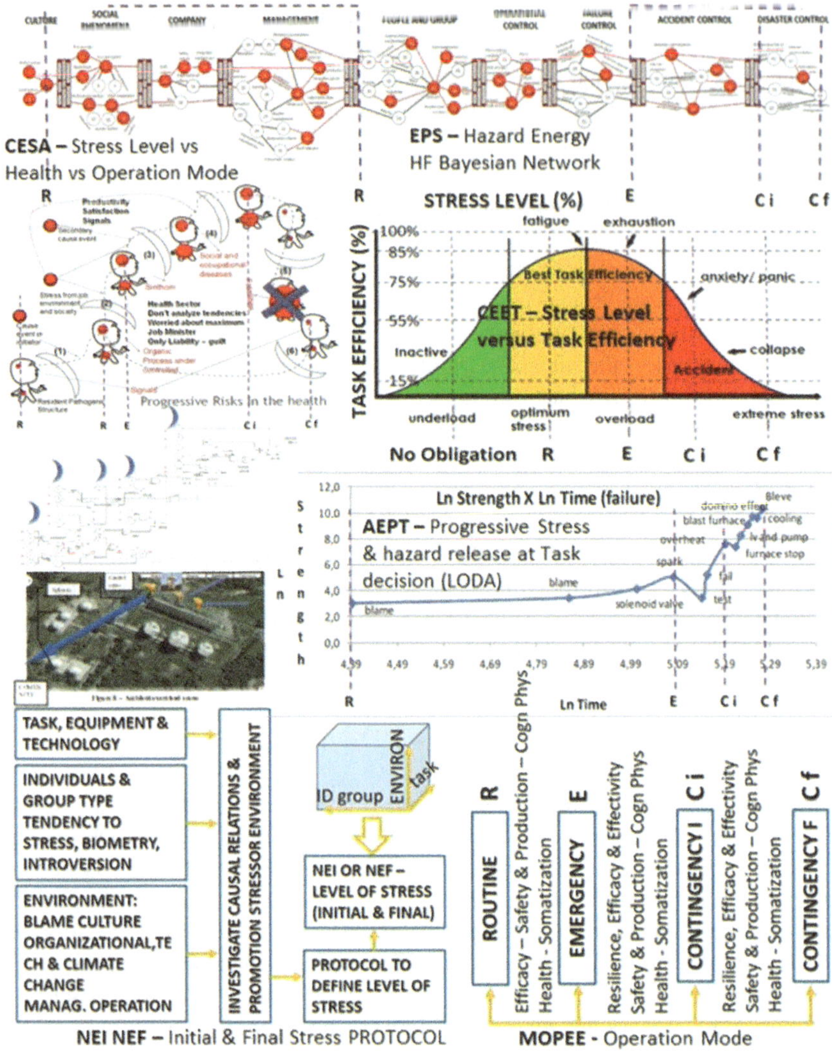

Fig. 1. Methodology characteristics

is related to human reliability in critical tasks or in tasks where the living environment is critical, which is the case of soldiers in war or the police team carrying out highly critical intervention. The result of this work will suggest human elements in the design of this mission to Mars. To discuss this subject, we initially considered the work of divers, exploration and production of oil, large projects with intensive work over long distances and finally the challenge of taking man to Mars, a mission of 3 years.

The human factors inherent to the task, technology, people and the group try to facilitate the discussion about the stressful environment and its consequences. An analysis of the level of release of the hazard energy, the type of flow of that energy indicates the best safeguards and surveillance design to avoid the onset of failure. A comparison of factors by type of activity of professionals will clarify important features.

3.1 Deepwater Diving [10] and Oil Platform Operator [14]

The diving service highly demanded by the oil industry has a low level of automation with some essential alarms for the execution of the task. The task seeks to maintain the stability of equipment structures and installations on the high seas (offshore). The risk of high pressure on the seabed demands specific requirements for the activity. This activity requires specific competence where cognitive gaps can cause human errors and low quality in the task.

The communication chain is intense in the activity to guarantee the safety and quality of the welding. Cooperation, commitment and group competence are important to support the task of diving. Physical challenges such as high pressure and low temperature are factors that must be controlled to avoid damage to the body and low quality of the task. On the other hand, the economic bias creates the conflict between security and production with the consequent reduction in investment in safeguards and routine control. Increased mental load impacts on stress and on the level of commitment and cooperation. The possibility of diseases arising from the activity is real. The workstation includes the supervision of the activity, the activity itself, welding of critical parts, in addition to the hyperbaric chamber. Physical movement is in depth and from the continent to the sea on the surface.

The project's stage of the project is that it requires more deep-water diver services. During the operation, there will be less capacity expansion services and the failure-incident phase will demand more. Stress research primarily tries to lessen the impact on health. Regarding the discussion on the mode of operation, the assembly is important because it is an activity for structuring the equipment and its interconnections, requiring greater care to avoid material and human fatigue.

Based on the proposed changes to increase the results in the activity of oil platform operator, we will analyze the stress factor and the need to correct or reduce the hazard energy to avoid the consequences on health and the task. The omissions in the activity can be a cultural or leadership characteristic, thus, the investigation of the cause requires greater effort by the organization. The repetition of cause diagnosed in the investigation of accidents as being low risk perception directs the organization to review procedures and training when the problem is in the acceptance of deviations due to human factors linked to social factors. Social inclusion & organizational integration are procedures that be revised to avoid feeling of being isolated from world.

3.2 Astronaut Service on Mars Mission [18]

This topic will discuss aspects that were not discussed in the activity of the platform operator or in the activity of the deep-sea diver. The identification and fulfillment of requirements for missions to Mars restricts the use of classic HRA methods that do not study human reactions to dangerous environments in this long-distance activity, very distant and with different levels of severity. We have to extrapolate the current knowledge about human errors based on less intense activities for these missions, which indicates the need to develop new tools to analyze teams-equipment-methods in carrying out these tasks.

The tasks and activities are intense in technology, high risk due to the unusual and involve complexity in the decision. Different activities will be carried out (launching, orbiting, mooring, landing activities) and there will be changes in the environment (isolation and microgravity) that can cause high psychological or physiological impacts. In this task, communication delays may occur, logistics problems for the supply of materials, durability of medicines and food, and non-availability of resources for emergencies. Nature is unknown and there are big gaps between the demands of the mission and what the methods of human reliability offer.

So, it is suggested: extrapolation of data to complex and ambiguous environments. Analysis of Effects on health and performance through studies and hot simulations. Investigate the priority performance factors for the mission, mainly the stress studied from the relationship of the qualitative description to quantitative data, the operator's speech in routine and emergency situations. We seek to standardize new situations and improve maintenance programs that rely on discipline and innovation during autonomous maintenance. In the risk analysis where work factors are compared in relation to stress as a cause, consequence or environment of influence, we can say that the risk of astronaut activity is 20% higher than that of the platform operator.

4 Conclusions

The factors that had high values in stress relationships in the case of the astronaut in relation to other professionals were: the planning and execution function of the operator, the analysis of hazard energy release and safeguards, analysis of the critical task.

References

1. Blackman, D.G.H., Smith, J.M.J.B.: The SPAR-H human reliability analysis method. Idaho National Laboratory Battelle Energy Alliance Idaho Falls, ID 83415 (2005)
2. CCPS: Center for Chemical Process Safety, Guidelines for Investigating Chemical Process Incidents, 2nd Ed. American Institute of Chemical Engineers. Wiley, New York (2003)
3. Lorenzo, D.K.: API770. A manager's guide to reducing human errors improving human performance in the process industries. American Petroleum Institute (2001)
4. Amorim, K.R.V.N.: A Percepção do Estresse dos Trabalhadores em uma Indústria de Revestimentos Cerâmicos do Nordeste: um Estudo de Caso. Master Degree Thesis at Federal University of Bahia, Salvador, BA (2016)
5. Watson, W.T.: Building a culture of health and well – being. London (2016)
6. Lipp, M.E.N.: Mecanismos neuropsicofiosiológicos do stress: teoria e aplicações clínicas. 3rd Ed, 227 p. Casa do Psicólogo, São Paulo (2010)
7. Smeltzer, S.C., Bare, B.G.: Brunner & Suddart: Tratado de enfermagem médicocirúrgica. 9th ed, 509 p. Guanabara Koogan, Rio de Janeiro (2002)

8. Gonçalves, G.G.: Influência do Estresse: Estudo de Caso Aplicado a Grupos de Estudantes de Engenharia na Escola Politécnica. Master Degree Thesis at Federal University of Bahia, Salvador, BA (2018)
9. Gonçalves, G.G., Nunes, K.R.V., da Silva, R.C.G., Ávila Filho S., Magalhães, R.: Protocol to evaluate the influence of stress on the task involving human types, behavior and health: an exercise proposal. In: Proceedings of European Safety and Reliability Conference (ESREL), Zurich (2015)
10. Silva, R.C.G.: Análise do estresse em mergulhadores profissionais, categoria mergulho profundo. Master degree thesis at Federal University of Bahia, Salvador, BA (2017)
11. Ávila Filho, S.: Dependent layer of operation decision analyzes (LODA) to calculate human factor, a simulated case with PLG event. In: 7th Global Congress on Process Safety, Chicago, Illinois (2011)
12. Avila, S.: LESHA – Multi-layer progressive stress & impact assessment on health & behavior. In: 9th Global Congress on Process Safety (2013)
13. Ávila, S.: Riscos na plataforma – Diretrizes sócio-humanas evitam acidentes e alteram padrões na produção de petróleo. Gestão de Riscos - Revista Proteção (2013)
14. Ávila, S., Mrugalska, B., Ahumada, C., Ávila, J.: Relationship between human-managerial and social-organizational factors for industry safeguards project: dynamic Bayesian Networks. In: 22nd Annual International Symposium Mary Kay O'Connor Process Safety Center. College Station, Texas (2019)
15. Ashrafi, M., Zadeh, S.A.: Lifecycle risk assessment of a technological system using dynamic Bayesian networks. Qual. Reliab. Eng. J. **33**, 2497–2520 (2017)
16. Reason, J.: Human Error. Cambridge University Press, Cambridge (2003)
17. Boring, R.L., Gertman, D.I., Ulrich, T.A.: Human reliability research needs for long-duration spaceflight. In: 10th International Conference on Applied Human Factors and Ergonomics (AHFE) and the Affiliated Conferences, Las Vegas. Conference Proceedings; International Conference on Applied Human Factors and Ergonomics (2019)

The Impact of Agile Project Management Model on the Performance of Technology Teams

Akif Onur([✉]) and Ismail Ekmekci

Industrial Engineering, Istanbul Commerce University, 34000 Kucukyali, Istanbul, Turkey
akif.onur34@gmail.com.tr, iekmekci@ticaret.edu.tr

Abstract. In the academic and business world, the number of leaders who are not familiar with the concept of an agile team has recently declined. Compared to traditional project management approaches, small but entrepreneurial teams, which are established close to the customers for the purpose of shortening the efficiency and time to market, are preferred in order to keep up with the rapidly changing conditions and competition The study prepared by taking into account the relationship between OECD countries and seven emerging countries, trying to show the difference between the share of research and development (R&D) expenditures from national income (GDP) and the number of patent applications and export in the Information and Communication Technologies (ICT) sector, and introduction-based literature review to provide information about the productivity and scalable evaluability of teams in engineering approaches, and the definition of the model, where the measures are, and the result stage.

Keywords: Agile team management · Agile project management · Team performance measurement

1 Introduction

Technological progress in classical economics was not directly calculated, and in modern growth theory it is clear that a well-educated workforce is an indispensable resource for the creation of technology. Because researchers are the driving force of technology, they need to be managed very differently from the usual workforce. An approach based on delivering needs iteratively and gradually called "Agile" is a preferred way of managing teams to end R&D with technology and innovation. Otherwise management would result in technology failure and the competition with technology creators would be shifted down and concluded as "creative destruction." Technologically leading countries are also well known as high-tech export performers.

Companies are unable to maintain their competitive edge in the market with conventional project management practices, and as a result, they are directed to teams managed under different frameworks under the program umbrella and want to update them with a new approach in areas where processes are inadequate [1]. It is stated that one of the most important factors hindering project and implementation success in

P. M. Arezes and R. L. Boring (Eds.): AHFE 2020, AISC 1204, pp. 282–288, 2020.
https://doi.org/10.1007/978-3-030-50946-0_38

traditional sectors is the lack of functional communication between stakeholders. However, in today's e-government applications there has been a rapid transition to a period where security is discussed, technologies are compared, customer satisfaction and product quality are more important than ever [2]. These applications such as blockchain, e-citizenship, and digital identity, developed with new methods and providing solutions, eliminate centralism [3].

2 Literature Review

R&D projects with the share allocated from the gross national product are encouraged, supported and followed by ministries, universities and R&D organizations. However, it is not possible to support to all projects since given the number, content, quality and limited resources of research and projects produced together, there is a scarce resource management requirement [4]. From this point out, traditional assessment software does not guarantee in the context of continuous improvement and learning, and research in light of more than 80 percent of these programs fail within the first 18 months [5]. This leads teams to present a running software to the requester in short circuits, moving forward with more frequent approval and feedback, and collecting tools to intuitively collect data as opposed to traditional methods. As a result of the above reasons, some questions may occur. The most important of these questions is how can companies easily measure performance when creating a team, but how can they measure and evaluate the performance of those teams when they create multiple teams? As we can see, while it is possible to diversify the questions, the need for high competition in today's uncertainty makes the idea of having teams that are fast, agile and compliant attractive. But institutions have difficulty deciding whether to meet the criteria to make such a strategy a reality. Especially in projects that require technology, situations that produce unnecessary results are often encountered despite the large costs incurred for a long time [6]. When the outputs of R&D expenditures such as company-level (micro) and country-level (macro) and export patents are analyzed, R&D expenditures sometimes do not lead to positive output as expected. R&D expenditures were taken as a share of GDP, exports and patents were examined at the micro level, with large amounts of R&D expenditures in Turkey, the largest exporters and 36 OECD countries and 7 emerging markets at the macro level.

R&D expenditures made by 36 OECD countries and 7 emerging market economies, the number of patents obtained between Information and Communication Technologies and their export to the Computer, Electronic and Optical sectors. In the table, since the R&D expenditures of some data should be concluded positively; This is because both patent applications and their exports show positive performance in the Information, Communication and Technology sector. Although some countries have high R&D expenditures, the number of patent applications is low and consequently the evaluation of the suitability of exporters in the high technology, information and communication technology sector (Tables 1 and 2).

While traditional agile formulations and recommendations highlight individuals and interactions on processes and tools, this analyze shows today's complex software systems and distributed teams. Management model that is provided by this analyzed

Table 1. Result of analyzed 241 companies in Turkey. Source: Turkish Exporters' Assembly

108 of 241 companies	Export Performance > R&D expenditures
44 of 241 companies	Export Performance = R&D expenditures
89 of 241 companies	Export Performance < R&D expenditures

system establishes the risk criteria used to prevent and reduce risk [7]. In Agile software development, not only is the interaction between individuals and each other more important than the process and the tools used, but the operation of the software is more important than the documentation in detail. Customer contributions are also more important than contracts and contracts, while responding to changes is more important than following a straight plan [8].

3 Agile Transitions Can Not Only Be the Job of Information Technologies

The digitalization we use as a concept, which is actually emptied of the content with the sentences we adorn, is similar to the youth talking about sexuality in adolescence: Actually, nobody knows exactly what it is, but it is always talked about.

Doing more and more the same things kills the talent within us and prevents us from seeing new ways and opportunities. Imagine a child learning to walk, ride a bike. While the child is doing this, there is no target for a year later, no planning, no success metrics. There is only courage and willingness to do so.

Agile transitions can not only be the job of information technologies.

What should we do, as the above assumption is unlikely to be true? We have to learn to look at it multi-dimensionally once we make decisions on such initiatives in our organizations. What are these dimensions?

Processes, organization, tools, infrastructure, finance, technology and of course human resources. When planning transition processes, we should definitely consider all these dimensions in an integrated way. The essence of the word, 'agile' transitions cannot be the business of organization and information technologies. It is necessary to take off these works with a team of all functions.

Especially the human dimension; Since human is the factor that will manage and develop the whole process, human resources must be prepared for this job much more.

Pre-measurement and analysis; It is possible to address the concept of agility, agility in seven dimensions in a person:

- Personal awareness
- Openness to feedback
- Environmental awareness
- Agility in change
- Agility in relationships
- Cognitive agility
- Determination to succeed

Table 2 The country's R&D expenditures, the number of patent applications in the field of Information and Communication Technology (ICT) and their exports in the Computer, Electronics and Optics sector. Source: OECD&WIPO

Country	ICT Patent Applications	R&D Expenditure as a share of GDP (%)	ICT Exports (Million$)
China	23.368	2,15	674.210
United States	17.831	2,79	199.303
Korea	5.317	4,55	166.757
Chinese Taipei	303	3,30	153.727
Singapore	240	1,95	134.847
Germany	2.850	3,04	130.316
Japan	11.840	3,21	96.572
Mexico	44	0,49	77.168
Netherlands	562	1,99	68.894
France	1.335	2,19	34.735
United Kingdom	1.459	1,66	33.294
Switzerland	356	3,37	31.415
Czech Republic	22	1,79	27.992
Poland	81	1,03	18.686
Italy	335	1,35	16.784
Ireland	190	1,04	16.102
Hungary	74	1,35	15.919
Slovak Republic	8	0,88	14.767
Canada	979	1,59	14.112
Belgium	200	2,70	13.551
Sweden	1.588	3,40	12.627
Israel	768	4,54	10.045
Austria	175	3,16	9.706
Spain	220	1,21	7.663
Denmark	137	3,05	7.055
Romania	19	0,50	4.449
Australia	417	1,79	4.333
Russia	277	1,11	3.801
Finland	541	2,76	3.686
Portugal	39	1,33	3.376
Turkey	133	0,96	2.516
Norway	90	2,09	2.396
Estonia	9	1,29	1.750

(*continued*)

Table 2 (*continued*)

Country	ICT Patent Applications	R&D Expenditure as a share of GDP (%)	ICT Exports (Million$)
Lithuania	5	0,90	1.627
South Africa	46	0,82	1.411
Latvia	1	0,51	1.334
Greece	20	1,13	1.200
Slovenia	6	1,86	1.177
New Zealand	35	1,37	605
Luxembourg	30	1,26	568
Chile	8	0,36	302
Argentina	3	0,54	144
Iceland	0	2,10	73

When you evaluate your human resources and leaders in these seven dimensions (with different measurement assessment tools and development center practices available in the market), you will actually see the adaptability of each individual and how they will react to the process. So how do you measure to achieve these results?

4 Agile Performance Methodology

Agile-Scrum model that is based on observers, developers, and repeats is framework of application development. Many of these models assume that modern software projects are quite complex, hard to coordinate team and it will be difficult to plan them all from the beginning. In order to reduce this confusion, six main criteria have been established. These six main criteria have 84 sub-criteria, so that in this framework there are 90 criteria.

First main criteria is agility index. It is the name given to a roof management for the management of agile-scrum complex projects to ensure that a product originally imagined and conformed to the design is produced at a fast, predictable cost and quality. The realization of the designed product is not carried out in the form of gradually realizing a list of requests prepared by the customer/user as detailed as possible [9]. Instead, the functions requested and defined by the customer/user are developed and revised within two or four-week periods called Sprint. This user-based requirement definition is described as a User Story and included in the job list. At the end of each Sprint, a functional piece of software is finished and can be delivered to the customer.

Second is the Performance measurement. Performance assessment one of the important step for decision making in competitive environment. The most important stage of the performance evaluation is the adjustment of the appropriate criteria. The performance measurement of teams is presented with 18 criteria as the most basic interpretable indicator, considering many criteria.

Third criteria measurement of planned iterations index which is designed to ensure that parts or functions of the product are delivered and evaluated on a regular basis. 16 measurement is designed to improve traceability by keeping progress and issues on a daily basis and ensuring they are localized for solving problems without receiving any complaints from the customer.

Forth one is agile goal index is defined as a specific goal to ensure that the team is gathered around a higher goal that exceeds all stakeholders' goals. Due to the nature of the work, the requirements of the product are not determined once in a while, but there is a target and definition to guide the team in each iteration to re-evaluate each delivery and make adjustments according to the situation.

Fifth criteria is customer satisfaction. Agile approaches focus on logical customer satisfaction. After all, the customer is the reason to develop the product in the first place [10]. In this section, 12 metrics are designed to participate in the basic key performance indicator process by obtaining net promoter score Calculation from the questions about determining Customer dissatisfaction to identify some Customer Satisfaction problems common in the project.

Last criteria is wellness of team. Discovering ways to motivate agile teams is positively correlated with output quality. Based on the work to be done, it is necessary to use the right resources to the teams and to control the team autonomously with a product accumulation and product owner in which the needs are handled correctly.

5 Importance and Drawback of Agile Modelling

Giant organizations have started to break up into tiny pieces. Because organizations growing as far as Titanic could not turn the rudder. Each piece sets its own rules, they set their own 1–2 week goals in an innovative environment, and they work only by drawing vision, without making medium and long term planning. In other words, we will witness that large companies will evolve into a new structure consisting of many start-ups. All the strict corporate working rules and ways of doing business in the old world will be left behind in the agile working culture. Institutions, on the other hand, will break down all their rules and change their educational perspectives in order to adapt their employees to this new world and to feed the new culture. Although the Agile model solves the problems completely, it has an effect on determining the method of dealing with the frequently encountered problems.

Agile model applications and being agile in communication has a high effect on describing the error and scoping performance of the team. This model lead to release your product and engineering management which can cause to improve your performance of product, team and work environment.

6 Conclusion

In order to catch the income level of developed countries, developing countries should invest in R&D in order to increase exports of goods and services. This will not only reduce the current account deficits, but also increase the welfare of its citizens. The team in R&D is very important for countries to achieve their goals. Therefore, the management of these teams should be carefully considered.

For teams that do not manage effectively, their inefficient teams, whose spending is far from producing an outflow, will result. Although inefficiency has various negative effects at both micro and macro levels, it can be understood from the analysis above, even though it causes time losses and productivity decreases. It is useful to use metrics in Scrum, to look at the product performance of the team, and to evaluate the maturity of a plan. Measurements made during the sprint run allow direct identification of problems while the team is working on jobs. Although Scrum prioritizes flexibility in the process, this measurement model built by us will allow things to be carefully planned, give full priority to customer requests, avoid risks, take into account all possibilities.

Our primary focus is to discover different ways to ensure the reliability of metrics, as this is the only factor for long-term use of metric programs that show strong potential in dictation.

References

1. Budayan, C.: Project portfolio management applications for Turkish Construction Industry in Istanbul Region. J. Polytech. **3**, 699–709 (2017)
2. Efe, A., Mühürdaroğlu, N.: Secure software development in agile development processes of e-government applications. J. Int. Sci. Res. **3**, 73–84 (2018)
3. Karahan, Ç., Tüfekci, A.: Blokzincir Teknolojisinin Dijital Kimlik Yönetiminde Kullanımı: Bir Sistematik Haritalama Çalışması. **3**, 73–84 (2018)
4. Ram, P., Rodriguez, P., Oivo, M., Martínez-Fernández, S.: Success factors for effective process metrics operationalization in agile software development: a multiple case study. In: 2019 IEEE/ACM International Conference on Software and System Processes (ICSSP), Montreal, QC, Canada, pp. 14–23 (2019)
5. Darrell, K.R., Jeff, S., Andy, N.: Scaling agile work. Harv. Bus. Rev. **5**, 60–69 (2018)
6. Ghane, K.: Quantitative planning and risk management of Agile Software Development. In: Proceedings of 2017 IEEE Technology & Engineering Management Conference (TEMSCON), Santa Clara, CA, pp. 109–112 (2017)
7. Top, Ö., Demirors, O.: Application of a software agility assessment model - AgilityMod in the field. Comput. Stand. Interfaces **62**, 1–16 (2019)
8. Cohn, M.: User Stories Applied: For Agile Software Development, 10th edn. Signature: Beck K., Addison-Wesley, Boston (2007)
9. Firdaus, M.B., Patulak, I.M., Tejawati, A., Bryantama, A., Putra, G.M., Pakpahan, H.S.: Agile-scrum software development monitoring system. In: 2019 International Conference on Electrical, Electronics and Information Engineering (ICEEIE), Denpasar, Bali, Indonesia, pp. 288–293 (2019)
10. PwC: Global Top 100 companies by market capitalisation. https://www.pwc.com/gx/en/audit-services/publications/assets/global-top-100-companies-2019.pdf. Accessed 24 Jan 2020

Cognitive Factors in Human Performance

Is There a Notable Comprehension Difference Between Abbreviations and Spelled-Out Words?

Tina M. Miyake(✉) and Katya Le Blanc

Idaho National Laboratory, PO Box 1625, Idaho Falls, ID 83415, USA
{tina.miyake, katya.leblanc}@inl.gov

Abstract. Nuclear power plant control rooms are information-rich environments. One source of information is the labels used to differentiate controls and alarms. Due to the multitude of alarms and controls, space is limited on the control boards. Consequently, labels are often abbreviated. Operators report abbreviations are faster to comprehend. However, these communications are anecdotal and in conflict with nuclear regulatory guidance. NUREG-0544 Revision 5 acknowledges that abbreviations are not ideal and are used for convenience. Moreover, operator anecdotes are also in conflict with decades of research, which have demonstrated that a word context allows for faster recognition of letters over a non-word context in adults. This paper outlines a coming study comparing abbreviations to spelled-out labels. Here, we discuss the potential in investigating this line of research.

Keywords: Human factors · Nuclear power · Comprehension · Semantic · Labels

1 Introduction

Nuclear power plant control rooms are information-rich environments. One source of information is the labels used to differentiate controls and alarms. Due to the multitude of alarms and controls, space is limited on the control boards, necessitating the use of abbreviated labels. The extensive use of abbreviations could potentially create comprehension difficulties.

U.S. Nuclear Regulatory Commission technical report designation (NUREG) 0544 Revision 5 [1] states that in documents abbreviations should be minimized in accordance with the Plain Writing Act of 2010. Plain writing primarily focuses on the understanding of the audience and not the convenience of the writer. Consequently, plain writing emphasizes avoiding or minimizing the use of abbreviations to facilitate comprehension. In the nuclear power plant control room, the audience is the crew of operators. Although the abbreviations are not in a document, comprehension of the terms is just as vital in this situation. The difficulty in the control room is that the use of the abbreviations is a function of a limit existing in the environment (i.e., space on the control boards) and cannot simply be overcome by increasing the number of pages in a

P. M. Arezes and R. L. Boring (Eds.): AHFE 2020, AISC 1204, pp. 291–295, 2020.
https://doi.org/10.1007/978-3-030-50946-0_39

document. Thus, the use of abbreviations is justifiable but might have a negative impact on comprehension of the labels.

Additionally, potential comprehension difficulties might be compounded by inconsistent labeling within a single system. For example, Ulrich and colleagues [2, pg. 32] noted that for the boration dilution display there were four different labels depending on the location:

- Boration Dilution Control from a dropdown menu at the top of the distributed control system window,
- Boration Dilution System/Control on the window title bar,
- Borate/Dilute Graphic Menu as a navigation button at the bottom of the display, and
- Control as a button on the Boration Dilution Navigation Menu

Ulrich and colleagues recommended that labels be consistent throughout a system and that the labels be spelled-out when possible. The latter recommendation was made, in part, because abbreviations were being used even when space on the control boards was available.

Curiously, some operators have reported a preference for abbreviations, suggesting not just a lack of a difference between abbreviations and spelled-out terms but a potential advantage for abbreviations. Operator reports could be dismissed as anecdotal evidence. However, operator feedback has been a vital part of control room development and research. Moreover, operators are the eventual users of a system, and, consequently, their preferences are important.

The focus of the proposed research is to investigate the potential influence of using abbreviations compared to spelled-out terms. Nuclear regulatory guidance [1] is clear. Moreover, decades of cognitive psychological research have demonstrated that a word context allows for faster recognition of letters over a non-word context in adults (e.g., word superiority effect [3–5]). Furthermore, the widely held concept of automatic semantic processing of words (e.g., [6]) suggests an advantage of spelled-out words over abbreviations. Thus, how can operator experience be explained? One possibility is that over time the abbreviations function in a similar fashion to spelled-out words (e.g., TV). Because the goal of our research is to develop a universal recommendation for labels in nuclear power plant control rooms based on empirical evidence, we are interested in any factors (e.g., repetition or orthographic construction) that could facilitate the comprehension of abbreviations.

2 Method

The purpose of the present study is to assess if there is a significant comprehension cost or benefit when using abbreviations as labels with an undergraduate population. The students serve as a representative sample that can be generalized to professional reactor operators. We are using a within-subjects design. Our independent variable is whether the term is abbreviated or spelled-out. Comprehension will be measured with reaction time and accuracy during a semantic sorting task. Additionally, we will ask participants to make familiarity judgments and complete a verbal ability test. These variables will be assessed as potential covariates. The study will be completed online.

2.1 Participants

Approximately 45–60 participants will be recruited from a small local university. The university's student population is primarily Caucasian. The surrounding community is ethnically homogenous and similar to the university student population. The university does have a larger non-traditional student population in terms of age than other universities. Additionally, many students are first-generation college students.

2.2 Materials

Study Session. Participants will be presented with abbreviations and spelled-out words. The term's respective category will be presented above. Participants will be instructed that they will need to study each term for a later task and that they will have as much time as they want to study each term. Before participants will be allowed to choose the study time for the terms, participants will make a quick familiarity judgment for each term. Familiarity judgments will be on a scale from 0 to 10. Zero indicates that the participant is not familiar at all with the word or abbreviation as a member of that category. A "10" indicates that the participant recognizes that word or abbreviation as a member of that category.

For example, if the participant saw the term below and they knew Bili is a member of the medical category, they would respond with a "10."

Medical

Bili

After all initial familiarity judgments are made, participants will be allowed as much time to study each term as they want. At the end of the study session, participants will complete another round of familiarity judgments.

Nelson-Denny Reading Test Form H. The Nelson-Denny [7] is a widely used verbal ability test (e.g., [8]). The Nelson-Denny [7] has a vocabulary section and a reading comprehension section. The vocabulary section consists of 80 vocabulary terms. Each question has 5 choices. For example, a participant would see the following, "To Repair is to: A) Destroy B) Finish C) Fix D) Work E) Show." Participants have 15 min to complete the vocabulary section. The reading comprehension section has several passages with accompanying questions. Each question has 5 choices. The entire section has a total of 38 items. Participants have 20 min to complete the reading comprehension section. The total Nelson-Denny score is calculated by doubling the total number of correct answers for the reading comprehension section and adding it to the total number of correct answers from the vocabulary section. We are modifying the Nelson-Denny in order to administer it online; consequently, we will not be measuring reading rate at the beginning of the reading comprehension section.

Semantic-Sorting Task. Participants will be presented with abbreviations and spelled-out terms. Participants will sort each term as quickly and accurately as they can into each term's respective categories (e.g., nuclear). If participants take longer than 7 s for a particular item, the item will be counted as a missed response, and the next item will appear.

2.3 Procedure

Participants will first read a consent script consisting of study information and describing their rights as a participant. Then, participants will study the terms that they will sort in the semantic sorting task later in the study. For this study session, participants will be making familiarity ratings. At the beginning of the study session, participants will be asked to make an initial quick assessment of their familiarity with each term. Familiarity ratings are being taken to ensure that participants consider the study terms at least once and to assess subjective initial familiarity of terms. Terms will be presented in random order with a label indicating their category membership. Once participants have completed their initial familiarity ratings, participants will be allowed to study the terms for as long as they like. Finally, participants will complete post-study familiarity ratings of the terms. After the study session is complete, participants will complete the modified Nelson-Denny reading test [7]. Next, the semantic sorting task will be administered. In this task, participants will see each term one at a time. Participants must sort each term as quickly and accurately as they can into their respective categories (e.g., nuclear category). Finally, participants will answer a few short demographic questions (e.g., gender, age, ethnicity).

We will compare the mean reaction time for abbreviations to the mean reaction time for spelled-out words controlling for vocabulary, reading comprehension, and familiarity ratings. We will analyze the accuracy data from the semantic sorting task by comparing abbreviations to spelled-out words in a similar fashion.

3 Prediction

If we find a difference between abbreviations and spelled-out words, this potential finding will be consistent with previous research on text-speak in a dual task [9]. Participants read a story either in text-speak or as correctly spelled out words. Additionally, participants had to monitor for a vibration on their torso while reading the story. In terms of reading comprehension, there was no difference in accuracy. However, there was a difference in the vibration detection task performance. Compared to the participants who only had to perform the vibration task, the dual-task conditions were slower and less accurate. More importantly, the participants who had to read the story in text-speak were the slowest and least accurate in the vibration detection task. This finding demonstrated that comprehension of material using abbreviations was more resource demanding. Moreover, this decrement was found even though the two tasks were in two separate modalities (i.e., visual and tactile). If comprehending abbreviations is more resource demanding, there could be potential consequences in how quickly an operator notices competing information. Head and colleagues' [9] findings suggest that we will find that abbreviations are more slowly categorized and potentially less accurately categorized than spelled-out terms.

4 Final Note

As mentioned in the introduction, this research is focused on an empirical basis for a universal recommendation for labels in the control room. At present, we are not focused on whether operators have developed strategies to coping with abbreviations in the control room. Our goal is to find support or lack of support as to which form is better. This study is a starting point and not an endpoint. Therefore, we view our proposed research as complementary to Ulrich and colleagues' [5] work. But, also noted in the introduction, the use of abbreviations is a practical one. At this point, it is not feasible to eliminate all abbreviations in the control room. Thus, future research should investigate how automatic processing factors could be manipulated to ease cognitive load induced by abbreviations [9].

Disclaimer. The opinions expressed in this paper are entirely those of the authors and do not represent official position. This work of authorship was prepared as an account of work sponsored by Idaho National Laboratory, an agency of the United States Government. Neither the United States Government, nor any agency thereof, nor any of their employees makes any warranty, express or implied, or assumes any legal liability or responsibility for the accuracy, completeness, or usefulness of any information, apparatus, product, or process disclosed, or represents that its use would not infringe privately-owned rights. Idaho National Laboratory is a multi-program laboratory operated by Battelle Energy Alliance LLC, for the United States Department of Energy under Contract DE-AC07-05ID14517.

References

1. United States Nuclear Regulatory Commission: Collection of Abbreviations. NUREG-0544, Rev. 5, pp. 1–57 (2016)
2. Ulrich, T., Boring, R., Phoenix, W., Dehority, E., Whiting, T., Morrell, J., Backstrom, R.: Applying human factors evaluation and design guidance to a nuclear power plant digital control system. INL/EXT-12-26787 Revision 0, pp. 1–62 (2012)
3. Cattell, J.M.: The time it takes to see and name objects. Mind **11**, 53–56 (1886)
4. Grainger, J., Bouttevin, S., Truc, C., Bastien, M., Ziegler, J.: Word superiority, pseudoword superiority, and learning to read: a comparison of dyslexic and normal readers. Brain Lang. **87**, 432–440 (2003)
5. Ripamonti, E., Traficante, D., Crippa, F., Luzzatti, C.: Word and pseudoword superiority effects in a shallow orthography language: the role of hemispheric lateralization. Percept. Motor Skills: Learn. Mem. **118**, 411–428 (2014)
6. Andrews, S., Lo, S., Xia, V.: Individual differences in automatic semantic priming. J. Exp. Psychol. Hum. Percept. Perform. **43**, 1025–1039 (2017)
7. Brown, J.I., Fishco, V.V., Hanna, G.: Nelson-Denny reading test (1993)
8. Turley-Ames, K.J., Whitfield, M.M.: Strategy training and working memory task performance. J. Mem. Lang. **49**, 446–468 (2003)
9. Head, J., Helton, W., Russell, P., Neumann, E.: Text-speak processing impairs tactile location. Acta Psychologica **141**, 48–53 (2012)

Operator's Human Error Features in Compensatory Tracking Task Based on Cognitive Process

Jintao Wu[1], Yan Lv[2], Weicai Tang[3], Qianxiang Zhou[1](✉), and Yi Xiao[3](✉)

[1] School of Biological Science and Medical Engineering, Beihang University, Beijing, China
wjtl9920317@163.com, zqxg@buaa.edu.cn
[2] Beijing Institute of Control and Electronic Technology, Beijing, China
[3] National Key Laboratory of Human Factors Engineering, China Astronaut Research and Training Center, Beijing, China

Abstract. To explore the cognitive characteristics of human error in different cognitive processes and the correlations between basic cognitive abilities and compensation tracking task, a simulated compensatory tracking task with four stages (perception task, judgment task, regulation task and comprehensive task, in which the comprehensive task include perception link, judgment link and regulation link) was performed. Four basic cognitive ability of 36 volunteers and the performance of the compensatory tracking task were obtained for statistical analysis. The results showed that the error rate of perception task/link was the lowest and that of regulation task/link was the highest. Moreover, the sustained attention ability was significantly associated with perception and regulation task. The working memory ability was significantly related to perception task. These observations indicated that human errors are most likely to occur in the regulation phase, and sustained attention and working memory are important abilities in compensatory tracking task.

Keywords: Human error · Compensatory tracking task · Cognitive process · Cognitive ability

1 Introduction

The complexity of the human-machine system leads to the high frequency of human error which has serious consequences. In serious space accidents, the number of factors related to human error can be as high as 80%–85% [1]. About 85% of the accidents in industrial around the world are directly or indirectly caused by human factors [2]. The proportion of accidents caused by human errors is gradually increasing, and the consequences of the accidents are becoming more and more serious. NASA put forward the guiding ideology of "human factors and equipment failure" and applied it to the human-system integration standard of the space shuttle and the International space station [3–6]. Human error analysis has been raised to a strategic level by the world's

P. M. Arezes and R. L. Boring (Eds.): AHFE 2020, AISC 1204, pp. 296–301, 2020.
https://doi.org/10.1007/978-3-030-50946-0_40

important space agencies. Therefore, it is necessary to study the characteristics and internal mechanism of human error, so as to reduce or even avoid human error.

At present, some methods are used to evaluate the impact of human error on equipment reliability. Through the combination of behavioral science and cognitive psychology, the cognitive reliability model was established according to the human cognitive process. The cognitive reliability analysis method was combined with the behavior analysis method to analyze the human error, such as CREAM (Cognitive Reliability and Error Analysis Method) proposed by Hollnage [7] and ATHEANA (A Technique for Human Error Analysis) proposed by Cooper [8]. Specifically, take CREAM, for example, which was based on a situational dependence cognitive model. It divided the human cognitive process into three functional stages: perception, judgment and decision execution, and analyzes the types of human errors that correspond to different cognitive stages. It emphasized the influence of environment on human cognitive activity and cognitive function failure, so it has a wide application prospect in the fields of accident analysis and risk assessment of complex man-machine system. However, the main limitation of this method is that it can only statically describe the result of human error, but cannot dynamically show its development process. With the development of computer technology, the dynamic analysis method of human reliability and human error based on simulation is derived, and the performance data can be obtained by simulating the human operation process in the real task through the virtual task scene or environment. Virtual task can reflect the characteristics of dynamic interaction between complex man-machine systems. Therefore, the analysis method of human reliability and human error based on cognitive model in dynamic operation is the trend and focus of future research.

At present, there are very few studies on the cognitive characteristics of human error in typical tracking task such as space rendezvous and docking, especially from the perspective of different cognitive stages and different cognitive characteristics. The traditional research based on human error mainly focuses on the static discontinuous monitoring operation tasks, such as nuclear power plant monitoring, air traffic monitoring. The results of these studies were mainly based on the result data, but lack of intermediate process data. In order to make up for this deficiency, present study simulates spatial rendezvous and docking by compensatory tracking task platform. Based on the cognitive stage theory, the idea of cognitive link segmentation was introduced to obtain the data of dynamic continuous cognitive process to explore the cognitive characteristics of human errors.

2 Methods

2.1 Participants

Thirty-four subjects were recruited from Beihang University. All the subjects were between 23–27 years old, right-handed, and had normal or corrected to normal vision. All subjects were paid for their participation and written informed consent was obtained.

2.2 Compensatory Tracking Task Experiment

Based on the cognitive process, four stages of compensatory tracking task experiments were designed, which were perception task, judgment task, regulation task and comprehensive task, and the comprehensive task included perception link, judgment link and regulation link.

In the task, the gray cross on the screen is the tracking target (as shown in the upper-left corner of Fig. 1(A)), and the white cross is the controllable tracking point (the middle of Fig. 1(A)). Subjects were asked to manipulate the joystick to control the tracker and to make the two crosses coincide within 40 s, that is, to complete the tracking task (Fig. 1(B)). In the perception task, there will be several speed changes in the tracking target, which requires the subjects to press "1" immediately after they perceive the speed change of the target. In the judgment task, after each speed change, it is necessary to judge how many times the changed speed is the initial speed (set to 3–5 times), and press the corresponding digital button of the control equipment. In the regulation task, the coolie cap should be moved to regulate the speed of the tracker after each change of speed until successful rendezvous and docking. In the comprehensive task, if the target speed changes, the subjects have to complete the perception, judgment and regulation sub-tasks within the limited time until successful rendezvous and docking.

In the whole experiment, each subject was asked to complete the perception task, judgment task, regulation task and comprehensive task, and the subjects were asked to complete 10 times in each task. In order to prevent the influence of learning effect, the task sequence is designed in balance.

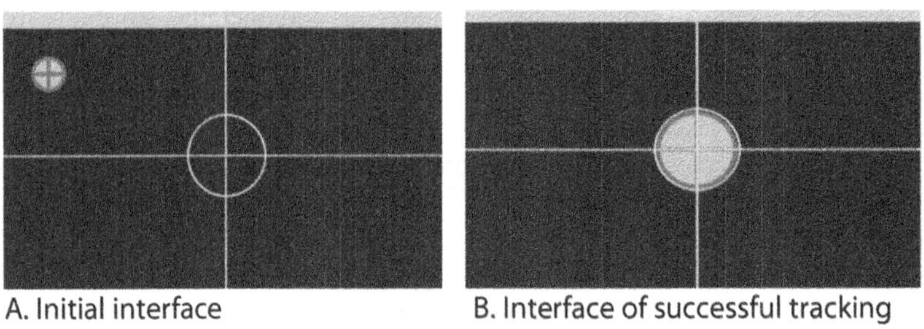

Fig. 1. Compensatory tracking task interfaces

2.3 Basic Cognitive Abilities Test

The current compensatory tracking task involves multi-cognitive process and cognitive ability. In the judgment task, the memory extraction of the speed before the mutation requires the ability of working memory. The attention and regulation of distance and speed in the task of perception, judgment and regulation requires sustained attention ability, attention distribution ability and motion control ability. Of course, due to the complexity of the task, there may be a lot of cognitive abilities involved, and in the end, above four key basic cognitive abilities were chose. The four abilities were respectively evaluated by "2-back" task, continuous performance task (CPT), visual dual task and motion control task.

For the "2-back" task, subjects were asked to identify whether the current number was the same as the second number preceding it, and then press the corresponding button. When doing CPT, one of the 0–9 numbers appeared randomly on the interface. Each number had one of three colors: red, green and blue. Numbers appeared at random anywhere on the screen. When red 3, green 5 and blue 9 appeared, the subjects need to press the corresponding number keys, and the rest do not react. When performing the visual dual task, there are two shapes on the left and right side of the screen, trapezoid on the left side and two circles on the right side. The orientation of the trapezoidal short side (left or right) and the position relationship of the two circles (separated, tangent, or intersecting) correspond to the keys of the different fingers of the left and right hand, respectively. During the motion control task, the ball on the screen is controlled by the rocker so that it moves along the set track to the end time limit of 60 s. In the process of moving, make sure that the ball does not collide with the edge of the track as much as possible. The correct response time (RT) of the first three tasks was recorded as working memory RT, sustained attention RT, attention allocation RT, respectively. The operation time of the fourth task was recorded as the motion control RTs. Both keystroke errors and omissions are counted as errors.

2.4 Procedure

First, the basic cognitive abilities of the subjects were tested, including working memory ability, sustained attention ability, attention distribution ability and motion control ability. Each test lasted about 15 min, and took a break of 5 min per test interval. The subjects entered the formal experiment after mastering the operation through practice.

3 Results

3.1 Comparison of Error Rates of Compensatory Tracking Tasks/Links

The error rates of the three tasks (perception, judgment, regulation) were significantly different ($F = 69.14$, $p < 0.01$). The error rate of perception task was significantly lower than that of judgment task and regulation task ($p < 0.01$), and the error rate of regulation task was the highest. The difference of the three links was significant ($F = 25.349$, $p < 0.05$), the error rate of perception link was significantly lower than that of judgment link and regulation link ($p < 0.05$), and the error rate of regulation link was the highest.

3.2 Pearson Correlation Between Basic Cognitive Abilities and Compensatory Tracking Task

The error rate of perception task was negatively correlated with the sustained attention RT ($r = -0.353$, $p < 0.05$), that is, the longer the sustained attention RT, the lower the error rate of perception task. The correct RT of perception task was negatively correlated with the correct rate of working memory RT ($r = 0.437$, $p < 0.05$). The better the ability of working memory, the shorter the correct RT of perception task. In addition, the correct rate of sustained attention task was negatively correlated with the correct RT of perception task ($r = -0.304$, $p < 0.1$, marginal significant), that is, the stronger the ability of sustained attention, the shorter the correct RT. The error rate of regulation task was negatively correlated with the correct rate of sustained attention task ($r = -0.372$, $p < 0.05$). The stronger the ability of sustained attention, the higher the success rate of regulation task.

4 Discussion

In the compensatory tracking task, the error rate of the perception stage was the lowest and that of the regulation stage was the highest, because the perception stage is the simplest compared with other stages and only involves stimuli perception. However, the judgment stage needs to recall the original speed and compare it with the current speed to identify the change. The regulation stage requires more memory, and the operation is more complex. In the process of compensatory tracking, the tracker should be accelerated first, and the tracker should be slowed down when approaching the tracking target. While regulating the speed, it is also necessary to pay attention to the distance between the two aircraft.

The correlation analysis between sub-tasks and basic cognitive abilities showed that the error rate of perception task was negatively correlated with the sustained attention RT. The longer the sustained attention RT, the more accurate the response to the perception task, which was in line with the speed-accuracy trade-off [9]. The correct rate of sustained attention task was negatively correlated with the correct RT of perception task, that is, the stronger the sustained attention ability, the shorter the correct RT, which indicated that the more attention is focused on the task, the easier it is to response quickly and correctly. Based on the results, working memory ability and sustained attention ability are the important cognitive abilities for the compensatory tracking task.

There is also limitation in the current research. The real compensatory tracking task is complicated, and many factors, such as relative distance, relative speed, relative attitude, relative position, fuel consumption, docking time and so on, need to be considered. The current simulation task is difficult to take into account all factors. However, even if we simplify the compensatory tracking task, the current research still has a certain reference value.

In conclusion, current studies showed that human errors are most likely to occur in the execution phase (regulation). Furthermore, the completion of compensatory tracking task requires a variety of basic cognitive abilities, among which sustained attention ability and working memory ability are particularly important.

References

1. NASA Committee: Report of the shuttle processing review team. R. 94N28178, NASA Kennedy Space Center, Florida (1993)
2. Baker, C.C., McCafferty, D.B.: Accident database review of human element concerns: what do the results mean for classification. Human Factors in Ship Design, Safety and Operation, London (2005)
3. NASA NASA-STD-3000: Man-Systems integration standards. NASA, Washington D.C. (1994)
4. NASA NASA-STD-3001: Space flight human-system standard. NASA, Washington D.C. (2011)
5. NASA NASA/SP-2010-3407: Human integration design handbook, NASA, Washington D.C. (2010)
6. O'Connor, B., Chief S.: Human-rating requirements for space systems. Report NASA/NPR, Washington D.C. (2011)
7. Hollnagel, E.: Cognitive Reliability and Error Analysis Method (CREAM). Elsevier, Amsterdam (1998)
8. Cooper, S.E., Wreathall, J., Thompson, C.M., et al.: Knowledge-base for the new human reliability analysis method, a technique for human error analysis (ATHEANA). Brookhaven National Lab, New York (1996)
9. Gueugneau, N., Pozzo, T., Darlot, C., et al.: Daily modulation of the speed–accuracy trade-off. J. Neurosci. **356**, 142–150 (2017)

Cognitive Performance After Repeated Exposure to Transcranial Direct Current Stimulation (tDCS) During Sleep Deprivation

Lindsey K. McIntire[1(✉)], R. Andy McKinley[2], Chuck Goodyear[1], John P. McIntire[2], and Justin M. Nelson[3]

[1] Infoscitex, Inc., 2027 Colonel Glenn Highway, Dayton, OH 45431, USA
{Lindsey.McIntire.ctr, Charles.Goodyear.1.ctr}@us.af.mil
[2] Air Force Research Laboratory, San Bernadino, USA
{Richard.McKinley.2, John.McIntire.1}@us.af.mil
[3] Consortium Research Fellow, 2510 Fifth Street, Building 840, WPAFB, Dayton, OH 45433, USA
Justin.Nelson.14@us.af.mil

Abstract. Previously, our lab demonstrated a 6 h performance benefit after a single 30-minute transcranial direct current stimulation (tDCS) session. Medical research with tDCS shows beneficial effects of repeated exposures; therefore, we hypothesize that repeated exposures will extend to healthy populations. Four groups of twelve participants in each received either active stimulation at 1800 and sham at 0400 h, sham at 1800 and active at 0400 h, active stimulation at both 1800 and 0400 h, or sham at 1800 and 0400 h during 36 h of continued wakefulness. Every two hours beginning at 1800 and ending at 1900 h the next day, participants completed 4 cognitive tasks along with subjective questionnaires. Contrary to our hypothesis, repeating stimulation did not have an additive benefit to cognitive performance. However, participants did report feeling significantly less fatigued on the subjective questionnaires compared to the other groups; suggesting possible additive effects for improving subjective mood.

Keywords: Transcranial Direct Current Stimulation (tDCS) · Fatigue · Sleep deprivation

1 Introduction

Long periods without sleep during off-circadian hours is common and necessary in many professions. For example, individuals in transportation, the military, or medicine regularly experience fatigue due to the nature of their work. Further, an estimated 25–30% of U.S. adults receive six hours or less of sleep per night [1]. Fatigue induced by sleep deprivation or working odd hours has repeatedly been shown in research to adversely affect accuracy, response times, attention, and multi-tasking ability which leads to forgetting or ignoring important aspects of the job [2]. These fatigue-related mishaps have resulted in an estimated $136 billion dollars annually cost from lost productivity [3]. A commonly used countermeasure to combat this type of sleep loss is

P. M. Arezes and R. L. Boring (Eds.): AHFE 2020, AISC 1204, pp. 302–313, 2020.
https://doi.org/10.1007/978-3-030-50946-0_41

caffeine. Caffeine has been shown to improve performance during insufficient sleep and sleep variation from circadian rhythm [4, 5]; however, the benefits of caffeine are short lived (approximately 2 h) and become less effective with chronic use [6, 7]. Therefore, it is of interest to investigate other possible forms of fatigue countermeasures that produce longer lasting effects with fewer side effects.

Recently, results from our lab have illustrated a robust effect of a single 30-minute session of transcranial direct current stimulation (tDCS) on aspects of cognition during sleep-deprivation induced fatigue stress [6, 8]. Specifically, tDCS led to a sustainment of vigilance/attention that persisted for approximately 6 h, faster response times in a simple reaction time test, improved working memory performance, and mood enhancements such as reduced fatigue and drowsiness and increased energy when compared to caffeine and no intervention [6]. When the tDCS treatment is given earlier in the sleep deprivation period (before onset of severe levels of fatigue), the effects on vigilance/attention were found to be consistent with our previous study (i.e., performance 'boost' lasting approximately 6 h). However, the effects on simple reaction times and mood remained elevated up to 22 h after the tDCS treatment [8]. Building upon our previous work, this study seeks to determine if a second tDCS treatment applied at the point where the effect on vigilance from the first treatment has perished could provide a longer period of protection from the fatigue stress; akin to having a second cup of coffee after the first cup's effects have worn off.

Several clinical studies have shown beneficial additive behavioral effects of tDCS resulting from repeated or cumulative tDCS sessions, especially when the tDCS treatment was paired with a behavioral/cognitive task [e.g., 9, 10]. In fact, the therapeutic benefits of tDCS appear to persist even longer with repeated or daily treatments (i.e., weeks or months rather than hours or days) in clinical patients [e.g., 11, 12]. Given that tDCS appears to induce long-term potentiation (LTP) [13], it is likely that such long term benefits are caused by lasting changes to synaptic connections in the stimulated tissue. Due to these previous findings that cumulative tDCS sessions appear to benefit clinical populations, it is reasonable to hypothesize that such cumulative tDCS effects might also extend to healthy populations by providing prolonged protection from fatigue or stress. The present study attempted to answer this question.

2 Materials and Methods

2.1 Equipment

tDCS Stimulator. The MagStim DC stimulator (Magstim Company Limited; Whitland, UK) was used to provide the tDCS stimulation. This battery-powered device was controlled with a microprocessor to ensure constant current at up to 5,000 μA. For safety, multistage monitoring of the output current and electrode/tissue impedance was included. The device automatically shut off if the impedance became greater than 50 k Ω to prevent electric shocks or burns. This device was investigational only (not FDA approved).

tDCS Electrodes. The electrodes were an array of 5 electroencephalographic (EEG) electrodes. Each electrode had an inner diameter of 1.6 cm yielding a contact area of 2.01 cm^2 for each electrode. At 2 mA of supplied current, there was an average current density of 0.199 mA/cm^2.

2.2 Subjects

Forty-eight active-duty military participants from Wright-Patterson Air Force Base completed this study. There were 14 female participants and 36 male with an average age of 26.6 $\pm$ 5. Participants were disqualified if they met any of the exclusion criteria described in McKinley et al. [14]. Fifty-seven total participants were enrolled. Three were disqualified based on the exclusion criteria. Disqualification criteria included: neurological or psychological diagnosis, currently pregnant, heart disease, frequent headaches or history of migraines, seizures, or learning difficulty. A further three participants did not get the required amount of sleep the night prior according to the activity watch and were asked to reschedule but never did. Three quit during their testing session because they no longer wished to continue.

2.3 Performance Tasks and Questionnaires

Mackworth Clock Test. The vigilance task was developed according to the description of the task used by Kilpaläinen, Huttunen, Lohi, and Lyytinen [15]. The task was an adapted version of the Mackworth clock test with parameters adapted from Teikari [16] and ran on a standard desktop computer. The participant was presented a visual display with 16 hole-like black circles arranged in a clock-like figure against a black background. Each circle changed from black to red for 0.525 s in turn, with each cycle lasting 3 s. The red light moved in a clockwise sequence by one step, which was considered the normal stimulus appearance. When the light moved twice the usual distance (i.e., skipping a circle), it was considered a critical signal and the participant was required to respond to this signal by pressing the spacebar as fast as possible on the keyboard with his or her preferred index finger.

Psychomotor Vigilance Task (PVT). Participants performed 10 min of the PVT each session during testing. The PVT192 (Ambulatory Monitoring, Inc.; Ardsley, NY) is a 8" $\times$ 4/5" $\times$ 2/4" handheld, battery-operated computerized test presentation and data capture system that records visual reaction times. The visual stimulus was presented on a small liquid crystal display (LCD) that displayed elapsed time counting up from zero by milliseconds. The stimulus presented for a maximum of 1 min (60,000 ms) before timing out. The task required the participant to detect whenever the PVT started counting up from zero, and to respond as quickly as possible by using a button press with the thumb. Once the participant depressed the microswitch, the device records the reaction time of the stimulus and resets to zero. The inter-stimulus interval times varied randomly from 2 to 12 s. Thus, the PVT requires sustained visual attention and rapid discrete motor responses.

Multi-Attribute Task Battery (MAT-B). This task was originally developed by the National Aeronautics and Space (NASA) to evaluate human performance metrics during a multi-tasking test paradigm [17]. MAT-B requires the operator to simultaneously monitor and respond to four separate tasks shown on a visual monitor, with each task in a separate quadrant of the display: a visual system alert monitoring task, a visuo-motor tracking task, an auditory communications monitoring task, and a visual resource management task. Participants performed a modified version of the MAT-B program for 20 min at a medium-high difficulty level.

Profile of Mood States (POMS) Questionnaire. The POMS is a 65-item questionnaire that measures subjective numerical reports of current mood using 6 categories: tension-anxiety, depression-dejection, anger-hostility, vigor-activity, fatigue-inertia, and confusion-bewilderment. Participants rated the intensity of their feelings about each item (example items: Tense, Vigor, Fatigue) on a scale of 1 to 5, with 1 being "not at all" and 5 being "extremely." The overall scores for each of the five categories were totaled, resulting in five factor scores. Each of the factor scores, except for the vigor-activity score, were summed. The vigor-activity score was reverse scaled, and so was instead subtracted from the total to produce a general mood disturbance score.

Visual Analog Scale (VAS). Subjective affect was also measured via the Visual Analog Scale (VAS) [18]. The VAS required participants to indicate the points on different lines that correspond to how he/she felt along the specified affect continuum at the time at which the test was taken. The adjectives included in the VAS were as follows: Alert/Able to Concentrate, Anxious, Energetic, Feel Confident, Irritable, Jittery/Nervous, Sleepy, and Talkative. The indicated magnitudes were later measured and converted into numerical scores for analysis.

Side Effects Questionnaire. Participants were asked to respond "none", "slight", "moderate", or "severe" to 33 items. Examples included: "Light Headed", "Headache", "Drowsiness", and "Drugged Feeling".

Mood Questionnaire. A 15-item mood questionnaire was administered where the participant checked a box (7 point scale) to indicate which one of the two moods they most identified with at the current moment. For example, "Fatigued vs. Energized", "Happy vs. Sad", and "Optimistic vs. Pessimistic" were items on this questionnaire. Depending on the boxes selected, the form outputs a numerical score from 1–7 to quantify the mood.

2.4 Procedures

Using a similar methodology described in McIntire, et al. [6], participants underwent 36 h of continuous wakefulness. Participants were randomly assigned to one of four groups (n = 12 for each group). In all groups, the anode was positioned over the left dorsolateral prefrontal cortex (lDLPFC) while the cathode was placed over the contralateral bicep. Active tDCS was administered for 30 min at 2 mA while sham tDCS was administered for 30 s at 2 mA, although both group's participants wore the tDCS instrumentation for a full 30 min regardless of active or sham stimulation, for proper blinding control. Group 1 (Early) received active tDCS at 1800 and sham tDCS at 0400 the next day. Group 2 (Late) received sham tDCS at 1800 and active tDCS at 0400. Group 3 (Early and Late) received active tDCS at both 1800 and 0400 while group 4 (Control) received sham tDCS at both time points. Group participation was randomized and condition was double-blinded using pre-programmed numbered codes for the stimulation parameters that were given out by a researcher not administering the stimulation. Stimulation was administered on-line during the beginning of their testing session.

Two days before their scheduled testing session and after consenting to participate in the study, participants were given an activity wrist monitor and instructed that their daily schedules should include a minimum of seven hours of sleep per night between the hours of 2300 and 0600. Participants also received training on all four performance tasks and subjective questionnaires utilized in the study.

On the day of their trial, participants were required to awaken at 0600 and perform their daily activities as normal. They were instructed to not consume any caffeine or central nervous system (CNS)-altering medications/substances on the experimental test day. Each participant arrived at the test facility at 1730 h. Upon arrival, their activity data was analyzed to ensure that proper sleep cycles were maintained. Starting at 1800 h, participants completed one session of the vigilance task (30 min), one session of the PVT task (10 min), one session of the Multi-attribute task battery (MATB) task (20 min), and answered the Profile of Moods States Questionnaire, Visual Analog Scale (VAS), mood questionnaire, and side–effects questionnaire. Afterwards, participants were provided a short break of approximately 45 min where they could talk, watch TV, walk, read, or play video games. The second session began at 2000 h. These procedures were repeated every two hours with the final session occurring at 1800 the following day (36 h continuous wakefulness). Both at 1800 on the first day and at 0400 on the next day, all participants were given either active or sham brain stimulation depending on their experimental group.

3 Results

Before 0400 there are essentially 2 large groups (Stim and Control). At 0400 and after, the Stim group effectively breaks into 2 new groups (Early, Early and Late), and the Control group subdivides into 2 new groups (Control, Late). The ANOVA results are for data collected at 0400 or later. A mixed-model ANOVA was used with ‘group’ as a between-factor (early, late, early and late, control) and ‘session’ as a within-factor (8 levels, 0400–1800). The dependent variable is then change from 1800 on the first day (session 1). This served as a baseline. An ANOVA determined at 1800 that there were no significant differences ($p \leq 0.05$) for all variables analyzed thus concluding that the starting baselines were similar among groups. For all plots, the values plotted are mean + or − 1.0 SE. Stars below a session indicate statistically significant paired comparison among the groups, where a single star (*) represents $0.01 < p \leq 0.05$, and a double-star (**) represents $p \leq 0.01$.

There were no statistically significant main effects (by group) for any of the performance tasks, when aggregated across time. Thus, there was no resounding finding to support our primary hypothesis of interest – cumulative effects of repeated tDCS stimulation when compared to early or late session stimulation, or when compared to controls. However, when the performance data is evaluated by each individual time epoch, some significant effects of tDCS stimulation were found and are worth noting; these findings will be discussed next.

For the Mackworth Clock Vigilance Task, examining percent hit changes from initial baseline at 1800 h (results shown in Fig. 1), we see that the three groups that received any stimulation (Early, Late, or Early and Late) outperformed the control group at the 1600 time epoch only (Early and Late vs. control, t = −2.19, p = .040; Early vs. control, t = −2.52, p = .020; Late vs. control, t = −2.43, p = .024). We can also see that one of the groups that received only early stimulation (at 1800) outperformed the control group at 2200 h (Early vs. control; t = 2.29, p = .033), but the other group that also received early stimulation (the Early and Late group vs. control) did not.

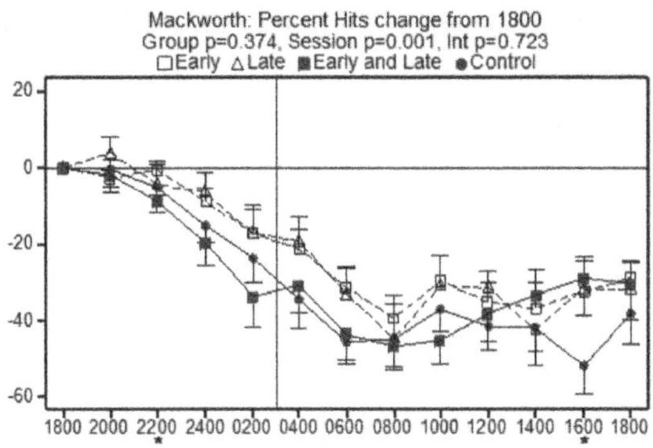

Fig. 1. Mackworth clock task percent hits change from baseline at 1800 h, by group. Significant differences detected via paired comparisons between groups at individual time epochs are indicated along the x-axis.

For the Psychomotor Vigilance Task, looking at mean response time changes from initial baseline at 1800 h (results shown in Fig. 2), we see that near the end of the study, at 1600 h only, the Late stimulation group and the group that received repeated stimulation (Early and Late) both were showing evidence of faster response times (Late vs. control; $t = -3.18$, $p = .005$; the Early and Late group vs. control; $t = -2.65$, $p = .015$).

For the Psychomotor Vigilance Task (PVT), looking at mean response time changes combined across the last four test sessions (when the participants had been continuously without sleep for 30–36 h) relative to the 1800 baseline, the two groups that received late stimulation (Late; Early and Late) show a non-significant trend towards faster response times than the control group. See Fig. 3 (left panel). A similar pattern was evidenced in the PVT lapses, as well as PVT errors, as shown in Fig. 3 (middle and right panels).

For the subjective ratings relating to mood and fatigue, we did find some possible evidence of a summation effect of repeated tDCS stimulation. On the Visual-Analog Scale (VAS), for the “Sleepy” measure, we found that at 1600 h, near the end of the

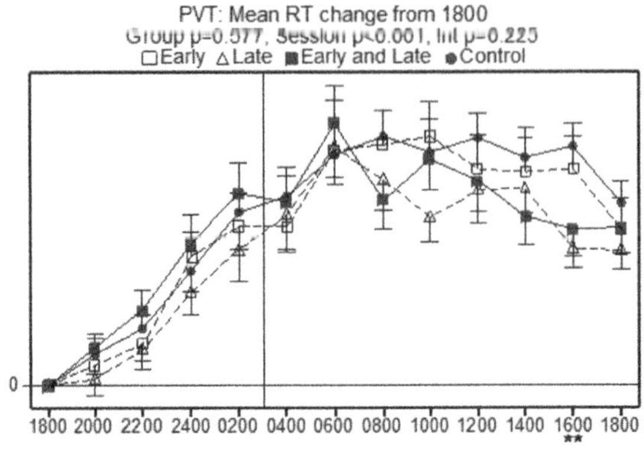

Fig. 2. Psychomotor vigilance task, mean response times change from baseline at 1800 h, by group. Significant differences detected via paired comparisons between groups at individual time epochs are indicated along the x-axis.

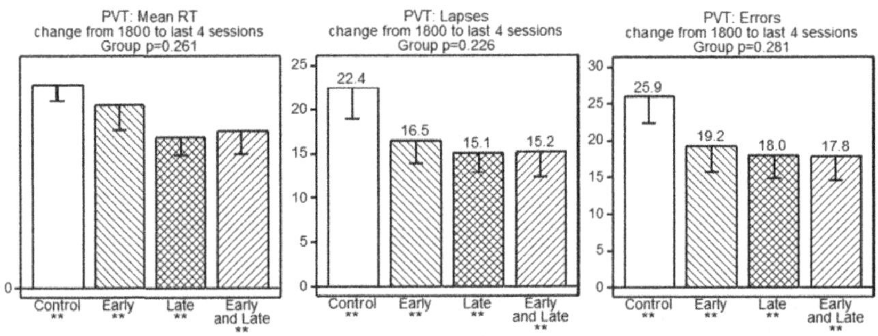

Fig. 3. Left panel: Psychomotor vigilance task, mean response times change from baseline at 1800 h, aggregated by group for the last 4 sessions. Middle panel: Psychomotor vigilance task, the number of lapses change from baseline at 1800 h, aggregated by group for the last 4 sessions. Right panel: Psychomotor vigilance task, number of errors change from baseline at 1800 h, aggregated by group for the last 4 sessions.

study, subjects in the repeated stimulation (Early and Late) group reported being less sleepy than the three other groups (see Fig. 4).

In the POMS subjective ratings data, on the "Anger/Hostility" item, we also found a benefit from repeated stimulation that was not evidenced in the other groups. In the earlier sessions, we see that the groups who received early stimulation (the Early group, and the Early and Late group) were reporting significantly less anger/hostility than the control group (Fig. 5, left panel). In the later sessions, the repeated stimulation group (Early and Late stimulation) were showing consistently less anger/hostility compared to the control group (Fig. 5, left panel). Taking only the last 4 sessions in aggregate, we see clearly that the repeated stimulation group had lower average ratings than the control group (Fig. 5, right panel); the Late group and the Early and Late group did not show change from baseline while the control group and early stimulation group did.

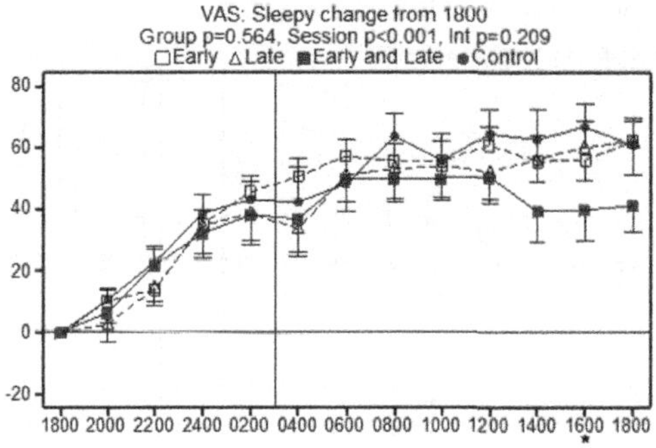

Fig. 4. Visual Analog Scale, 'Sleepy' ratings change from baseline at 1800 h, by group. Significant differences detected via paired comparisons between groups at individual time epochs are indicated along the x-axis.

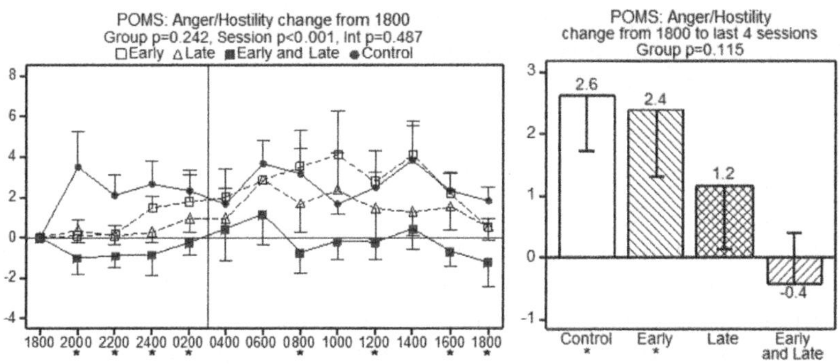

Fig. 5. Left panel: Visual Analog Scale, 'Anger/Hostility' ratings change from baseline at 1800 h, by group. Significant differences detected via paired comparisons between groups at individual time epochs are indicated along the x-axis. Right panel: Visual Analog Scale, 'Anger/Hostility' ratings change from baseline at 1800 h, aggregated last 4 sessions, by group.

In the Mood Quesionnaire self report data, we found that providing any form of stimulation (early, late, or repeated) tended to improve ratings of "Contentment/Satisfaction" when compared to the control group, especially in the later part of the study (see Fig. 6, left and right panels).

On the Side Effects Questionaire, we did find a main interaction for ratings of 'Fatigue' ($F(21, 294) = 1.98$, $p = .007$). *T*-tests revealed a significant difference between the Control group and the group getting stimulation both Early and Late at times 1400, 1600, and 1800 h the next day (see Fig. 7). There was also a significant difference between the Early and Late group and the Early stimulation group at 1800 the next day. At 1800 h, the group who received stimulation twice (Early and Late) reported feeling 44% less fatigued than the control group and 47% less fatigued than the group who received stimulation 24 h prior. Similarly, providing stimulation late only, or combining with early and late stimulation, tended to result in smaller ratings of

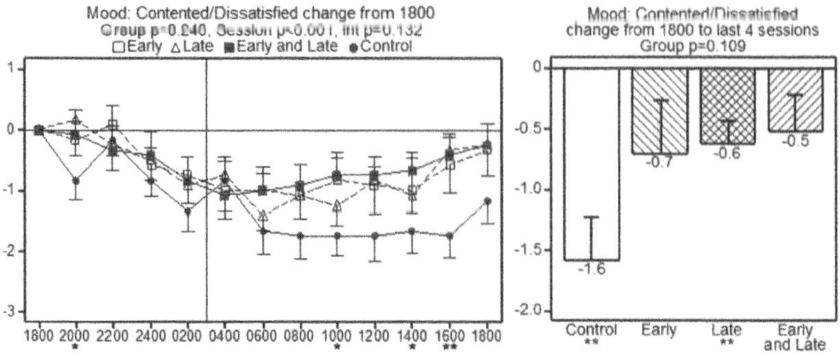

Fig. 6. Left panel: Mood Questionnaire, 'Content/Satisfaction' ratings change from baseline at 1800 h, by group. Significant differences detected via paired comparisons between groups at individual time epochs are indicated along the x-axis. Right panel: Mood Questionnaire, 'Anger/Hostility' ratings change from baseline at 1800 h, aggregated last 4 sessions, by group.

"Trouble Staying Awake" especially toward the later phase of the experiment (see Fig. 7). This same pattern of results was found for ratings of "Difficulty Staying Awake" (see Fig. 7).

4 Discussion

This study examined the effects of anodal tDCS over the left dorsolateral prefrontal cortex as a fatigue countermeasure during 36 h of sustained wakefulness. Specifically, the study measured the effects of two 30-minute stimulation times versus one to help mitigate declines in mood and performance as a result of fatigue induced by sleep deprivation. The stimulation was given to the assigned groups at either 1800 h (Early; at the start of testing), later into the sleep deprivation vigil at 0400 h (Late), or at 1800 and 0400 h (Early and Late). Contrary to our hypothesis, the overall results suggest that repeating the stimulation for a second time during the sleep deprivation vigil did not provide much of an added or extended performance benefit. However, providing both early and late stimulation did improve subjective mood ratings and perceived fatigue at later time periods, as compared to the relevant control groups. We also found further supporting evidence to suggest that receiving any type of tDCS stimulation, regardless if administered early, late, or early and late (repeated), can improve some performance metrics as well as some subjective metrics of mood and fatigue. In other instances, we found that late tDCS stimulation, regardless if administered for the second time in the vigil (repeated), performed better and reported less fatigue than the control group and the early only stimulation group; possibly suggesting that any positive effects of early stimulation had worn off, and that the late stimulation session served to provide a needed "boost."

Therefore, it appears that repeating tDCS multiple times during a vigil (in this case, two stimulations 10 h apart) simply may not have an additive cognitive performance benefit as hypothesized. Interestingly, in contrast to our findings, several clinical studies have found beneficial effects of repeated tDCS sessions involving: stroke

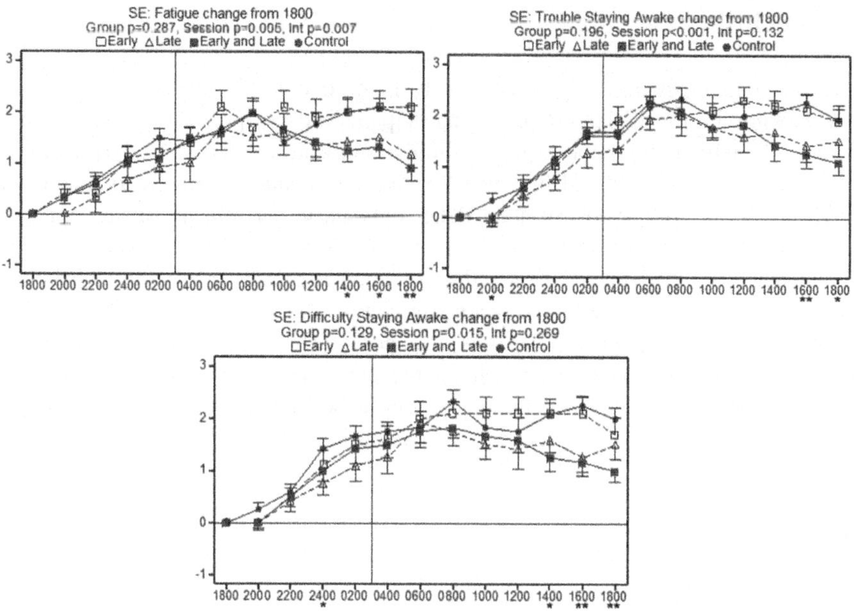

Fig. 7. Side Effects Questionnaire, 'Fatigue', 'Trouble Staying Awake', and 'Difficulty Staying Awake' ratings change from baseline at 1800 h, by group. Significant differences detected via paired comparisons between groups at individual time epochs are indicated along the x-axis.

rehabilitation [19], depression [11], and chronic migraines [20], to name a few. However, most of these studies repeat stimulation 20–24 h later, with normal sleep periods in between, which is when neuronal LTP is primarily suspected to occur in support of memory consolidation. Also, other research has found that repeating tDCS during a period in which the aftereffects of the first stimulation were still present seemed to enhanced the effectiveness of the stimulation in rodent models [21, 22] and in humans [9]. It is possible that 10 h of separation between repeated stimulation, with no sleep, is too small a time window for beneficial performance effects to accrue, at least under the present paradigm of extended wakefulness. Alternatively, is also possible that the present study delivered the second dose too late in the vigil to provide any cumulative effects.

Although in this study we found no distinct *performance* benefits of repeating tDCS stimulation 10 h after the first stimulation during sleep deprivation, when compared to single-session tDCS, we did find a beneficial effect on *mood*. People who received the second dosage reported feeling less fatigued than both the control group (sham stimulations at 1800 and then 0400) and the group who received active stimulation only at 1800. There is a large amount of psychiatric clinical literature that shows repeating stimulation for 5 consecutive days with stimulations spaced as often as twice a day can have cumulative improvements in mood for these patients when compared to controls [e.g. 11, 19, 23] and the effects appear long lasting, e.g., 1 month [e.g. 23–25].

5 Conclusions

In conclusion, repeating tDCS twice 10 h apart to enhance attention, arousal, and multi-tasking ability in sleep deprived participants did not provide an enhanced performance benefit compared to a single stimulation event (early or late) during the 36-h sleep deprivation vigil. It did however provide an enhancing benefit to mood when compared to the group who received tDCS only once, early in the sleep deprivation watch (at 1800). Further research should examine other timings for a repeated trial(s) as it is unclear if we delivered the second stimulation too early or too late, or whether a sleep session may be needed in-between stimulations for any additive effects to accrue via memory consolidation (LTP). It is also possible that repeating stimulation may simply not have a cumulative benefit on the particular types of cognitive tasks we utilized in this study (despite showing additive beneficial mood effects), so future research could consider testing other cognitive abilities or other brain areas of interest.

References

1. Gottlieb, D., Ellenbogen, J., Bianchi, M., Czeisler, C.: Sleep deficiency and motor vehicle crash risk in the general population: a prospective cohort study. BMC Med. **16**(1), 44 (2018). https://doi.org/10.1186/s12916-018-1025-7
2. Perry, I.C. (ed.): Helicopter aircrew fatigue. AGARD (Advisory Rep. No. 69). Advisory Group for Aerospace Research and Development, Neuilly sur Seine, France (1974)
3. Alger, S., Brager, A., Capaldi, V.: Challenging the stigma of workplace napping. Sleep **42** (8), 97 (2019). https://doi.org/10.1093/sleep/zsz097
4. Tharion, W.J., Shukitt-Hale, B., Lieberman, H.R.: Caffeine effects on marksmanship during high-stress military training with 72 hour sleep deprivation. Aviat. Space Environ. Med. **74** (4), 309–314 (2003)
5. McLellan, T.M., Kamimori, G.H., Voss, D.M., Bell, D.G., Cole, K.G., Johnson, D.: Caffeine maintains vigilance and improves run times during night operations for special forces. Aviat. Space Environ. Med. **76**, 647–654 (2005)
6. McIntire, L.K., McKinley, R.A., Nelson, J.M., Goodyear, C.: A comparison of the effects of transcranial direct current stimulation and caffeine on vigilance and cognitive performance during extended wakefulness. Brain Stimul. (2014). https://doi.org/10.1016/j.brs.2014.04.008
7. Miller, N.L., Matsangas, P., Shattuck, L.G.: Fatigue and its effect on performance in military environments (Report No. 0704-0188). Naval Postgraduate School Operations Research Department, Monterey, CA (2007)
8. McIntire, L.K., McKinley, R.A., Nelson, J.M., Goodyear, C.: Transcranial direct current stimulation versus caffeine as a fatigue countermeasure. Brain Stimul. **10**, 1070–1078 (2017). https://doi.org/10.1016/j.brs.2017.08.005
9. Monte-Silva, K., Kuo, M.F., Hessenthaler, S., Fresnoza, S., Liebetanz, D., Paulus, W., Nitsche, M.A.: Induction of late LTP-like plasticity in the human motor cortex by repeated non-invasive brain stimulation. Brain Stimul. **6**(3), 424–432 (2013)
10. Tippett, D.C., Hillis, A.E., Tsapkini, K.: Treatment of primary progressive aphasia. Curr Treat Options Neurol. **17**(8), 362 (2015)

11. Fregni, F., Boggio, P.S., Nitsche, M., Bermpohl, F., Antal, A., Feredoes, E., et al.: Anodal transcranial direct current stimulation of prefrontal cortex enhances working memory. Exp. Brain Res. **166**(1), 23–30 (2005)
12. Valle, A., Roizenblatt, S., Botte, S., Zaghi, S., Riberto, M., Tufik, S., et al.: Efficacy of anodal transcranial direct current stimulation (tDCS) for the treatment of fibromyalgia: results of a randomized, sham controlled longitudinal clinical trial. J. Pain Manag. **2**(3), 353–361 (2009)
13. Rohan, J., Carhuantanta, K.A., McInturf, S.M., Miklasevich, M., Jankord, R.: Modulating hippocampal plasticity with in vivo brain stimulation. J. Neurosci. **35**(37), 12824–12832 (2015)
14. McKinley, R.A., Nelson, J., Bridges, N., Walters, C.: Modulating the brain at work using noninvasive transcranial stimulation. NeuroImage **59**(1), 129–137 (2012)
15. Kilpeläinen, A.A., Huttunen, K.H., Lohi, J.J., Lyytinen, H.: Effect of caffeine on vigilance and cognitive performance during extended wakefulness. J Aviat. Psychol. **20**(2), 144–159 (2010). https://doi.org/10.1080/10508411003617847
16. Teikari, V.: Vigilanssi-ilmiön mittaamisesta ja selitysmahdollisuuksista [On measurement and explanation of vigilance] (Jyväskylä Studies in Education, Psychology and Social Research No. 35). University of Jyväskylä, Jyväskylä, Finland (1977)
17. Comstock, J.R., Arnegard, R.J.: The multi-attribute task battery for human operator workload and strategic behavior research. NASA Technical Report (No. NASA-TM-104174) (1992)
18. Penetar, D., McCann, U., Thorne, D., Kamimori, G., Galinski, C., Sing, H., Thomas, M., Belenky, G.: Caffeine reversal of sleep deprivation effects on alertness and mood. Psychopharm **112**, 359–365 (1993)
19. Boggio, P.S., Nunes, A., Rigonatti, S.P., Nitsche, M.A., Pascual-Leone, A., Fregni, F.: Repeated sessions of noninvasive brain DC stimulation is associated with motor function improvement in stroke patients. Rest Neuro Neurosci. **9**(54), 1–7 (2006)
20. DaSilva, A.F., Mendonca, M.E., Zaghi, S., Lopes, M., DosSantos, M.F., Spierings, E.L., et al.: tDCS-induced analgesia and electrical fields in pain-related neural networks in chronic migraine. Headache: J. Head Face Pain **52**(8), 1283–1295 (2012)
21. Bindman, L.J., Lippold, O.C.J., Redfearn, J.W.T.: The action of brief polarizing currents on the cerebral cortex of the rat during current flow and in the production of long-lasting after-effects. J. Physiol. **172**, 369–382 (1964)
22. Gartside, I.B.: Mechanisms of sustained increases of firing rate of neurons in the rat cerebral cortex after polarization: role of protein synthesis. Nature **220**, 383–384 (1968)
23. Brunoni, A.R., Ferrucci, R., Bortolomasi, M., Vergari, M., Tadini, L., Boggio, P.S., et al.: Transcranial direct current stimulation (tDCS) in unipolar vs. bipolar depressive disorder. Progr. Neuro-Psycopharm Biol. Psychiatry **35**(1), 96–101 (2011)
24. Ferrucci, R., Bortolomasi, M., Vergari, M., Tadini, L., Salvoro, B., Giacopuzzi, M., et al.: Transcranial direct current stimulation in severe, drug-resistant major depression. J. Eff. Disord. **118**, 215–219 (2009)
25. Loo, C.K., Alonzo, A., Martin, D., Mitchell, P.B., Galvez, V., Sachdev, P.: Transcranial direct current stimulation for depression: 3-week, randomised, sham-controlled trial. Br. J. Psychiatry **200**, 52–59 (2012)

A Fuzzy Logic Model for Quantifying the Likelihood of Human Decision-Making in Nuclear Emergency Situations

Young A Suh(✉) and Jaewhan Kim

Risk and Reliability Assessment Research Team, Korea Atomic Energy Research Institute, 111 Daedeok-daero 989, Yuseong-gu, Daejeon 34057, Republic of Korea
{sya, jhkim4}@kaeri.re.kr

Abstract. The purpose of this paper is to develop a fuzzy logic model for quantifying the likelihood of decision-making actions as one of the human reliability analysis (HRA). Especially, under emergency situations in Nuclear Power Plants (NPPs), the reliability of a decision while following a beyond design basis accident (after core damage situation), i.e. Fukushima Accident in 2011, is important part of HRA. In this paper, the fuzzy model is proposed for quantifying the likelihood of human decision-making under Severe Accident Management Guideline (SAMG) implementation corresponding to the nuclear emergency situations. Trial and error techniques were applied to identify suitable input parameters, fuzzification, the inference system, and defuzzification to develop the fuzzy model of SAMG decision actions. The results of fuzzy model show that ambiguous language expression of decision-making by operators can be quantified. In addition, we demonstrates the developed model is feasible to quantify the decision action with comparing the result of expert judgement. This study shows that the fuzzy model can be applied to evaluating human decision likelihood or error in overall HRA as well as SAMG HRA.

Keywords: Human decision-making · Fuzzy logic model · Human reliability analysis · Quantitative method · Severe accident management guidelines

1 Introduction

In case of nuclear emergency situations like 2011 Fukushima accidents, severe accident management guideline (SAMG) program has been required to mitigate the accident. In SAMG situation, technical support center (TSC) has authority of managing severe accidents and has similar role on nuclear power plant (NPP) operators. The TSC should anticipate and evaluate negative impacts associated with each of SAM strategies and compare the positive impacts and the negative impacts when the strategy is implemented. During these SAMG process, the decision-making on whether to implement the strategy or not is highly complex and requires a higher level of cognitive process. Thus, decision-making is important process during SAMG implementation [1].

Conventional human reliability analysis (HRA) methods have not dealt with a decision-making part of human activities in a serious way because most of the actions

P. M. Arezes and R. L. Boring (Eds.): AHFE 2020, AISC 1204, pp. 314–321, 2020.
https://doi.org/10.1007/978-3-030-50946-0_42

required in emergency operating procedures (EOP) belong to rule-based actions [2]. EOP do not require a serious decision-making function of human operators in Nuclear Power Plants (NPPs). However, the decision-making activity by the TSC plays a crucial role in implementing SAMG, thus a new technique dealing with decision-making actions in SAMG HRA could be developed.

In this paper, fuzzy logic model (FLM) is used for expressing SAMG decision-making process quantitatively. The three reasons why we chose the FLM are followings. The FL is a very useful tool for modeling complex systems qualitative, inexact, or uncertain information [3]. Second, FL has had many HRA applications [3–7] in the last years. Lastly, FL resembles the way humans make inference and take decisions [3]. It also accommodates ambiguities of real world human language and logic [8, 9]. Thus, this study used the fuzzy model to quantitatively evaluate the decision action in SAMG as well as optimize decision-making.

The purpose of this study is to develop a fuzzy logic model for quantifying the likelihood of decision-making actions as one of the HRA. This paper is organized into two parts as modeling approach and practical application. This study first focused modeling to identify suitable FLM for imitating TSC's SAMG decision actions. Trial and error techniques are applied to identify suitable fuzzification, the inference system, and defuzzification. A case dealing with SAMG decision situation is applied into the developed FLM to confirm practicality of this model. This result may be used in quantitative HRA with reflecting decision-making actions.

2 Fuzzy Logic Model for Quantifying SAMG Decision-Making

Fuzzy Logic Model (FLM) can be categorized into three steps: fuzzification, fuzzy inference process, and defuzzification. This study used Mandani's fuzzy inference process for developing FLM. Fuzzification is the process of transformation from crisp data to fuzzy set using fuzzy linguistic term (i.e. Low, Medium, High). Fuzzy operators (AND or OR) and fuzzy rules (if-then rules) are applied in the FLM. Logical operators represent that AND/OR as selecting a minimum/maximum value among input values, respectively. Fuzzy rules are expressed in human words, and usually used in form of If-then rules. For example, this format is 'If input 1 is Low and/or input 2 is High, then output is Medium'. Membership Functions (MFs) as parts of fuzzification in addition to fuzzy rules are identified by expert judgement since this is based on knowledge. Finally, the result of output is defuzzified using the centroid method to get a crisp value. These steps can be done by using the fuzzy logic toolbox simulator of MATLAB [10].

A SAMG decision-making action can narrow down as follows. Before decision process, requisite information can be gathered from critical function monitoring system (CFMS), main control room (MCR) and field staffs. Based on the information, the TSC evaluated the feasibility of SAGs. In detail, TSC identify the positive effects associated with the strategy. Then, they identify negative impacts associated with the strategy, and evaluate mitigative actions against negative impacts. After that, the TSC compare the positive and negative impacts of the mitigative action and then decided whether to implement the strategy or not. This is the decision-making process of SAMG. This process can be determined as fuzzy sets of inputs and output variables in Table 1.

Table 1. Overview of parameters

Parameters		Fuzzy sets
Input 1	Likelihood of negative impacts	3 (L, M, H)
Input 2	Evaluation complexity of negative impacts	3 (L, M, H)
Input 3	Implementation feasibility of mitigative actions	5 (VL, L, M, H, VH)
Input 4	Decision burden from the consequences of mitigative actions	5 (VL, L, M, H, VH)
Output	Final decision on a strategy	5 (VL, L, M, H, VH)

Based on selected four inputs and outputs, the FLM, which can quantify a SAMG decision action, is developed. Figure 1 shows a screen capture of the FLM for the influence on the final decision output under pressure of NIs using the MATLAB simulator, including the rule editor and results. Results can give a crisp value, which is a single scalar quantity, through defuzzification. Defuzzification using centroid approach extracts a precise value from the range of fuzzy sets as an output variable.

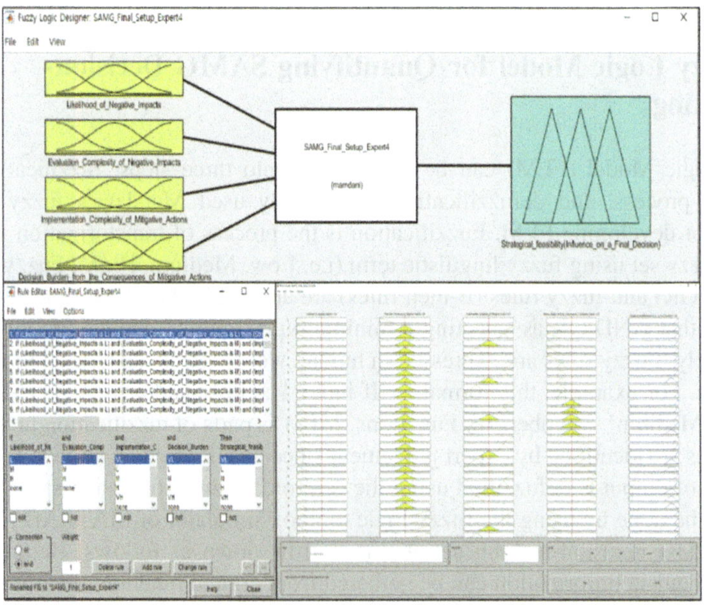

Fig. 1. Fuzzy Logic Model (screen capture) including rule editor and results

3 Application of the FLM into Nuclear Emergency Situations

To validate the developed FLM, it was applied to a case study to obtain quantifiable data. The Steam Generator Tube Rupture (SGTR) accident was selected, for which an SAMG expert performed an evaluation of SAG implementation. Their result was then compared with the quantified result from the developed FLM, which confirmed the ability of the FLM to quantify decision-making actions.

3.1 SGTR Scenario

Steam Generator Tube Rupture (SGTR) was selected for our case study. The SGTR accident sequences initiated by the rupture of one or more helical tubes in one SG, causing primary coolant leakage to the secondary side at the rate exceeding the makeup capacity of the charging pumps. This accident leads to core damage by the continuous leakage of the reactor coolant through the ruptured steam generator tube. Details of case scenario is following. SGTR is initiated with success of reactor trip under normal secondary heat removal using the turbine-driven auxiliary feedwater (TDAFW) system. However, there was the loss of multiple safety systems including all the trains of safety injection pumps and containment spray pumps following the reactor trip due to SGTR. For detail assumption, we categorized into case A and B. Case A is the success of isolation of SG with about 100 h after the accident, while Case B is the fail of isolation of SG due to stuck open status of Main Steam Safety Valve (MSSV) with 4 h after the accident. Figure 2 describes the SGTR event tree for expressing accident scenario sequences.

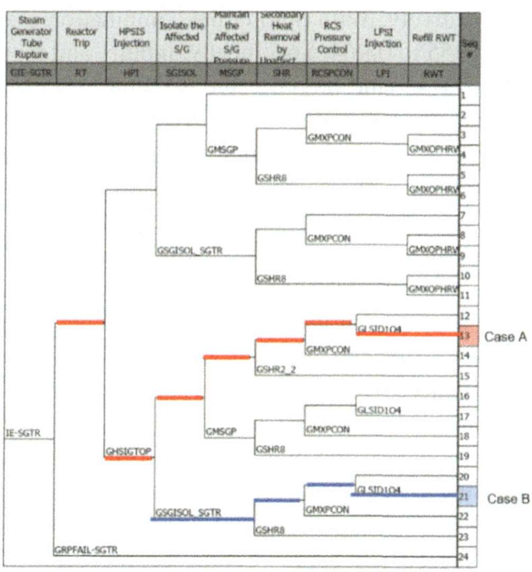

Fig. 2. Case scenarios through SGTR PDS-ET

Based on the scenario assumptions, first, an expert identified candidate of Accident Management (AM) strategies (SAGs) for Level 2 PSA. Based on Diagnosis Flow Chart in SAMG, SAG-01, SAG-02, SAG-03 and SAG-06 among seven SAGs were evaluated in that case of SGTR. According to the above assumptions, all SG levels is not enough to satisfy the standard of SG level, thus injection into SG (SAG-01) should be considered in case of SGTR. It might be possible to use portable pump for injecting into SGs. In this scenario, SAG-02 might be considered for mitigating SGTR. Since Core Exit Thermocouple (CET) is above 371.1 °C (CET_T), injection into RCS (SAG-03) using portable pump will be considered for candidate AM action. There is no means for injection into CNMT (SAG-04), which might not be implemented. In case of SAG-05, this is not applicable due to the no Fission Production (FP) release. Containment pressure might be considered for the mitigation of TLOCCW, thus SAG-06 is one of candidate of AM strategies. CTMT spray will be acted using the high capacity portable pump. SAG-07 is also not applicable due to the assumption that hydrogen is assumed to be under Normal Control. Each individual SAM strategy applicable to the SGTR scenario was evaluated for the following aspects: (1) the likelihood of negative impacts (Input 1), (2) the evaluation complexity for negative impacts (Input 2), (3) the implementation feasibility of mitigative actions (Input 3), (4) decision burden from the consequences of mitigative actions (Input4), and (5) final decision on a strategy (Output).

3.2 Application

Table 2 describes the result from an expert judgement on SAMG in cases of SGTR-A and B with comparing the result of FLM. A well-trained TSC staff evaluated the four inputs and output. Since the expert evaluated the linguistic value (represented by fuzzy set), the output of crisp value has the range depending on the range of fuzzy set. For example, a participant evaluates input 1 and input 2 as Low(L) which has the interval [0, 0.2] and [0, 0.3] respectively. With putting the combination cases into our FIS, the output has values from 0.955 to 0.964. Specifically, input 1 and input 2 have zero value in the range of 'Low' fuzzy set and output value is 0.964; whereas input 1 and input2 have 0.2 and 0.3, respectively and output value is 0.955.

In the SGTR scenario, if the environmental release of the fission product through the containment bypass release path is high, the TSC is expected to use 'SAG-5: fission product release control' as a priority of order. In order to control the release via the ruptured SG, the SAG-5 guides the TSC to use 'SAG-1: SG injection' to consider the SG injection strategy to scrub the released fission product. But in the given scenario, the TSC may have a decision burden in considering the SG injection using the portable pump because this strategy requests depressurization of the ruptured SG as a prior action, which may release more fission products than before implementing the strategy. On the other hand, if the environmental release of the fission product through the containment bypass release path is not high, the TSC is expected to enter the 'SAG-1: SG injection.' The negative impacts associated with the SG injection using the portable pump are 'thermal shock of the intact SG' and 'creep rupture of the intact SG tube'. The decision burden of the TSC in association of this strategy is expected to be high because only one SG remains intact and the implementation of the strategy have the potential to induce an additional damage to the intact SG.

Table 2. Comparison of the results from expert judgement and the developed FLM on the SAMG in case of SGTR-Case A[a)] & Case B[b)]

Related SAGs	Negative Impact (NI)	Mitigative Actions against NI	Input 1	Input 2	Input 3	Input 4	Output from expert judgement	Fuzzy output (Crisp value)	Corresponding Fuzzy set
SAG-01: Inject into SGs_Robust SGs	Fission product release from leaking SG tubes	SG water boiling	L	L	–	–	VH	0.955–0.964	VH
		RCS depressurization							
		Steam removal from condenser/SG depressurization							
	Thermal shock of SG	Limit flow to SG to xxx gpm for the first xx minutes of injection	H	L	M	H	M	0.7–0.725	M-H
		Begin feeding only one SG at a time to minimize consequences of SG tube failure until minimum wide range SG level is indicated	H	L	VH	H	M	0.85	H
		Feed only isolatable SGs to minimize consequences of SG tube failure	H	L	VH	H	M	0.85	H
	Creep rupture of SG tubes	Depressurize only one hot, dry SG at a time to minimize consequences of SG tube failure	M	M	VH	H	M	0.7–0.725	M-H
		Establish feed flow as soon as possible once SG pressure is below the shutoff head of the feed source. Limit flow to SG to xxx gpm for the first xx minutes of injection	M	M	M	H	M	0.7–0.725	M-H
		Depressurize the RCS (use SAG-2)	M	M	M	H	M	0.7–0.725	M-H
SAG-02: RCS Depressurization	Fission product release from containment building		L	L	–	–	VH	0.955–0.964	VH
	Containment severe challenge from a hydrogen burn		L	L			VH		
	Containment severe challenge from over pressurization (from corium concrete interaction (CCI))		L	L			VH		
	Fission product release from leaking SG tubes		L	L			VH		
	Loss of SG inventory		L	L			VH		
SAG-03: Inject into RCS	Containment severe challenge from a hydrogen burn when refilling water into core		L	M	–	–	VH	0.855–0.874	H-VH

(continued)

Table 2. *(continued)*

Related SAGs	Negative Impact (NI)	Mitigative Actions against NI	Input 1	Input 2	Input 3	Input 4	Output from expert judgement	Fuzzy output (Crisp value)	Corresponding Fuzzy set
	Creep rupture of SG tubes		L	L			VH	0.955–0.964	VH
	RCP seal damage		L	L			VH	0.955–0.964	VH
	Containment severe challenge from over pressurization (from CCI)		L	M			VH	0.855–0.874	H-VH
	Containment Flooding		L	M			VH	0.855–0.874	H-VH
	Habitability of auxiliary building		L	L			VH	0.955–0.954	VH
	Containment severe challenge from a hydrogen burn when acting containment spray		L	L			VH	0.955–0.954	VH
	Fission product release		L	L			VH	0.955–0.954	VH
SAG-06: Control Containment Pressure	Containment severe challenge from a hydrogen burn		L	L	–		VH	0.955–0.964	VH
	Damage threat of safety related equipment due to frame acceleration in containment fan cooling duct		L	L			VH	0.955–0.964	VH
	Insufficient spray source for accident mitigation		L	L			VH	0.955–0.964	VH
	Containment Flooding	Throttle spray flow to minimize rate of containment water level increase	M[a] H[b]	M	H	L[a] M[b]	VH[a] M[b]	0.85[a] 0.5–0.55[b]	H[a] M[b]
		Use the spray pumps in recirculation mode (N/A under the given scenario)	M[a] H[b]	M	VL	L[a] M[b]	VH[a] M[b]	0.5–0.55[a] 0.15–0.3[b]	M[a] VL[b]
		Use the fan coolers (N/A under the given scenario)	M[a] H[b]	M	VL	L[a] M[b]	VH[a] M[b]	0.5–0.55[a] 0.15–0.3[b]	M[a] VL[b]
	Containment severe challenge from over pressurization (from CCI)		L	M	VH	L	VH	0.955–0.964	VH

4 Conclusion

The objective of this paper is to develop a reasonable FLM for quantitatively express SAMG decision-making. This result showed the possibility of quantification of SAMG decision action in the decision complexity. This result can be applied to perform detailed task analysis for SAMG. Moreover, it will support modeling of SAMG actions into Level 2 PSA to adequately evaluate the effect of SAM strategy on the risk of an NPP.

Acknowledgments. This research was supported by a Nuclear Research & Development Program of the National Research Foundation of Korea (NRF) grant, funded by the Ministry of Science and ICT (MSIT) (Grant Code: 2017M2A8A4015291).

References

1. Kim, J., Cho, J.: Technical challenges in modeling human and organizational actions under severe accident conditions for Level 2 PSA. Reliab. Eng. Syst. Saf. **194**, 106239 (2018)
2. Pyy, P.: An approach for assessing human decision reliability. Reliab. Eng. Syst. Saf. **68**(1), 17–28 (2000)
3. Konstandinidou, M., Nivolianitou, Z., Kiranoudis, C., Markatos, N.: A fuzzy modeling application of CREAM methodology for human reliability analysis. Reliab. Eng. Syst. Saf. **91**(6), 706–716 (2006)
4. Baziuk, P.A., Rivera, S.S., Nunez Mc Leod, J.: Fuzzy human reliability analysis: applications and contributions review. Adv. Fuzzy Syst. **2016**, 9 (2016)
5. Zio, E., Baraldi, P., Librizzi, M., Podofillini, L., Dang, V.N.: A fuzzy set-based approach for modeling dependence among human errors. Fuzzy Sets Syst. **160**(13), 1947–1964 (2009)
6. Ung, S.T., Shen, W.M.: A novel human error probability assessment using fuzzy modeling. Risk Anal.: Int. J. **31**(5), 745–757 (2011)
7. Li, P.C., Chen, G.H., Dai, L.C., Li, Z.: Fuzzy logic-based approach for identifying the risk importance of human error. Saf. Sci. **48**(7), 902–913 (2010)
8. Sheridan, T.B.: Telerobotics, Automation, and Human Supervisory Control. MIT Press, Cambridge (1992)
9. Richei, A., Hauptmanns, U., Unger, H.: The human error rate assessment and optimizing system HEROS—a new procedure for evaluating and optimizing the man-machine interface in PSA. Reliab. Eng. Syst. Saf. **72**(2), 153–164 (2001)
10. Mathworks. Fuzzy Logic Toolbox User's Guide, MATLAB. Mathworks, inc. (2019)

Human Response Characteristics According to the Location of Visual Stimuli

Yejin Lee[1,2(✉)], Kwangtae Jung[1], and Hyunchul Lee[2]

[1] Department of Industrial Design Engineering, KOREATECH, Cheonan 31253, South Korea
{Yejin3210,ktjung}@koreatech.ac.kr
[2] Nuclear ICT Research Division, KAERI, Daejeon 34057, South Korea
leehc@kaeri.re.kr

Abstract. The correct perception of visual information in the design of human-machine systems is one of the most important factors to consider for improving the usability and safety of systems. Humans rely on sight for more than 80% of the information that is processed in their work performance and daily lives. However, the recognition of visual stimuli is not always accurate. Human beings do not recognize the subtle changes in visual stimuli. Such characteristics should be considered in the visual information design of human-machine systems. The study was conducted on the assumption that the layout of visual stimuli could affect human cognitive performance. The effect of signal stimulus location on the visual stimulus array was investigated through the experiments. Three black circles were used as an experimental stimulus. The reference stimulus was a 30 mm diameter circle and the signal stimulus was a 30.8 mm diameter circle. The signal stimulus was the size that 90% of the subjects correctly determined in the JND measurement experiment. The experimental equipment was a tachistoscope and 20 college students participated as subjects. The subjects judged whether the three circles presented in the horizontal or vertical direction were the same size. They pressed the left button if they were the same and pressed the right button for the other. As a result, the location of the signal stimulus in the horizontal arrangement of visual stimuli did not significantly affect the error rate and response time of the stimulus judgment. However, the location of the signal stimulus in the vertical arrangement had a significant effect on the error rate and response time, and the error rate and response time were the smallest when the signal stimulus was located at the center. These results should be considered in the visual information design or interface design.

Keywords: Visual stimuli · Just noticeable difference · Tachistoscope · Recognition error · Response time

1 Introduction

Humans visually receive more than 80% of their information for work on human-machine systems. Therefore, it is important to be able to correctly transmit visual information to humans for proper work performance. However, the perception of visual stimuli is not always accurate. Humans do not recognize subtle changes in visual

P. M. Arezes and R. L. Boring (Eds.): AHFE 2020, AISC 1204, pp. 322–327, 2020.
https://doi.org/10.1007/978-3-030-50946-0_43

stimuli [1]. The smallest stimulus difference that humans can recognize is a just noticeable difference (JND), a concept from psychophysics [2].

Several factors affect the perception of visual information. Park (2013) found experimentally that inadvertent disturbances can significantly affect the identification of targets [3]. Kim (2008) identified that the periphery, interior angle, and orientation of the plane were the main determinants of size perception [4]. Kwon and Shin (2002) studied the effect of the shape and location of targets on the performance of the visual search task [5]. Ghuntla et al. (2014) found that the visual reaction time(VRT) became smaller after practice for both simple and choice VRT tasks [6]. They found that the reaction time decreased with practice. Smith (1969) studied the effect of shape on the perception of area [7].

In addition to the factors considered in the above studies, it is not difficult to predict that the human judgment ability can vary according to the location of the signal stimulus in the arrangement of visual stimuli. The location of visual stimulus is an important factor to be considered in visual information and interface design. Therefore, the purpose of this study was to investigate how the location of the signal stimulus in the arrangement of visual stimuli influences human response to the judgment of stimulus.

2 Method

This study was performed under the hypothesis that the performance of human stimulus judgments will vary according to the location of the signal stimulus in the visual stimuli. Three circles were used as experimental stimuli to validate this hypothesis. The three circles were arranged at the same intervals and included one circle of a different size. The experiment was conducted on a horizontal and vertical arrangement. The reason why the circle was used for the experiment was that an icon or a control device often had the shape of a circle. In order to determine the size of the signal stimulus, JND measurements of the size of the circle were performed.

2.1 Just Noticeable Difference of Circle Size

JND is the smallest difference value that can detect any changes between two stimuli by relative judgment. Typically, JNDs are determined by asking subjects to make forced choices between two simultaneously presented stimuli of different magnitude, as the interstimulus difference is varied over a series of trials. The 50% point of the resulting discrimination probability curve may then be taken as the JND, although other values can be used to suit the needs of the individual researcher [8]. In this experiment, JND was obtained for the size of a black circle with a diameter of 30 mm. 30 subjects participated in the experiment. As a result, the JND for the size of the black circle with a diameter of 30 mm was 0.45 mm. The Weber ratio was 0.016 from the result. This was about the same as previous studies that showed that the Weber ratio for vision was 0.017 [1]. Based on the experimental results, 0.8 mm, which was the difference value of the stimulus correctly judged by 90% of the subjects, was used for the experiment (Fig. 1).

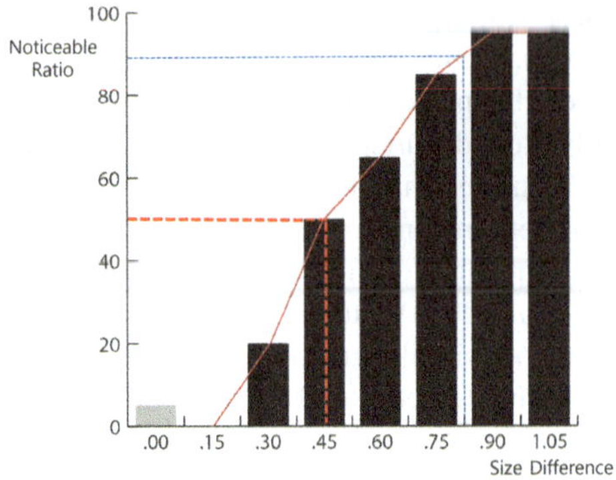

Fig. 1. JND for circle size

2.2 Experiment Method

The purpose of the experiment was to determine if the array of three circles included a circle of different size. A tachistoscope, which is a device that displays an image for a specific amount of time, was used in the experiment (Fig. 2). 20 subjects participated in the experiment: 10 males and 10 females. The subjects were all college students with normal vision. The mean age of the subjects was 23.1 years. The subjects judged whether the sizes of the three circles presented on the screen were the same. They then pressed the left button if they were the same size. Otherwise they pressed the right button. The experimental stimuli were presented in the center of the screen. Figure 2 shows the experimental scene.

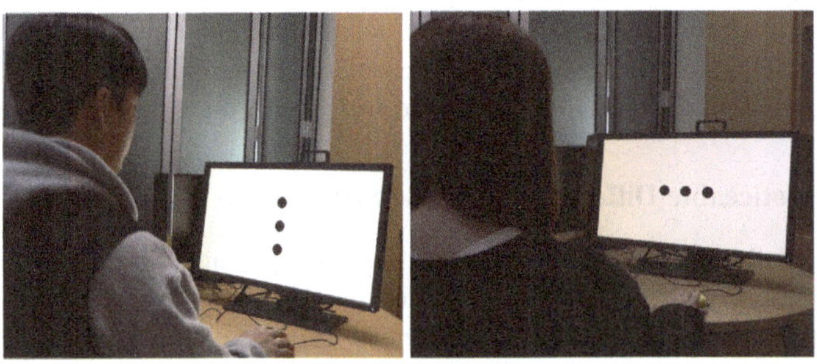

Fig. 2. Experimental scene

In the experimental stimulus, the size of the reference circle was a black circle with a diameter of 30 mm, and the distance between the circles was 30 mm. The experimental stimulus consisted of three circles. For the signal stimulus, a circle with a diameter of 30.8 mm was used. The size of the signal stimulus was based on the results of the JND experiment (Fig. 1). The presentation time of the visual stimulus was 2 s, and the time from the disappearance of one stimulus to the appearance of the next stimulus was 2 s. During the time, the landscape image was presented on the screen to remove the afterimage of the previous stimulus.

The number of presentations of experimental stimuli was 40 for both the horizontal and vertical arrangements. 40 experimental stimuli consisted of ten stimuli consisting of the same sized circle, ten stimuli containing a larger circle in the first location, ten stimuli containing a larger circle in the second location, and ten stimuli containing a larger circle in the third location. The order of presentation of each stimulus was randomized. The subjects learned the experimental method through the explanation of the experiment and a preliminary experiment.

3 Results

3.1 Response Characteristics in Horizontal Arrangement

Analysis of Recognition Errors. In this study, the recognition error is an error of not recognizing that there is a signal stimulus (i.e., a circle with a different size) in the visual stimulus consisting of three circles. We analyzed whether there was a significant difference in the recognition error according to the location of the signal stimulus in the horizontal array of visual stimuli. First, we analyzed whether there was a gender difference in the recognition error of the stimulus judgment. There was no gender difference in the recognition error ($p = 0.686 > 0.05$). Therefore, the effects of the location of the signal stimulus were analyzed without distinguishing the gender of the subject. As a result, the location of the signal stimulus in the horizontal arrangement did not significantly affect the recognition error ($p = 0.261 > 005$).

Analysis of Response Time. Response time is the time from the moment a visual stimuli are presented on the screen until the subject presses a button. We analyzed whether the response time of the subject had a significant effect on the location of the signal stimulus in the horizontal arrangement of the three circles. First, we analyzed whether there was gender difference. There was no gender difference in response time ($p = 0.265 > 0.05$). Therefore, the effects of the location of the signal stimulus were analyzed without discriminating gender. The results showed that the location of the signal stimulus did not significantly affect the response time for the horizontal arrangement ($p = 0.155 > 005$).

3.2 Response Characteristics in Vertical Arrangement

Analysis of Recognition Errors. We analyzed the difference in recognition error according to the location of the signal stimulus in the vertical arrangement of the three circles. First, we analyzed whether there was gender difference in the recognition error of stimulus judgment. There was no gender effect on the subject's recognition error ($p = 0.445 > 0.05$). Therefore, the effects of the location of the signal stimulus were analyzed without discriminating gender. As a result, the location of the signal stimulus in the vertical arrangement significantly influenced the recognition error (p = 0.030 < 005). In addition, the recognition error was the lowest when the signal stimulus was located in the middle (Fig. 3).

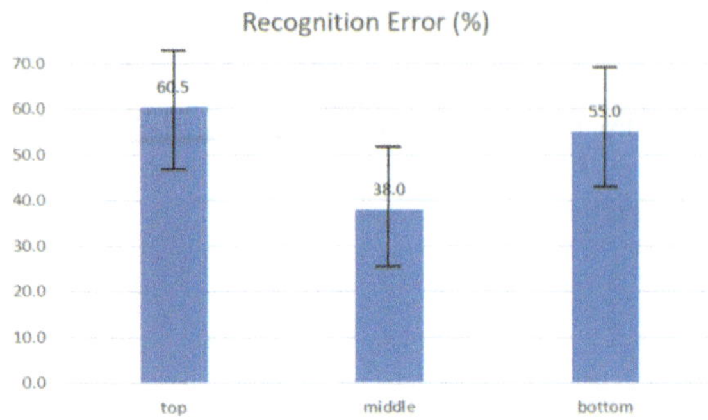

Fig. 3. Recognition error according to the location of the signal stimulus in vertical arrangement

Analysis of Response Time. We analyzed whether there was a significant difference in the response time according to the location of the signal stimulus in the vertical arrangement of three circles. First, we analyzed whether there was gender difference. There was no gender difference in the response time (p = 0.120 > 0.05). Therefore, the effects of the location of the signal stimulus were analyzed without discriminating gender. As a result, the location of the signal stimulus had a significant effect on the response time of the subject (p = 0.047 < 0.05). Looking at the average of the response times, the response time was the shortest when the signal stimulus was in the middle, which was similar to the recognition error (Fig. 4).

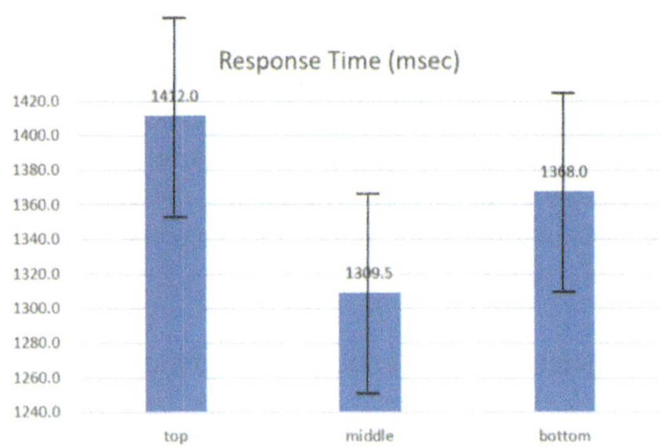

Fig. 4. Response time according to the location of the signal stimulus in vertical arrangement

4 Conclusion

When presenting visual information, the location of the signal stimulus can have a significant effect on the recognition of correct information. This study was carried out to investigate how the human response characteristics differ according to the location of the signal stimulus in the arrangement of visual stimuli. Experimental studies have shown that the location of the signal stimulus can have a significant effect on the human response characteristics depending on whether the stimulus array is horizontal or vertical.

For horizontal stimulus, the location of the signal stimulus did not have a significant effect on the recognition error and response time in judging the stimulus. However, for the visual stimulus in the vertical array, the location of the signal stimulus had a significant effect on the response time and recognition error of judging the stimulus. This means that when designing visual information or an interface, the location of the component does not need to be considered important in the horizontal arrangement, but the location of the component needs to be considered in the vertical arrangement. In addition, when the signal stimulus was located in the center of the vertical arrangement, the accuracy was the highest and the response time was also found to be fast. This means that important information or components in visual information or interface design should be placed in the center of the array.

Since this study was only conducted with a group of college students, it will be necessary to conduct future studies on a wider range of age groups in the future. The results of this study can be used in the layout of user interface and the design of visual information.

References

1. Sanders, M.S., McCormick, E.J.: Human Factors in Engineering and Design. McGraw-Hill Education, New York (1993)
2. Aron, Y., Grauman, K.: Just noticeable differences in visual attributes. In: Proceedings of the IEEE International Conference on Computer Vision (ICCV) (2015)
3. Park, C.: The influence of unattended distractors on the identification of targets. Korean J. Cogn. Sci. **24**(4), 365–391 (2013)
4. Kim, H.: The effect of configuration characteristics of geometrical figure on size perception. J. Integr. Des. Res. **7**(2), 23-3 (2008)
5. Kwon, O., Shin, H.: Effects of target types and locations on visual detection performance. Korean J. Exp. Cogn. Psychol. **14**(2), 127–143 (2002)
6. Ghuntla, T.P., Mehta, H.B., Gokhale, P.A., Shah, C.J.: Influence of practice on visual reaction time. J. Mahatma Gandhi Inst. Med. Sci. **19**(2), 119–122 (2014)
7. Smith, J.P.: The effects of figural shape on the perception of area. Percept. Psychophys. **5**(1), 49–52 (1969)
8. Papas, E.B., Keay, L., Golebiowski, B.: Estimating a just-noticeable difference for ocular comfort in contact lens wearers. Invest. Ophthalmol. Vis. Sci. **52**(7), 4390–4394 (2011)

New Findings and Methods in Human Performance

How Do Pilots and Controllers Manage Routine Contingencies During RNAV Arrivals?

Jon Holbrook[1], Lawrence J. Prinzel III[1(✉)], Michael J. Stewart[2], and Daniel Kiggins[3]

[1] NASA Langley Research Center, Hampton, VA, USA
{jon.holbrook, lawrence.j.prinzel}@nasa.gov
[2] NASA Ames Research Center, Moffett Field, Mountain View, CA, USA
michael.j.stewart@nasa.gov
[3] San Jose State University, San Jose, CA, USA
daniel.kiggins@nasa.gov

Abstract. Traditional approaches to risk and safety management have focused on collection and analysis of data describing unwanted outcomes, such as accidents and incidents. These approaches are ill-equipped to enable learning from events in which procedural non-adherences do not result in reported undesired outcomes, nor to explore whether these non-adherences may, in fact, reflect desired behaviors. This study investigated how operators responded to expected and unexpected disturbances during RNAV arrivals into Charlotte Douglas International Airport. Pilots from mainline and regional airlines and terminal radar approach control air traffic controllers were interviewed with regard to how they anticipate, monitor for, respond to, and learn from these routine disturbances. In addition, event reports submitted to NASA's Aviation Safety Reporting System were also examined. Results of these analyses were used to identify behaviors and strategies that support the everyday resilient performance.

Keywords: Resilience · Human error · Safety · RNAV · STARs · OPD

1 Introduction

Area navigation standard terminal arrival route (RNAV STAR) procedures used at major airports are intended to increase predictability and efficiency. These procedures provide vertical, lateral, and speed profiles for aircraft to follow as they descend toward an airport. Analyzing aircraft flight track data for more than 10 million flights into 32 domestic airports revealed that only 12.4% of flights fully complied with the vertical and lateral profiles in the RNAV STARs [1]. This study [1] provides an example in which published procedures were frequently misaligned with normal operations. Questions remain, however, with regard to the reasons for the misalignment, and thus, how to interpret this finding. Traditional approaches to risk and safety management have focused on what can go wrong or especially, what did go wrong through a forensic analysis of unwanted outcomes, such as accidents and incidents. In this view,

P. M. Arezes and R. L. Boring (Eds.): AHFE 2020, AISC 1204, pp. 331–338, 2020.
https://doi.org/10.1007/978-3-030-50946-0_44

safety is both defined and measured by its absence, namely the lack of safety. While these approaches have proven utility, they are ill-equipped to enable learning from events such as those described in [1], in which procedural non-adherences do not result in reported undesired outcomes, nor to explore whether these non-adherences may, in fact, reflect desired behaviors.

An alternative and complementary approach to risk and safety management is to focus on what goes right, and identify how successes can also contribute to safety management and assurance. Focusing on the rare cases of failures attributed to human error provides little information about how or why human performance routinely results in safe, successful outcomes. Therefore, a complete definition of safety should reflect not only "avoiding things that go wrong", but also "ensuring that things go right". Hollnagel [2] posits that things go right because people continuously adjust their work to match their operating conditions, especially as systems grow in complexity. Monitoring everyday work can help to efficiently and proactively identify new strategies that work and conditions under which existing strategies break down. Few mechanisms to monitor everyday work currently exist in the aviation domain, which limits opportunities to learn how designs function in reality.

Rather than focus on rare events in which things went wrong, the current study focused on frequent events in which operators adjusted their work to ensure things went right. Namely, this study investigated how operators responded to expected and unexpected disturbances during RNAV arrivals. Aircraft flight track data for arrivals into Charlotte Douglas International Airport (KCLT) were isolated from the dataset collected by [1]. There are seven (7) RNAV STARs with vertical profiles into KCLT. Data collection ranged from May 26, 2016 to December 7, 2017 (560 days). The dataset included 349,555 arriving flights equipped to fly RNAV arrivals, of which, 323,194 flew some portion of a STAR. In addition, 52,295 flights (16%) complied fully vertically and laterally to the STAR. The total number of non-adherence events was 231,766, of which 49% were misses below published altitudes, 9% were misses above published altitudes, and 42% involved lateral non-adherences.

For the current study, pilots from mainline and regional airlines and terminal radar approach control (TRACON) air traffic controllers were interviewed with regard to how they anticipate, monitor for, respond to, and learn from routine disturbances during RNAV arrivals into Charlotte Douglas International Airport (KCLT). In addition, event reports submitted to NASA's Aviation Safety Reporting System (ASRS) that referenced one or more of the KCLT RNAV STARs were also examined. The ASRS database is the world's largest repository of voluntary, confidential safety information provided by frontline aviation personnel. In addition to the narratives submitted by reporters, the ASRS database contains coded assessments by expert analysts, including contributing factors/situations and human factors associated with each report.

Results of these analyses were used to identify behaviors and strategies that support the everyday resilient performance of human operators in the National Airspace System (NAS). Better understanding of operators' resilient behaviors and strategies can inform development of procedures, policies, training, and technologies based on frequently occurring successes, not just rarely occurring failures.

2 Method

2.1 Operator Interviews

Twenty-four (24) air transport pilots (ATP) flying for an airline conducting routine operations into KCLT, and 6 TRACON controllers at KCLT were recruited to participate. All pilot participants were employed by a major airline (12 pilots, averaging 18,000 h) or regional airline (12 pilots, averaging 3600 h) operating under Federal Aviation Regulations part 121. The average estimated number of KCLT RNAV arrivals flown by each pilot over their career was 542 for the major airline pilots and 402 for regional pilots. All TRACON controller participants were highly experienced, with an average of 16 years of operation experience and with an average of over 4600 KCLT RNAV arrivals worked by each in their careers. All interviews were conducted under approval from NASA's Institutional Review Board.

Participants were interviewed individually, using a semi-structured protocol designed to elicit specific instances of unplanned or unexpected events experienced during RNAV arrivals into KCLT, as well as their goals, motivations, pressures, and knowledge at the time of the described actions [3, 4]. The pilot interviews resulted in 15,353 transcribed words specific to this research focus collected from the approximate 8 h of interview data (average = 22 min per interview). The controller interviews generated 12,100 transcribed words over approximately four hours of data collection (average = 38 min per interview). All participants also completed a written questionnaire, in which they estimated the frequency of behaviors associated with resilient performance, and other tailored questionnaire items specific to operator role.

2.2 ASRS Event Reports

NASA's Aviation Safety Reporting System (ASRS) database was searched for event reports within the date range of November 2015 to June 2019 that referenced one or more of the KCLT RNAV STARs. The database search returned 32 ASRS event reports, of which 29 involved air carrier operations on one of the STARs. Those 29 event reports included 39 narratives, with 32 filed by flight crew and 7 by air traffic controllers. Those 39 narratives were examined to identify statements describing resilient performance using the Resilience Analysis Grid (RAG) framework, which identifies four basic capabilities of resilience performance: anticipating, monitoring for, responding to, or learning from disruptions [5]. For reports that described a non-adherence event, information was collected on whether that non-adherence was intentional as well as the "type" of non-adherence (i.e., vertical, lateral, and speed).

3 Results

3.1 Interviews

The RAG framework was used to extrapolate strategies for resilient performance from the events and behaviors that participants described in the interviews. These strategies are shown in Table 1. This analysis extended the taxonomy of resilient performance strategies initially identified in [3].

Table 1. Identified resilient performance strategies employed in routine aviation contexts.

	Strategy	Description
Anticipate	Anticipate procedure limits	Predict when current context inhibits normal use of a procedure, regulation, policy, norm
	Anticipate knowledge gaps	Predict whether crew member or other actor lacks required knowledge or information
	Anticipate resource gaps	Compare level of available resources (e.g., time, fuel, workload) to perceived resource needs
	Prepare alternate plan and identify triggering conditions	Have an actionable plan ready within the time available
	Conduct pre-action briefing	Discuss planned action and identify variables that might affect that plan
Monitor	Monitor environment for cues signaling change from normal operations	Identify triggering variables that signal something has changed from what was expected
	Monitor environment for cues signaling need to adjust or deviate from current plan	Identify triggering variables that signal something will not continue to work as planned
	Monitor own internal state	Self-assess physiological state, emotional state, workload, knowledge
Respond	Adjust current plan to accommodate others	Help others in system by changing timing or other action
	Adjust or deviate from current plan based on risk assessment	Change plan based on monitoring of triggers associated with safety boundaries
	Negotiate adjustment or deviation from current plan	Work with others to accommodate competing goals and come to mutually acceptable solution
	Defer adjusting/deviating from plan to collect information	Continue with current plan because acting without critical information may worsen situation
	Manage available resources	Preserve finite resources by adjusting controllable aspects of the situation
	Recruit additional resources	Obtain resources locally or externally
	Manage priorities	Change goals, task order, task content, or pace of operation to accommodate resource limitations
Learn	Leverage experience and learning to modify or deviate from plan	Compare formal expectations and experience to current situation to develop real-time assessment of acceptability or risk
	Understand formal expectations	Understand applicability of laws, procedures, policies, and cultural norms

(*continued*)

Table 1. *(continued)*

	Strategy	Description
	Facilitate others' learning	Share information with others to increase their immediate understanding and long-term learning
	Conduct after-action debriefing	Discuss performance after mission has concluded to foster understanding and identify opportunities for improving future performance

3.2 ASRS

Pilots and controllers reported 37 non-adherences to published RNAV STAR procedures across the 29 event reports. Types of non-adherences reported included 13 lateral non-adherences, 17 altitude non-adherences (9 above and 8 below published limits), and 7 speed non-adherences (5 above and 2 below published limits). Of these non-adherences, 21 were unintentional (e.g., "I recognized at 33,200 feet that the autopilot was not going to capture FL [flight level] 330. Descent rate was approximately 2,000 fpm [feet per minute] and not decreasing." ASRS Report #1309485) and 16 were intentional (e.g., "While descending into [K]CLT on the CHSLY 3 RNAV arrival we asked for 10 degrees right for TCU [towering cumulus] near WHIZE." ASRS Report #1545309).

Analysis of the 39 narratives revealed 99 statements describing resilient behaviors or strategies by the flight crew (62 statements), air traffic control (34 statements), airline operations (2 statements), and air traffic control operations (1 statement). At least one statement was identified for each of the 19 resilient performance strategies listed in Table 1.

Coding performed by ASRS analysts identified "human factors" as associated with 26 of the 29 events, and as a contributing factor in 14 of the events. Multiple codes for different human factors could be applied to a single event; across all event reports, there were 63 instances in which the following human factors codes were applied: Situational awareness, confusion, human-machine interface, communication breakdown, distraction, workload, time pressure, troubleshooting, and training/qualification.

3.3 Questionnaire

Pilots estimated that the RNAV arrivals that they have flown into KCLT went as expected or planned 45% (mainline) and 55% (regional) of the time (range = 20%–90%). Furthermore, pilots estimated that they fully adhered (whether planned or unplanned) to published KCLT RNAV STARs 48% (mainline) and 37% (regional) of the time. TRACON controllers, by contrast, estimated that 30% of KCLT RNAV STAR operations fully adhered to the published procedure as designed.

Resilience was defined for study participants as "the ability to adjust functioning prior to, during, or following changes and disturbances, so that required operations can be sustained under both expected and unexpected conditions." On average, pilots estimated that 53% (mainline) and 46% (regional) of RNAV arrival operations into KCLT required resiliency, with more than 50% of pilots reporting that they exhibited resilient performance daily. The controllers estimated that approximately 80% of these RNAV STAR operations required resiliency, with 100% of controllers reporting that they exhibited resilient performance at least one per day, and 66% of controllers estimating "more than once per shift".

Table 2 presents the percentage of time that participants estimated the following factors contributed to non-adherence to RNAV arrivals into KCLT.

Table 2. Frequency estimates (% of operations) by pilots and TRACON controllers of factors associated with non-adherences to RNAV arrivals into KCLT.

Factors *describing* RNAV STAR non-adherence	Mainline pilots	Regional pilots	TRACON
Involved lateral deviation from published arrival	15%	47%	85%
Involved vertical deviation from published arrival	20%	39%	24%
Involved speed deviation from published arrival	43%	58%	60%
Non-adherence was initiated by ATC	47%	67%	70%
Non-adherence was initiated by flight crew	15%	21%	30%
Non-adherence was unplanned or unexpected	18%	32%	15%
Factors *contributing to* RNAV STAR non-adherence	Mainline pilots	Regional pilots	TRACON
Weather	23%	24%	65%
Unexpected aircraft performance	17%	08%	25%
ATC-Flight deck communication issues	13%	10%	10%
Flight deck automation issues (e.g., FMS)	23%	09%	40%
ATC actions (e.g., traffic, runway change)	23%	46%	10%
Flight crew requests	10%	14%	15%
RNAV procedure design	28%	07%	35%
Flight crew errors	12%	15%	25%

Note: Factors can occur in combination, therefore estimates can sum to greater than 100%.

4 Discussion

Pilot and controller interviews, questionnaires, and operational event reports were examined to elicit specific examples of resilient performance during RNAV arrivals to KCLT. The research objective was to explore how pilots and controllers manage contingencies during routine, everyday operations. Using Hollnagel's RAG framework, 19 resilient performance strategies were identified from operator interviews and event reports. While operational data have been used to create detailed taxonomies of human error (e.g., [6]), rarely have these data been used to catalog human success, in particular, behaviors that represent safety-producing performance.

Building upon previous research [1], this study highlights the high frequency of non-adherences to RNAV STAR procedures and the factors that contribute to them. It is noteworthy that different data sources resulted in different estimates of the frequency of RNAV STAR non-adherences at KCLT, ranging from 30% (TRACON controller estimate) to 43% (pilot estimate) to 84% (estimate based on flight track data specific to KCLT arrivals collected for [1]). It is possible that TRACON controllers' estimate were lower because these controllers work only the last segment of the RNAV arrival procedure, and thus may not have experience with additional non-adherences that occur earlier in the arrival. For both pilots and controllers, it is possible that their estimates of non-adherences did not fully consider routine intentional non-adherences (e.g., approved course deviations to circumvent convective weather). Over 40% of the non-adherences reported in the analyzed sample of ASRS events described intentional non-adherences to published course, altitude, and speed, which may not be adequately represented in pilot or controller estimates of non-adherence frequency. While objective data can provide more accurate and confident estimates of event frequency, subjective data are needed to explain why those events occur. It is noteworthy that many procedural non-adherences are intentional, and are, in fact, examples of resilient performance.

An average of 3.4 statements describing resilient safety-producing performance were identified in each ASRS narrative (99 resilient performance statements across 29 event reports). This is a notable finding, given that ASRS reports are primarily filed to describe things going wrong. While human failures are systematically coded (e.g., loss of situation awareness, communication breakdowns, etc.), human successes are not. The prevalence of safety-producing behaviors in the ASRS database highlights an opportunity to systematically explore this understudied aspect of human performance.

While the study of human error is clearly important, it presents an incomplete picture of human performance which includes both failures and successes. The resilient performance strategies and methods for identifying them described in this report can serve as a starting place for broader systematic exploration of how humans contribute to safety. Additional study and analysis are needed to create methods to improve procedures, policies, training, and technologies based upon thorough study of human performance, including both successes and failures.

Acknowledgements. The authors would like to thank Mr. John Koelling, Dr. Misty Davies, and Dr. Kyle Ellis (NASA's System-Wide Safety Project), and Ms. Vanessa Aubuchon and Mr. Chris Teubert (NASA's Transformational Tools and Technologies Autonomous Systems Subproject) for providing funding to support this work. The authors would also like to thank Bryan Matthews for providing KCLT flight track data from [1].

References

1. Stewart, M., Matthews, B., Avrekh, I., Janakiraman, V.: Variables influencing RNAV STAR adherence. IEEE/AIAA. In: Proceedings of the 37th Digital Avionics Systems Conference (DASC). London, UK (2018)
2. Hollnagel, E.: Safety-I and Safety-II: The Past and Future of Safety Management. Ashgate, Farnham (2014)

3. Holbrook, J.B., Stewart, M.J., Smith, B.E., Prinzel, L.J., Matthews, B.L., Avrekh, I., Cardoza, C.T., Ammann, O.C., Adduru, V., Null, C.H.: Human Performance Contributions to Safety in Commercial Aviation. NASA-TM-2019-220417 (2019). https://hsi.arc.nasagov/publications/NASA-TM-2019-220417-comp.pdf
4. Holbrook, J.B., Prinzel, L.J., Stewart, M.J., Smith, B.E., Matthews, B.L.: Resilience and safety for in-time monitoring prediction, and mitigation of emergent risks in commercial aviation. In: International Symposium on Aviation Psychology. Dayton, OH: Wright State University, 7–10 May 2019 (2019)
5. Hollnagel, E.: RAG – the resilience analysis grid. In: Hollnagel, E., Pariès, J., Woods, D.D., Wreathall, J. (eds.) Resilience Engineering in Practice. A Guidebook. Ashgate, Farnham (2011)
6. Wiegmann, D.A., Shappell, S.A.: A Human Error Approach to Aviation Accident Analysis: The Human Factors Analysis and Classification System. Ashgate, London (2003)

Validating a Human Performance Model Without a Complete System

Richard Steinberg[1(✉)], Alice Diggs[2], and Jade Driggs[3]

[1] Raytheon IDS, Huntsville, AL, USA
richard.k.steinberg@raytheon.com
[2] Raytheon IIS, Sterling, VA, USA
alice.c.diggs@raytheon.com
[3] United States Air Force, Colorado Springs, CO, USA
jade.driggs@us.af.mil

Abstract. Human performance modeling (HPM) has been effective for determining crew designs. Crew design includes determining the number of operators needed, the role of automation, and member task responsibilities required to operate a system. Crew models are built for target systems under development to better inform the design and the needs of each operator's user interface. This is a Verification and Validation (V&V) challenge—it is difficult to obtain metrics to validate a model when the completed system does not yet exist for obtaining sample measures. This paper addresses the challenge with an implementation of the V&V approach applying it to an HPM that successfully received V&V Accreditation for a large United States Air Force (USAF) command and control system. This paper then compares the results of the validated model with NASA TLX workload metrics and discusses the V&V effectiveness. Finally, it recommends better methods for V&V for human performance models.

Keywords: Human Performance Modeling · Workload · Human systems integration

1 Introduction

Human Performance Modeling (HPM) can be used to help design systems and allow trades to be performed between different crew sizes, task allocation and distribution, and levels of automation. However, a model is only effective if it reflects real world behaviors closely enough to inform decision making analysis. This problem is exacerbated by the complexity of estimating workload by means of computer models. While the notion that human operators have capacity limitations and exceeding them degrades performance is widely accepted [1, 2], the thresholds for defining overload are less certain and vary across many factors [3]. Additionally, human workload models are complex since the goal of the crew design is not merely a matter of minimizing workload, but striving to balance each operator's workload so that it is neither too high nor too low. Yerkes and Dodson [4] found that insufficient or low workload also has a negative effect on operator performance. Furthermore, cognitive tasks with low workload (e.g., monitoring tasks), have deleterious impacts on operator situation

P. M. Arezes and R. L. Boring (Eds.): AHFE 2020, AISC 1204, pp. 339–345, 2020.
https://doi.org/10.1007/978-3-030-50946-0_45

awareness [5]. In addition to this delicate balance of operator workload, we also must overlay the complexities of implementing automation and decision support tools, when they are needed. The challenge with high workload is not merely a matter of automating as many tasks as possible, but rather, defining the specific tasks, or task steps, that will improve human-system team performance, at an acceptable workload level. Evidence shows that automating tasks can increase workload [6] due to the increased need for monitoring the health and status of the processes, a task for which operators are not well suited [7].

Even with the stated challenges of building and validating a HPM, it can be extremely cost effective since building a complete system or prototype for evaluation can be prohibitive. Computer models can evaluate a wide range of scenarios and combinations of events far more cost effectively and exhaustively than with direct observation of events. Using Monte Carlo simulations affords the opportunity to perform a myriad of combinations of events enabling a comprehensive analysis of workload. The apparent advantages of a computer model to evaluate workload is accepted, but the question remains: Does the model have high enough accuracy to provide supportive evidence for design decisions? Without confidence in the results of the model, decision makers will likely dismiss the outputs and the model will fail to impact the design of the system. However, effective means for validating and verifying HPMs exists. This paper describes one such method and demonstrates a case study for how implementation.

2 Verification and Validation (V&V)

Validation and Verification are analyses that fulfill somewhat reciprocal roles in establishing the credibility of a model for its intended uses. Verification determines if we are effectively representing the system in the implementation of the model. Validation determines if we are doing the right things by using the model. Validation further bifurcates into two analyses: Structural Validation and Output Validation. Output validation is also referred to as results validation. Validation is establishing that the model matches the real world. It involves ensuring the model reflects the operator's tactics, techniques, and procedures. It answers the question, "Does the model output match real-world behaviors with accurate predictions of timing of events and external triggers?" Verification is confirmation that the data is not corrupted during the tool's operation. Verification defines whether the model logic and actions match the concept model. According to Air Force Instruction (AFI) 16-1001 [8], verification is the process of determining that a model implementation and its associated data accurately represent the developer's conceptual description and specifications. It means answering the question, did we build the model correctly? V&V data is typically much easier to obtain for computer models of real world systems than for human performance models for proposed systems. For example, a V&V practitioner obtains output data from real world measurements and then performs a statistical T-test to ensure the mean observed data is not statistically different from that resulting from the model output for a given confidence interval. Human performance models are much more difficult to V&V due to several factors including the following:

1. Variability among human actors on the system.
2. Gathering empirical data to validate the model depends upon having the right user and large enough sample sizes in the right environment. For example, data on astronauts will be different from users in a low-stress office environment. Modeling actions in military environments will be different from those in low-stress civilian environments.
3. Data observation depends on having a representative system of the real world to obtain sufficient data fidelity.

These challenges must be addressed within the context of recognized process for V&V. One well accepted process for V&V is described in the AFI 16-1001 [8] and defines three pillars of V&V: structural validation, verification, and output validation (results validation). Structural validation is made up of two components, Conceptual Model Validation and Design Verification. The three steps are summarized in Table 1, Components of V&V. Each of the sections in the table will be discussed and how the challenges unique to HPMs are addressed.

Table 1. Components of V&V.

AF 16-1001	V&V function	Question to be answered
Structural validation	A. Conceptual Model Validation	*Does the model reflect operational tactics, techniques & procedures?*
	B. Design Verification	*Does the design reflect conceptual model behavior?*
Verification	C. Implementation Model Verification	*Does the implementation of the model execute the Design?*
Results validation	D. Results Validation	*Do results model real world behaviors?*

At first glance, it appears that performing verification and validation using the steps above would be rather difficult. For example, getting enough data for results validation without a complete system for obtaining it is challenging before considering the confidence level needed to justify a design decision. However, one of the characteristics of this process is the cumulative effect of the three components of structural, verification, and results validation. The idea is that the cumulative case analysis – with the probability of passing a deficient tool through any one step in the V&V being 1/20–yields four nines confidence in the M&S for a tool which has passed all three analyses —structural validation, verification and output validation [9].

Conceptual Model Validation. The model was constructed, and a list of assumptions and constraints developed. These included a baseline crew size, operator crew positions, operator goals, and the tasks performed to complete the goals. Nominal and off-nominal workflows were constructed, representing the functional series of tasks needed to execute the crew mission. Peer reviews were conducted with domain experts to confirm the right assumptions and operational concept. Peer reviews involved five experts with at least five years' experience on the system. The challenge in building a conceptual model for a system under development is that the design concept is evolving as it is developed. One of the lessons learned is to ensure flexibility in adding and removing operator tasks by using an object oriented design in the development process. This enabled the model to be more easily adapted as the system design matured.

Design Verification. This was the most time consuming step in our model validation, but also the most important foundation enabling a successful V&V process. Thirty-two workflows were constructed in Step A to define the set of functions and procedures that would involve operators in an operational mission. The workflows were constructed of lower level tasks that were included in the model. The low level tasks became common blocks or objects in the computer model. These representative tasks include items such as: a) Simple point and click, b) Simple strategy decision, c) Select an option from a menu, d) Complex strategy decision, e) Respond to an alert.

Because the workflows are constructed from low level tasks, the model can be validated based common tasks performed on similar C2 systems. Decomposing the workflows into small, common tasks, reduces the need to validate larger workflows, while still showing that the small tasks are valid and that the collection of tasks into the larger workflows provides a good representation of the operator procedures. Furthermore, the low task level objects were the types of tasks that more easily afforded data collection for performance measures. The model developers can obtain sufficient data on each type of sub-task that represents the larger workflows and thereby gather estimates on the full set of mission workflows.

For example, during an early release of our product, the team was able to gather timing data on three operational tasks using operators in a laboratory. Using the HPM for this Case Study, our team was able to match the timing for the larger sequences of low level tasks as shown in Table 2. This provides evidence for that for this Case study, the approach for building from low level tasks can successfully construct a model representative of the real world system. While initially this does not seem significant, it enables V&V to be performed on the low level tasks rather than the larger sequences of tasks which depends on a completed system for accurate data.

Table 2. Compare time in lab with model

	Time to perform task in lab (sec)	Time to perform task with model (sec)	Difference (sec)
Add scheduled task	210.8	214.15	<4 s
Modify scheduled maintenance task	43.7	44.41	<1 s
Manual override	38.65	39.83	<1.5 s

Given this approach, determining whether the model has the right workflows and sub-tasks becomes crucial. The workflows are Peer Reviewed by an operational Subject Matter Expert to assess each step to ensure the workflow contained all the required sub-tasks, and reviewed the task timing estimates for accuracy. SMEs ensured that the steps with the operational flows matched real world tactics, techniques, and procedures, and provided external task and timing estimates. It became important that the sub-tasks were correct prior to Results Validation in step D.

Implementation Model Verification. This step involved selecting reasonableness tests based upon expected behaviors. It means determining relationships indicative of a HPM. For example, a heavy load scenario should increase workload when compared to a low load scenario. Adding additional operators to distribute tasks should diminish workload. For the this step in the V&V process, our team came to the conclusion that it is better to emphasize relationships and data trends rather than statistics. If the test follows the trends then it passes. The tests then become pass/fail enabling statistics to be analyzed using a binomial experiment. This method provides the benefits of being testable, measurable, and repeatable. It also provides the ability to use a binomial distribution for evaluating the data.

Results Validation. This is the most challenging, yet the most important of all step in the V&V process. It involves a comparison between expected behaviors and real world results. For this step, our team constructed a simulator in visual studio of the target system. Since it would be cost prohibitive to emulate every operator tasks performed in the model. Scenarios were specifically selected that contained each type of sub-task employed in the model. This enabled persons doing the model V&V effort to obtain comparison of timing data for all types of sub-tasks in the Model.

We constructed a simulation in a rapid prototyping tool, Visual Studio. The team built a part task simulation selecting scenarios that covered all the types of low level sub-tasks an operator would perform, but also reflected a broad spectrum of the common workflows that would be most utilized. One of the risks for this approach was to verify that the simulator reflected the system design for the target system.

Six operators with current experience on the existing system performed thirteen of the thirty two scenarios. 2,076 mouse clicks were logged and incorporated into the data table.

Sample Case Study. For our case study, we have a crew of three operators constructed using the Extend Sim modeling library. The workload occupancy was measured after each simulation run. Five hundred Monte Carlo runs simulated the operator's eight hour shift were modeled and produced the following average results.

2.1 Workload Model Outputs

Utilization of Time, or occupancy, is defined as the percentage of time during a scenario that an operator is performing tasking. Utilization of time is calculated as the time an operator is doing active tasking, more than passive monitoring, divided by the total time of their shift. To ensure that the reproduction of your illustrations is of a reasonable quality, we advise against the use of shading. The contrast should be as pronounced as possible. If screenshots are necessary, please make sure that you are happy with the print quality before you send the files (Table 3).

Table 3. Workload occupancy per crew member and NASA TLX values

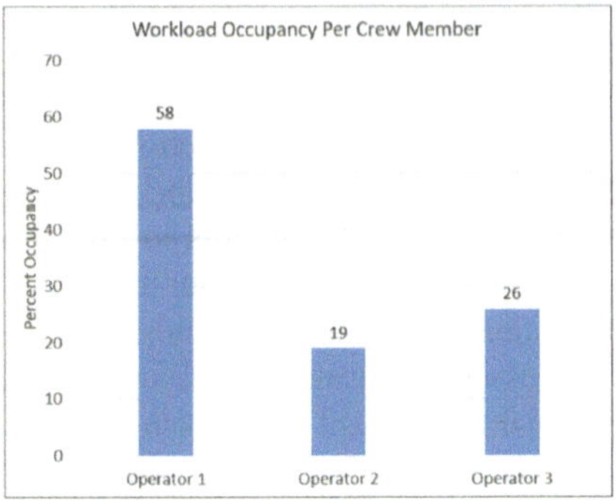

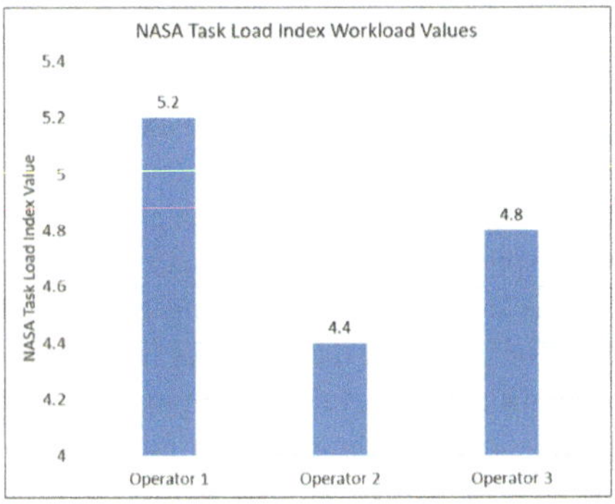

A usability test performed on this system, which included NASA TLX survey data found a similar relationship between workload among crew members. The relationships among the modeled workload occupancy data compared with the NASA TLX data have the same relative trends. Discussions with other human performance modeling experts at the Human Factors and Ergonomics Society Human Performance Modeling Interest group suggests that comparing this type of value to value data is akin to comparing apples and oranges. However, the relative data trend gave evaluators improved confidence in the model results.

Lessons Learned. The experience in V&V on the HPM revealed several recommendations to the process. First of all, design the model by testable objects. Validation is easier to obtain at the lowest or object level. Model designers should build a model based upon sub-tasks required to compete each mission workflow, and sub-tasks that can be simulated for obtaining validation data. Secondly, pre-define and perform reasonableness tests. Prior to building the model, define the reasonableness tests and the anticipated results. These can be used during the development process of the model. When the target system is not fully designed, building a model by objects enables modelers to revise the model as the system design matures. Finally, multiple V&V tests by evaluating patterns of reasonableness should be used rather than solely relying on statistics. This enabled the test results to be combined statistically improving the confidence of the end-to-end V&V process results with relatively lower confidence test data for the individual steps.

3 Conclusion

The challenges for V&V for a HPM are unique and require different procedures than performing V&V on physics based computer models. Performing V&V of a HPM is critical since the model must be of sufficient fidelity to ensure the data products will provide meaningful information that useful for design decisions. Our demonstrated method built the model using object low-level tasks, which allowed for data collection in support of model validation. Additionally, implementation validation of the model focused on data trends to perform reasonableness tests rather than focusing on solely on confidence intervals that may not be obtainable. In conclusion, V&V for a specified HPM was successfully performed for an Air Force Space Program.

References

1. Wickens, C., Goh, J., Helleberg, J., Horrey, W., Talleur, D.: Attentional models of multitask pilot performance using advanced display technology. Hum. Factors **45**, 360–380 (2003)
2. Wickens, C., Hollands, J.: Engineering Psychology and Human Performance, 3rd edn. Pearson Publishing, New York (1999)
3. Armstrong, S., Brewer, W., Steinberg, R.: Usability testing. In: O' Brien and Charlton (eds.) Handbook of Human Factors Engineering Methods (2001)
4. Yerkes, R., Dodson, J.: The relation of strength of stimulus to rapidity of habit formation. J. Comparative Neurol. Psychol. **18**, 459–482 (1908)
5. Matthews, G., Reineman-Jones, L.: Workload Assessment: How to Diagnose Workload Issues and Enhance Performance, Santa Monica (2017)
6. Parasuraman, R., Mouloua, M.: Automation and Human Performance. Erlbaum, Mahwah (1994)
7. Endsley, M.R., Kiris, E.O.: The out-of the loop performance problem and level of control in automation. Hum. Factors **37**(2), 381–394 (1995)
8. United States Air Force Instruction 16-1001. Verification, Validation and Accreditation (VV&A), June 2016
9. Plotkin, N., Imhoff, S., Gililland, W.: VV&A of Models and Simulations: The Power of Independent Cumulative Analysis. Raytheon Company IIS Publications (2017)

Dynamic Modeling of Field Operators in Human Reliability Analysis: An EMRALD and GOMS-HRA Dynamic Model of FLEX Operator Actions

Thomas A. Ulrich(✉), Torrey Mortenson, Ronald L. Boring, and Steven Prescott

Idaho National Laboratory, Idaho Falls, ID, USA
{thomas.ulrich, torrey.mortenson, ronald.boring, steven.prescott}@inl.gov

Abstract. In response to the Fukushima Dai-ichi accident, the Nuclear Energy Institute developed guidance for establishing flexible coping strategies for beyond-design-basis external events (BDBEE). These coping strategies, termed FLEX, consist of additional equipment and procedures to safeguard the plant in the event of a BDBEE. Little research has been performed to model the FLEX activities with dynamic human reliability analysis (HRA). The research described here sought to build an HRA model of FLEX activities for a station blackout event, add dynamics to the model using the GOMS-HRA method, and evaluate the feasibility and ease of building the dynamic model within an existing dynamic PRA tool called Event Modeling Risk Assessment using Linked Diagrams (EMRALD). The modelling effort was successful in building a dynamic HRA model of FLEX activities within the existing dynamic PRA tool and highlighted several potential improvements to enable EMRALD to better incorporate HRA more effectively within its framework.

Keywords: Human reliability analysis · GOMS-HRA · FLEX · Nuclear process control · EMRALD

1 Introduction

In response to the Fukushima Dai-ichi accident, the Nuclear Energy Institute developed guidance for establishing flexible coping strategies for beyond-design-basis external events (BDBEE) [1]. These coping strategies, termed FLEX, are intended to allow for the quick addition of supplementary equipment that will increase the overall system defense-in-depth and ensure that the reactor operations are not endangered during such an event. The relevancy of the acronym FLEX is linked to the fact that these strategies are deemed to be flexible in implementation and procedure, which is a change from the highly prescriptive nature of most activities in nuclear energy. The specific nature of these particular strategies is that they exist outside of the planned resilient design of nuclear power plants, particularly design basis accidents. For a nuclear utility to license a power plant they must prove to the regulator that the plant can survive a number of

P. M. Arezes and R. L. Boring (Eds.): AHFE 2020, AISC 1204, pp. 346–352, 2020.
https://doi.org/10.1007/978-3-030-50946-0_46

design basis events that, while unlikely, could result in a release of radioactive material. FLEX processes operate outside of these known design basis event scenarios and are therefore intended to be more malleable to accommodate any situation and could be implemented across an unlimited number of potential events. Rather than attempting to expand the probabilistic risk assessment (PRA) models to include significantly more design basis accidents, the industry instead adopted a set of implementations that would enable their operators to deploy emergency equipment as needed to protect the plant.

Current considerations of human reliability analysis (HRA) consider existing methods for defining operator responses to BDBEE where FLEX equipment would be deployed [2–4]. These approaches acknowledge that BDBEE scenarios are not adequately modeled in existing PRA models. They largely use existing HRA methods, but these methods were designed primarily for design basis events within the main control room. Recent guidance by Electric Power Research Institute (EPRI) has provided example analyses using existing HRA methods for the most common deployment scenarios of portable equipment [5]. Many questions remain about the suitability of legacy methods to address FLEX events. To circumvent shortcomings associated with generalizing HRA methods beyond their intended scope, the U.S. Nuclear Regulatory Commission (U.S. NRC) has used expert elicitation methods to arrive at human error probabilities (HEPs) [6].

These approaches to anchor HRA for FLEX with existing HRA methods are challenged by the wide variety of circumstances that can underlie a BDBEE scenario. In this paper, we propose a dynamic simulation approach to help account for the contextual uncertainties that complicate the determination of HEPs for FLEX deployment. This dynamic simulation is presented as a proof of concept, and further research is necessary prior to its use in FLEX HRA. However, the method explored here shows great promise for modeling error uncertainty in BDBEEs.

2 Method

2.1 FLEX Scenario

In a FLEX situation there are modified roles for control room and field operators as the primary goals of each change to respond to the emergency. Control room operators move to a mitigation focus and away from the primary goal of efficient power generation for the customer base. Field operators are the primary actors in a FLEX scenario and perform many of the prescriptive steps in procedures as well as respond to novel circumstances. These modified roles must be carefully considered as HRA models interpret potential and likely actions taken by the operators in the event of a FLEX deployment. There is a level of variability and unpredictability that will need to be explored in order to adequately capture its effects in a quantitative model.

For this initial effort, the authors began by assuming a BDBEE induced extended loss of AC power (ELAP) scenario. This entry condition was assumed in order to limit the model considerations to responses of an ELAP rather than also considering the probabilities and ranges of entry for an ELAP occurrence. The original intent was to model the steps from the ELAP declaration to response and deployment. Therefore, a scenario that allowed for power restoration but required emergency core cooling capabilities was selected to demonstrate the processes that operators step through to demonstrate several different procedures. The generic initial ELAP procedure guides operators to establishing makeup feed flow to steam generators, which necessitates the deployment of an auxiliary FLEX pump. To deploy this FLEX pump, the procedures direct operators to the FLEX staging process and installation hard cards (i.e., field procedures) for the individual pieces of equipment including the pump itself, hosing, electrical connections, and a diesel generator to provide power.

2.2 GOMS-HRA

To begin modeling operator actions for the FLEX staging process in a realistic way we began by assigning normal distributions around the Goals, Operators, Methods, and Selection (GOMS) task level primitives used in the GOMS-HRA method [7–9]. The overall GOMS model assumes all tasks can be decomposed into subtasks at more granular levels. Subtask modelling allows a direct understanding of the specific cognitive demands that are placed on the operator by the tasking. Even the simplest of tasks are complex combinations of different cognition functions that can quickly lead to incredibly complex modeling instances. Conceptually, the GOMS model can be described based on its elements which include: goals specify the desired end state of the task that the operator is acting toward; operators are specific actions available to the operator; methods are the groups of operators that construct the steps needed in the task; and selection identifies the heuristics or rules that guide the operator choices when multiple actions for the same goal are possible.

2.3 EMRALD

The Event Modeling Risk Assessment using Linked Diagrams (EMRALD) [10] tool was developed to support the increasing need for dynamic PRA models that can respond to evolving plant conditions during the simulation. Historically, PRA models are completed in a static fashion in which a fixed plant state is defined, and models are based on those unchanged characteristics throughout its duration, where different severity of events or timing must be binned into groups are different sequences. EMRALD is one of a growing number of tools to support plant state variations through simulation to better model realistic event progressions in time. Many of the dynamic PRA tools in use today are scripting methods that are difficult to use and debug and do not possess a graphical user interface (GUI). EMRALD is a robust platform that can generate probabilistic findings, process variables, and timing of events for analysis by using a drag-and-drop GUI to enhance the usability and ease of model creation. EMRALD adopts an iterative method of building models and allows for continued improvement and modification of modeling structures.

In assessing the use cases of HRA deployment into a FLEX situation, there are some key characteristics that mandate a more dynamic tool than the more common static HRA methods. As mentioned previously, the entire purpose of FLEX processes is to respond to events that are, by definition, beyond the design basis of the current plant infrastructure. The few serious event occurrences within the nuclear industry, historically, were unpredicted and therefore unplanned. The variability of a FLEX deployment is such that a dynamic tool is required to give the model any sense of realism in its application to an actual event. Many of the procedures require consecutive actions, with each having a large uncertainty distribution. With most static modeling, a failure in any of the actions causes a conservative overall failure, while dynamic modeling can credit faster completion time to other slower or repeated actions, providing a more realistic time sequence. EMRALD's GUI-based interface and quick spin-up time enables a more efficient demonstration and testing of ideas surrounding the actual interaction of FLEX procedures and human reliability.

Since the activities for FLEX occur outside the control room, it is important to adopt a tool that supports performing the actual simulation. EMRALD achieves this through events and their timing to form a structure for the simulation. Traditional activities are often coupled to thermohydraulic simulations such as RELAP [11]; however, for activities outside the control room these simulations are less applicable. Though the time thresholds are based on values derived from simulations such as RELAP, the thermohydraulic code itself is not necessary. A self-contained simulation alternative is needed. EMRALD supports this structure and therefore is suitable for dynamic HRA.

2.4 Model

The EMRALD tool was used to build the logic for the fault trees to analyze the FLEX deployment variables. The scenario modelling is intended as a proof-of-concept demonstration of how operator FLEX activities can be dynamically modeled to more accurately account for human risk contributors to overall plant safety. The model itself is comprised of three primary hierarchical types of elements. The main model diagram defines the overall simulation. The main model ties the underlying models together and initializes the simulation state, e.g., the state of the plant and environment in which the FLEX activities are performed. During each simulation run, only one of the three possible staging areas is deemed usable for FLEX deployment, but the path to the one available staging area must still be cleared of debris. Furthermore, the highest priority staging area is assumed to be the closest, the second priority is the second closest, and the third is the farthest away. The distance magnitude then factors into the amount of time the debris clearing action requires for each of the different staging areas, such that the third priority location requires the longest clearing time. Another critical function of the main model is tracking the overall simulation time against the allowed 74-min to deploy the FLEX pump to maintain sufficient water in the steam generators and support the necessary heat removal to keep the core cool. Exceeding this 74-min timespan comprises a system failure due to an inappropriate amount of time elapsing to perform the FLEX activities.

Since this is the initial effort to examine and demonstrate the use of EMRALD for HRA FLEX activities, there are many simplifications made to the model. During an ELAP scenario resulting from a loss of offsite power, the operators perform a number of activities to restore their instrumentation within the main control room. For the sake of simplicity, this modelling effort assumed those activities occurred successfully and simply assigned a time distribution to the length of time required to perform them. This impacts the model by adding a time delay before the actual FLEX deployment activities, which occur within the FLEX procedure, take place. Specifically, this time delay reduces the amount of time the operators have to perform the actual FLEX deployment.

Once the FLEX response is initiated from the main model, the operator model is invoked (see Fig. 1). The operator model is comprised of the critical procedure steps required for the operator to make the decision to deploy the FLEX equipment due to the unavailability of the auxiliary feedwater system and then begin clearing debris from the route to the staging location (see Fig. 2). The operator model performs two critical functions. First, it tracks the sequence of procedure steps and ensures that the operator performs them in the correct order. The second function serves to track when the operator is engaged performing a step and then when the operator becomes available to perform a step. This is important to ensure that only one procedure step is performed at a time.

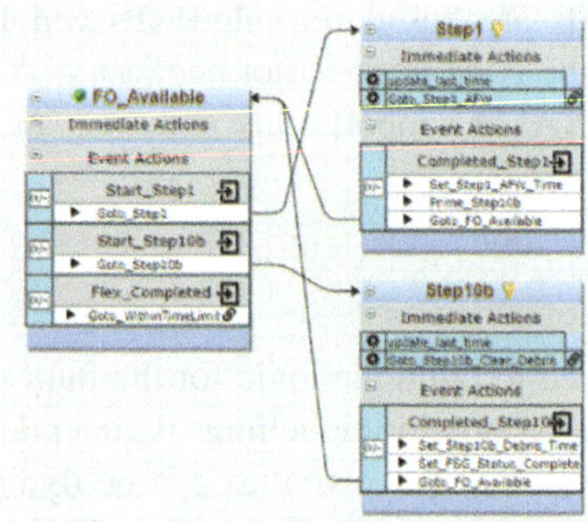

Fig. 1. Model of the operators' actions comprised of procedure steps that must be executed to complete the flex deployment. Operators are either available or engaged in performing a specific procedure step.

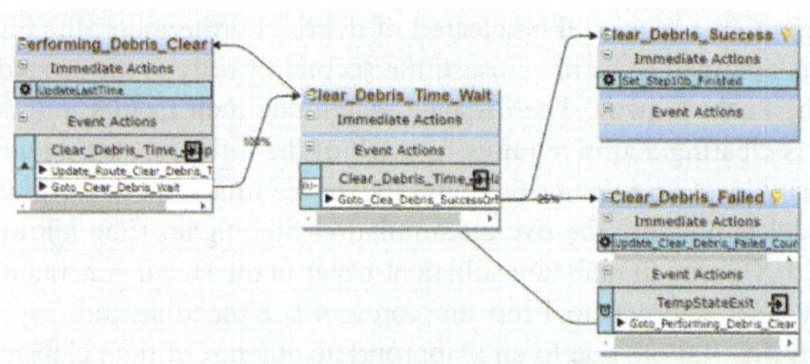

Fig. 2. Individual procedure step model depicting the operator clearing debris.

As the operator performs a procedure step, the model invokes an individual procedure model associated with that specific procedure step (see Fig. 2). The individual procedure step model contains the GOMS-HRA elements, which include the probability of successfully completing the procedure steps based on the relevant subtask primitive and performance shaping factors in addition to the timing distribution for the time required to perform the subtask primitives. For a given procedure step, if the step is failed, the operator repeats the step until a successful outcome is achieved. Each repetition adds to the total time taken for the procedure step execution. Within this first modelling effort, only one GOMS-HRA primitive was included. During the scenario, the operator performs a check to determine if auxiliary feedwater is available. If the check succeeds, the operator proceeds to the next task, or it can fail, in which case the operator checks the status again. Currently no performance shaping factors are integrated into the model when sampling the probability estimate for the check primitive subtask. Future work will investigate the best practices to integrate performance shaping factors into the model.

3 Discussion

The current project demonstrated the feasibility to model human operator FLEX activities dynamically with EMRALD. Static HRA provides an approximation of human activities in using FLEX equipment during BDBEEs. Dynamic HRA such as the current EMRALD modeling example allow simulation of a range of possible outcomes to reflect the complex nature of BDBEEs. A final use of dynamic FLEX HRA simulation is to allow consideration of the benefits of FLEX equipment for non-BDBEE situations, such as using FLEX equipment to enhance safety margins during outage work. Creating a realistic simulation of deployment of FLEX equipment allows modeling of these scenarios that goes beyond current PRA models.

These initial efforts were purely for proof of concept and require additional fidelity to accurately capture realistic conditions during a FLEX event. EMRALD was designed for PRA for hardware systems, and as a result it was not initially clear whether it would prove feasible to model human activities. This demonstration shows that it is feasible, but there are still challenges that must be overcome to better enable unique aspects of human actions.

For example, EMRALD conceptualizes components as individual objects within the model. Each component can be instantiated in only one instance within the model such that any component exists only once within the model. It can be referenced from different elements within the model; however, it is only a single instance. This works well for PRA, since the physical world represents the function of a component as one entity. Human actions, however, may be used repeatedly at different times in the simulation and on different components that might be easier or more challenging to assess. The assessed action therefore exists as separate instantiations of that particular human action type. For example, a human action may consist of assessing the state of a valve multiple times during the simulation. The assessed action is not the same across time spans, since the state of the operator might change due to performance shaping factors of the context at that time. Therefore, the conceptualization of a generic action type is needed to eliminate the tedious process of manually redefining the action each time it is invoked in a particular context. Instead, a generic action component could be invoked and populated with the context of its use, including the performance shaping factors that govern the human operator's ability to perform the assessment. Since humans repeatedly perform various actions, this would greatly reduce the amount of modelling time required to accommodate the unique human operator aspects when they are performing basic human actions. Furthermore, it would be useful to chain together individual human action components into more sophisticated generic actions to further reduce the level of effort required for capturing repeatedly used and more complicated human actions.

Acknowledgments. This work of authorship was prepared as an account of work sponsored by Idaho National Laboratory, an agency of the United States Government. Neither the United States Government, nor any agency thereof, nor any of their employees makes any warranty, express or implied, or assumes any legal liability or responsibility for the accuracy, completeness, or usefulness of any information, apparatus, product, or process disclosed, or represents that its use would not infringe privately-owned rights. Idaho National Laboratory is a multi-program laboratory operated by Battelle Energy Alliance LLC, for the United States Department of Energy under Contract DE-AC07-05ID14517.

References

1. NEI: Diverse and Flexible Coping Strategies (FLEX) Implementation Guide. NEI 12-06, Rev. 4 (2016)
2. Boring, R.L., St. Germain, S., Banaseanu, G., Chatri, H., Akl, Y.: Applicability of simplified human reliability analysis methods for severe accidents. In: 7th International Conference on Modelling and Simulation in Nuclear Science and Engineering (7ICMSNSE), Paper 98, pp. 1–11 (2015)
3. Park, J., Jeon, H., Kim, J., Kim, N., Park, S.K., Lee, S., Lee, Y.S.: Remaining and emerging issues pertaining to the human reliability analysis of domestic nuclear power plants. Nuclear Eng. Technol. **51**, 1297–1306 (2019)
4. Reid, M.: Human reliability assessment for 'Flex' equipment. In: Proceedings of the Probabilistic Safety Assessment and Management Conference (2018)
5. EPRI: Human Reliability Analysis (HRA) for Diverse and Flexible Mitigation Strategies (FLEX) and Use of Portable Equipment: Examples and Guidance, EPRI 3002013018. EPRI, Palo Alton (2018)
6. Kichline, M.: Human reliability analysis for using portable equipment. In: EPRI FLEX Workshop (2018)
7. Boring, R.L., Rasmussen, M.: GOMS-HRA: a method for treating subtasks in dynamic human reliability analysis. Risk, reliability and safety: innovating theory and practice. In: Proceedings of the 2016 European Safety and Reliability Conference, pp. 956–963 (2016)
8. Boring, R.L., Rasmussen, M., Ulrich, T., Ewing, S., Mandelli, D.: Task and procedure level primitives for modeling human error. Adv. Intell. Syst. Comput. **589**, 30–40 (2017)
9. Ulrich, T., Boring, R.L., Ewing, S., Rasmussen, M.: Operator timing of task level primitives for use in computation-based human reliability analysis. Adv. Intell. Syst. Comput. **589**, 41–49 (2017)
10. Prescott, S., Smith, C., Vang, L.: EMRALD, dynamic PRA for the traditional modeler. In: Proceedings of the 14th International Probabilistic Safety Assessment and Management Conference. Los Angeles, CA (2018)
11. RELAP5-3D Code Development Team: RELAP5-3D Code Manual Volume I. Idaho National Laboratory, Idaho Falls (2018)

Credible Evidence Continues to Surface Regarding a Likely "Friendly Fire" Incident Along the Sesame Street and Shrine Corridor Area on June 30, 2013

Fred J. Schoeffler[1(✉)], Lance Honda[2], and Joy A. Collura[3]

[1] Sheff LLC Pine, Pine City, USA
dougfir777@yahoo.com
[2] Prineville, USA
[3] Yarnell Hill Fire Revelations, Congress, Yavapai County, USA

Abstract. On June 30, 2013, nineteen Granite Mountain Hot Shots (GMHS) perished on the Yarnell Hill Fire. The Serious Accident Investigation Team - Report conclusion states: "no indication of negligence, reckless actions, or violations of policy or protocol." Was an unfeasible firing operation dismissed? Video-audio and Hearsay-Exception evidence indicated uncollaborated independent action while GMHS hiked downhill through unburned fuels. Was there a concurrent rogue firing operation? Indications of an unfeasible goal pursuit continued with everything contradicting a sound plan. Contemplated failure led to more entrenched behaviors. Weather deterioration and increased fire behavior were interpreted unrealistically. Both visual and auditory stimuli decreased significantly under stress; listening to cues weakened vision - intense visual cues diminished hearing triggering tunnel vision and auditory exclusion. Fixated goal setting, non-critical thinking, indecision-making, single-mindedness, and leadership dysfunctions concealing possible dire consequences resulted in disaster. "Friendly Fire" decisions and actions are discussed for lessons to reduce similar tragedies.

Keywords: Wildland fire · Hearsay Exception · Tunnel vision · Auditory exclusion · Destructive goal pursuit · Friendly Fire

1 Introduction

On June 30, 2013, nineteen Prescott F.D. Granite Mountain Hot Shots (GMHS) were entrapped and killed on the Yarnell Hill Fire. The Serious Accident Investigation Team (SAIT) conclusion published in their Report (SAIR) (Sept. 2013) *questionably* stated they found "*no indication of negligence, reckless actions, or violations of policy or protocol*" (emphasis added) [1]. A firing operation along old roads in the Sesame Street - Shrine Corridor improved with a dozer to backfire from was unfeasible [1]. This paper posits evidence of several independent WF-FF groups likely firing off the Corridor area - *without* collaborating their actions - while the GMHS left their Safety Zone (S/Z) and hiked downhill toward a supposed "*at-risk*" Boulder Springs Ranch (BSR) [1–5].

P. M. Arezes and R. L. Boring (Eds.): AHFE 2020, AISC 1204, pp. 353–361, 2020.
https://doi.org/10.1007/978-3-030-50946-0_47

Destructive goal pursuit is deemed the single basic error credited to all disasters [6]. Seemingly unrelated elements interacting with others when their side effects and outcomes are unassessed, soon become problems never considered possible. Research indicates the greater insecurity a group feels of their chance of achieving goals, the harder they try [6]. Despite mounting evidence that it was unfeasible, why would the GMHS continue to pursue their BSR attempt? Surely, they observed adverse fire weather and fire behavior – yet, they unrealistically interpreted those conditions, seeking more evidence implying failure as likely. These signs would ultimately influence them to put even more effort into their fatal BSR goal pursuit that surely seemed counterintuitive [6]. Being so totally leader-dependent was detrimental to the GMHS.

The GMHS - dehydrated, tired, stressed, distracted – somehow felt the need to vacate their S/Z. Research indicates that under stressful conditions while focused intently on visual tasks, auditory stimuli decreases significantly [7]. Conversely, when focused on spoken messages, the visual image activity diminishes. So, a WF intently *listening* to audible cues, (e.g. radio or cell phone), could have diminished visual performance; a WF intently *visually* focused could have diminished hearing. Notable hastiness in the face of a desperately pursued goal, they focused on the BSR. They knew better, ignored their training, logic, and gut feelings because of destructive goal pursuit [6]. Hence, tunnel vision, auditory exclusion, and the bizarre destructive goal pursuit are bona fide, potentially critical threats that pose unusual hazards for wildland firefighting [6, 7].

2 Wildfire Rules - Firing Operations - Friendly Fire - Hearsay Rules

Former U.S. Forest Service (USFS) Fire Director Jerry Williams stated: "The Ten Standard Firefighting Orders must be firm rules of engagement. They cannot be simple guidelines, and they cannot be 'bargained.' They are the result of hard-learned lessons. Compromising one or more of them is a common denominator of all tragedy fires. On the Dude, South Canyon, and Thirtymile Fires, the Fire Orders were ignored, overlooked, or otherwise compromised" [8].

"Wildland fire is a high-risk, high-consequence business. …[often] surrounded by uncertainty and danger. … [T]he tragedies at Dude, South Canyon, and Thirtymile and the accident at Cerro Grande remind us of the danger that is always present in our world. *Entrapment avoidance* must be our primary emphasis and our measure of professional operational success" [8]. (link here)

This "*Yarnell Hill Fire - 2013*" (4-30-19 - WTKTT) video reveals three GMHS performing a minor firing operation near their "*lunch spot*" from a photo taken from near Deertrack Drive at 1036 to "*get the fire squared up with the two-track road*" [1]. The three GMHS, (left - Steed in red hardhat) working as a firing group began an '*indirect*' burnout Sunday. The one on the right is pointing with his arm outstretched or possibly using a flare gun (https://youtu.be/b3KShvWbbl8).

Escape Routes are the paths WFs/FFs take from unsafe present locations to safer ones; it is the most elusive safety prong due to its ever-changing status [9], performed by the abstract GMHS fatal escape route. A recent WF study found key portions of entrapment potential lies in human factors, prior WF entrapment investigations have similar reviews and proposals, weak entrapment investigation process and reporting systems, and a likely sizable under-reporting of entrapments [10]. *Why* did the GMHS skillfully pursue a true *E/R to get to a S/Z*, *pervert that term*, leave their viable S/Z - and then fatally hike downhill in chimneys/chutes of unburned chaparral [1, 2]?

2.1 Wildland Firing Operations Tactics and Strategy - Fighting Fire with Fire

Burning out and backfiring are dissimilar firing operations, often misunderstood and substituted. *Burning Out* is setting fire inside a control line to consume fuel between the fire edge and the control line strengthening and straightening them by eliminating fuel between the fire edge and the control line; considered direct attack [11]. A Single Resource Boss (CRWB, ENGB, etc.) has authority to initiate burnouts with Division Supervisor approval, usually an on-going part of line construction. *Backfiring* is a fire set along the inner edge of a fireline to consume the fuel in the path of a fire or to change the fire's direction, approved by the Incident Commander (IC) or the Operations Section Chief (OPS) and put into effect at the Division level [11]. Adhering to basic wildland firefighting rules and guidelines with *informing adjoining forces as critical* [11]. Some groups justly question whether "*extensive backburning*" (slang for firing operations) is "a cause or consequence" of large wildfires posting greater WFs/FFs risks than usually sensed. "It is easy to understand … why suppression firing may be the dominant form of fire use today" [12].

2.2 Wildland Firing Operations and Friendly Fire

The notion of *Friendly Fire*, (*aka fratricide* or *amicicide*) is atypical, presented here *from* a unique wildland fire perspective. Consider now a short account of the phrase based in historic military passages and quotes to propose as an analog for wildland fire incidents. According to retired Army Colonel turned academic, Scott Snook extensively researched a friendly fire incident in Iraq when two U.S. Air Force fighters by mistake shot down two Army Blackhawk helicopters [13]. "Friendly fire-casualties unintentionally inflicted on one's own forces-is not a new problem in the history of warfare. However, until quite recently, little explicit attention has been paid to studying its causes and possible solutions" (footnote omitted) with certain cases frequently revealing that the fratricide was the final link in a chain of mistakes [13]. Retired Army Lt. Colonel Charles Shrader refers to it as "*amicicide.*" He wrote "… in the '*fog of war'* friendly fire casualties are inevitable … 'fog of war' is an oft-mentioned, if imperfectly understood, …" [14]. With few sources, researchers are left with scattered, cryptic, notes found in general operational histories or official combat records [14]. The authors can duly relate to the notion of *few sources*.

Rare wildland "friendly fire" incidents are most often fatal as documented and recounted by many experienced WFs and human factors researchers that firmly believe that the Mann Gulch (MT-1949), Loop (CA-1966), Battlement Creek (CO-1976), Mackenzie (AZ-1994), Cedar (CA- 2003), and Yarnell Hill Fires (AZ-2013) were the result of likely "friendly fire" [5, 15]. These are discussed in some detail on InvestigativeMEDIA (IM) [15] and *incompletely* recorded (read "*official*" records) in the Wildland Fire Lessons Learned Center Incident Reviews. Likewise, the WF brother of a deceased GMHS that was unsuccessful getting a GMHS position, acknowledged and agreed that his brother had died "*from fratricide*" on June 30, 2013. He conceded that to cope with losing his brother, he accepted they were adults with many options that day. Contrary to the SAIT-SAIR, groupthink poor choices caught up with them, resulting in the fatal outcomes [2, 3].

Consider the AZ State Forestry (ASF) video (6-30-13), YH Fire with a USFS Type 3 Engine and two WFs with torches, steadily firing off one of the spur roads in the Peeples Valley area [15] on the fringes of the Sesame Street and Shrine Corridor area verifies that Engine Crews decided to perform fairly assertive burnouts despite the intense conditions. This Engine *correctly* follows *behind* the lighters, compared to the naive Municipal FFs firing operation with the Engine *in front of* the lighters [4]. The relevant video segment is from *00:00 to 01:25*. Later segments reveal separate and distinct smoke columns, indicative of a firing operation, from about the *1:30 to 4:17* timeline with some notable, aggressive backdrop fire behavior, including some clear, relevant radio transmissions throughout the video (https://youtu.be/7UVL8pxSBJc) [16]. The authors contend that these videos are foundational to undergirding the Corridor firing operation prospect.

2.3 Sesame Street/Shrine Road Corridor Area Likely Firing Operations Evidence

The Sesame Street to Shrine Corridor area likely firing operation is sustained by at best seven (7) suggestive proofs: (1) three SAIT-SAIR excerpts [1]; (2) photos that the YHFR site posted, based on separate and distinct smoke columns [2, 3, 5]; (3) over twenty WFs, FFs, citizens saw firing operation videos in several places [3–5]; (4) mentions of burnt fusees and "*accelerants*" [3–5]; (5) InvestigativeMEDIA posts about a "*back burn*" [5, 15]; (6) the ASF/IFE video noted above; and (7) WF hearsay evidence regarding at least three separate Corridor firing operations [4, 15].

SAIT-SAIR: [1, 4] "[BRHS] … clear out the two-track road … [for access] and … *possible backfire* … (an old fuel break) between *Sesame Street and Shrine Road.* … dozer to open the line. [BRHS] move their crew carriers toward the *Shrine of St. Joseph (the Shrine)* and a youth camp area around 1500 and then start preparing for *burnout along the dozer line*." (pp. 18, 21) "… SPGS1 … [asks] if they still have the option to burn out from the dozer line. [BRHS] tells him no. … DIVS A, … agrees and says … the fire is almost as far as the [GMHS] vehicles. … DIVS A says, "*... we're going to make our way to our escape route*." (p. 24) "A task force is … expecting … [the] *need to burn off this line* … indirect line connects to the dozer line between *Shrine Road and Sesame Street.* (p. 25) (emphasis added) *Indeed, a burnout was planned for.*

2.4 Corridor Firing Operation Evidence and Arizona Rule 803. Exceptions to the Rule Against Hearsay - Regardless Whether the Declarant Is Available. Article VIII

Consider now the existence of a Sesame Street and Shrine Corridor firing operation, using legal literature to establish standing, for legal credibility, while utilizing the Arizona Rule 803 [18] to qualify statements for "*particular probative value*" (evidence useful to prove something in a trial).

'A *firsthand account* is based on one's personal experience; a *secondhand account* is based on an one's research, rather than personal experience, generally considered to be circumstantial. Hearsay is a statement made in court by someone other than a party to prove the truth of matter.'

'Evaluating factual matter, an expert is allowed to rely on anything normally relied upon by other experts in his field, anything that has gained "*general acceptance*" in arriving at opinions. The material upon which *expert opinion* is based need not be admissible; it may be hearsay' [19].

Accepting a veteran WF as an "*expert witness*," granting credibility is permissible. Tight groups, (i.e. WFs), at times share unreliable beliefs. Consider whether the WF provider's experience base is amply extensive to support what is presented as bona fide, albeit second- or third-hand, plus expert opinion. Deferring to the AZ Rule 803, discounting opinion(s) lacking ample basis ensuring reliable inferences fairly drawn from experience are presented to you [20].

This component of the probable firing operation - planned and discussed - was likely fired off by a mixture of Federal, State, and *local* Municipal Fire personnel [4, 5, 15]. It is a widely held belief among WFs/FFs, explicitly those involved in the Sesame Street and Shrine Corridor area likely firing operation, that the increased outflow winds and resultant fire behavior created was akin to an exponentially violent fire storm that overtook the GMHS as they *hiked downhill into the lit fire surging uphill* toward them, with no lookout [1, 4, 5, 15]. *See Fig. 1 & Fig. 3* in Ref [4] (Fig. 1).

Fig. 1. June 30, 2013, **1624** (4:24 PM) (L); **1629** (4:29 PM) (CL); **1631** (4:31 PM) (CR); **1649** (4:49 PM) (R). Aggressive fire behavior; separate & distinct smoke columns in the Sesame Street and Shrine Corridor. **Source:** Sun City FD, Lauber, News12, Tham; WTKTT, Google Earth [17]

The context to follow is a mixture of *'first-person' – 'second hand' – 'third-person'* perspective anecdotes and comments of *three separate firing operations*: (1) a training academy where the students are engaged in a group discussion about fire behavior on the June 30, 2013, YH Fire, and eventually human factors [4, 5]; (2) a separate (*circa* 1432) firing operation in a spur road area by a Single Resource Agency Type 3 Engine [4, 5, 16]; (3), a municipal PFD FF perspective at the BSR [4, 5]; and (4) a 25-year veteran WF, using the avatar "*Downhillndirect*" (DND), that spoke with each of these individuals about their respective likely firing operations details [4, 5].

Quotes and paraphrases from DND's proximate dialogue notes with the respective likely firing operations personnel follow [4, 5, 15]. Excerpts from military literature on the historic notion of "*friendly fire*" are assessed and then applied to the wildland fire equivalent [13, 14]. Wildland "friendly fire" incidents originate mostly from veteran WFs and FFs anecdotes and experiences, and from several InvestigativeMEDIA website posts on this vital contentious issue [4, 5, 15].

'The training academy FF student (SFF) spoke of "*firing out multiple spur roads while trying to keep it square so it wouldn't jump the main road. I asked several times if the area was clear of FFs and civilians before they did any firing.*' A supervisor told them: "*Sometimes we do things we have to, not because we want to.*" He thought the Firing Boss was *'part of the IMT ... like a Structure Protection Specialist'* ... *'the closer we got to the Sesame Street and Shrine area ... a feeling of absolute panic as we pulled out.'* ...*'we could hear everything those guys were saying. Marsh said that the [BSR] isn't that far from where he thought the guys are now ... we knew Marsh went down that ridge a few times before [the guys deployed]'* [4, 5, 15, 17].

'We were part of a firing operation off one of the spur roads .. close to where everybody piled out [Sesame Street to Shrine Road near Miner Rd] about twenty (20) min *before hearing all the ruckus on the radio about the GMHS, Air Attack, yelling ... we had the Engine out in front. No spots, absolute miracle because the wind was howling parallel with the road they were on and every now and then gust over the road and throw embers, but no spots. The Engine was going fast; we had a hard time keeping up with it'* [1, 4, 5, 17]. (Engines out front suggest inexperience).

The SFF noted: the *'urgency and fear in the eyes of everyone was unnatural. It was burning in so fast ... it was ripping right in* [17]. *We tied into a corner, a bunch of vehicles came out including the BRHS and the GMHS buggies* ... [he started crying again]. ... *we felt much better lighting to the corner of this road intersection, because ... we figured they [GMHS] were out of the area.'*

'The fire was gone, half mile or mile away from us within 5–10 min *of firing, uphill and gone. We all staged in our rigs listening to the TAC channels and then we heard the GMHS A/G chaos. The [Engine] Crew, including our Captain, felt like we couldn't breathe. He said he felt like he was going to have a panic attack; then Captain said: "we just f**king killed people.*" The SFF stated they were later spoken to by several higher ranking WF/FF Supervisors in Yarnell telling them that *'everything was gonna be alright'* and *'you saved a lot of structures ...'* [4, 5, 17].

Indeed, there is unsettled evidence of a Sesame Street and Shrine Fuel/Fire Break Corridor firing operation. Consider the legal literature on standing for legal credibility utilizing the Arizona Rule 803. Exceptions to the Rule Against Hearsay [18] to qualify WF DND's accounts as having "*particular probative value*" (evidence sufficiently useful to prove something important in trial). Instructors were sternly directed on '*YH Fire discussions to get preapproved USFS Regional and Washington Office lesson plans' and 'no disrespecting the dead and no YH Fire talk here'* [5, 15].

3 Human Factors - Errors - Failures - Mindfulness

Consider some novel *Human Failure* values unexplored/unrecorded by investigators of how and why WFs and FFs believed their actions as acceptable risks along their *drift into failure* [2–5, 15].

3.1 Tunnel Vision and Auditory Exclusion Vulnerabilities and Perils

John Hopkins University researchers found, 'By narrowing attention, … attention shifts from vision to audition caused increased activity in auditory cortex and decreased activity in visual cortex and vice versa, reflecting the effects of attention on sensory representations. The experiment was designed to create tunnel vision, but a completely unexpected event occurred. While vision was being tunneled, performance of the audible control center decreased' [7]. The researchers further found that tunneled vision leads to diminished hearing. Tunneled hearing led to diminished vision. The researchers concluded that a person intently listening to audible cues - like a radio or cell phone - could have diminished visual performance. In some cases, when the stress is severe enough, the hearing receptors in the brain may shut off completely, referred to as *auditory exclusion* [7], that can directly restrict wildland firefighting situational awareness [2, 3].

Surely, the GMHS would '*see*' the noticeable changing weather and increased fire behavior, but their stressed brains maybe sabotaged or delayed their threat perception and reaction away from those focus areas [7, 17]. Several cell phone and radio conversations likely diverted considering emerging hazards, and their failure to reassess priorities resulted in a drift into failure [2, 7, 17].

3.2 Destructive Goal Pursuit Relevance from Mountaineering to Wildland Firefighting

As per Kayes, destructive goal pursuit, [6] the single basic error attributed to all catastrophes, warrants noting three *counterintuitive* points: (1) the limits of goals and their pursuits, (2) the need for less attention on goals and more on group dynamics, and (3) problems brought about by leadership. Organizations driven by goal attainment are inclined to continue in that same direction rather than adjust course, even when that course of action leads to failure. And when new and contradictory information continues to suggest that the goal cannot be safely attained, inspired action directed toward their goal - not to potential environmental changes - (*i.e. adverse fire weather and increased fire behavior*), leaders choose to ignore those. The GMHS leadership failed to appreciate potential contributions from their concerned WFs, and so as a group they (un)willingly accepted their leaders' decision(s) to abandon their S/Z in pursuit of the BSR goal [2, 5, 6]. Stressed and distracted, *how* and *why* did they succumb to the folly to vacate their S/Z?

4 Organizational Culture

4.1 Municipal Fire Department (Hybrid) Attitudes, Influxes, and Influences

'*The Prescott Way. Just one of those things that happened, an accident. Fire Order #10* (Fight fire aggressively having provided for safety first) *is hillbilly. We're smarter. We're a lot smarter*' [5]. These are strong markers of the conflicting causal factors of confusion and ambiguity; latent human failures on any wildfire, mainly those with threatened structures like the YHF. These will continue unless WFs truly learn "*complete*" lessons from these lethal recurring tragedies [2, 15]. *Demanding* to include "*adjoining forces*" into Watch Out #7 should lessen friendly fire incidents.

5 Conclusion

The authors concede to scarce evidence of the Sesame Street and Shrine Corridor area likely firing operation due, largely, to fears of exposing the truth. In the coming years, the authors will justly seek, examine, and release scores of current and future likely firing operation records. One enduring inquiry remains - when will those WFs, FFs, and citizens possessing anecdotes, photos, videos, do what's right and courageously come forward?

Acknowledgements. Dr. Ted Putnam; YH Fire Hikers; Mike Williams; Woodsman; Gary Olson; WantsToKnowTheTruth; Norb; Downhillndirect; those who lost family, friends, and loved ones; 19 young men that inspire our writing for healing and complete lessons learned.

References

1. Yarnell Hill Fire Serious Accident Investigation Team (SAIT): Yarnell Hill Fire Serious Accident Investigation Report (SAIR), 122 pp. (2013). https://sites.google.com/site/yarnellreport/
2. Schoeffler, F.J., Honda, L.: Epic human failure on June 30, 2013. In: International Conference on Applied Human Factors and Ergonomics (AHFE) Advances in Human Error, Reliability, Resilience, and Performance (HERRP) (2017). https://philpapers.org/archive/SCHEHF.pdf
3. Schoeffler, F.J., Honda, L.: It Could Not Be Seen Because It Could Not Be Believed on June 30, 2013. AHFE (2018)
4. Schoeffler, F.J., Honda, L., Collura, J.A.: Formerly Unrevealed Public Records Should Change the Account of What Occurred on June 30, 2013. AHFE, HERRP (2019)
5. Yarnell Hill Fire Revelations (YHFR) website (2019). www.yarnellhillfirerevelations.com
6. Kayes, D.C.: Destructive Goal Pursuit. Mt. Everest Disaster. Palgrave MacMillan, New York (2006)
7. Shomstein, S., Yantis, S.: Control of attention shifts between vision and audition in human cortex. J. Neurosci. **24**, 10702–11070 (2004)
8. Williams, J.: Next steps in wildland fire mgmt. Fire Mgmt. Today **62**, 31–35 (2002)
9. Natl. Wildfire Coordinating Group (NWCG) 6-Minutes for Safety. Escape Routes 2 (2019)

10. Jolly, M., Butler, B., Page, W., Freeborn, P.: An Assessment of Research Needs Related to Wildland Firefighter Safety. JFSP #18-S-011 (2019)
11. NWCG: Firing Operations NFES courses. S-234 Ignition Ops and S-219 Firing Ops (2014)
12. Ingalsbee, T.: Managing Large Wildfires by Design. FUSEE. Eugene, OR (2014)
13. Snook, S.A.: Friendly Fire. Princeton University Press, Princeton (2000)
14. Shrader, C.: Amicide: the problem of friendly fire in war. J. Army War College 23 (2005)
15. Dougherty, J.: InvestigativeMEDIA blog. Relevant posts - Chapters VI to XXX (2013–2020). https://www.investigativemedia.com/yarnell-hill-fire-2/ (Use "*Ctrl F*" w/ key words to search)
16. The Institution of Fire Engineers - YH Fire - Storyful News video - (11-10-14) YH Fire (2013) 1036 AM GMHS firing op cross-fade video WTKTT (2019). https://youtu.be/b3KShvWbbl8
17. Thompson, J.: June 30, 2013, 1624 (4:24 PM); Lauber, B.: 1629 (4:29 PM); ABC News 12 photo(s) 1629; Tham, R. (RiP) 1649 (4:49 PM), also included in reference [4] at Figure 3 (2013)
18. Westlaw: AZ Rules of Evidence, Art. VII, Rule 803. Exceptions to the Rule Against Hearsay - Regardless of Whether the Declarant is Available as a Witness. Arizona Court Rules (2015)
19. Freedom School: Objections to Evidence and Testimony - Federal Rules of Civil Procedure http://freedom-school.com/objections_to_evidence_and_testimony.pdf. Accessed Jan 2020
20. Poulin, A.B.: Experience-based opinion testimony. Pepperdine Law Rev. 39 (2011)

Identification of Collectible Items in the Rancor Microworld Simulator Compared to Full-Scope Studies

Jooyoung Park[1], Thomas A. Ulrich[1], Ronald L. Boring[1(✉)], Sungheon Lee[2], and Jonghyun Kim[2]

[1] Human Factors and Reliability Department, Idaho National Laboratory, 2525 Fremont Avenue, Idaho Falls, ID 83415, USA
{Jooyoung.Park, Thomas.Ulrich, Ronald.Boring}@inl.gov
[2] Department of Nuclear Engineering, Chosun University, 309 Pilmun-Daero, Dong-Gu, Gwang-ju 61452, Republic of Korea
leesh0867@chosun.kr, Jonghyun.Kim@chosun.ac.kr

Abstract. Most studies on collecting reliability data in human reliability analysis (HRA) have concentrated on research using full-scope simulators and actual operators. However, researchers have discovered challenges in collecting various items and amount of data needed to support HRA. As an opposite and complementing approach to a full-scope study, Idaho National Laboratory (INL) developed Rancor Microworld, a simplified simulator, to generate HRA data from student subjects. This paper aims to identify collectible items in Rancor Microworld versus those in full-scope simulators. In this paper, collectible items and their varying analysis levels in Rancor Microworld are identified in comparison with full-scope simulators. The appropriate method for treating the items is suggested through experiment directions for the future.

Keywords: Human reliability analysis · Data collection · Simulator study · Simplified simulator · Microworld

1 Introduction

Most studies on collecting reliability data in human reliability analysis (HRA) have concentrated on research using full-scope simulators and actual licensed operators [1–4]. This is primarily because these simulators largely replicate conditions in actual reactors. The biggest, most recent efforts in this regard are led by the U.S. Nuclear Regulatory Commission and the Korea Atomic Energy Research Institute (KAERI), which collect data from full-scope simulators using the Scenario Authoring, Characterization, and Debriefing Application (SACADA) database [3] and the Human Reliability data Extraction (HuREX) [2] framework, respectively.

However, researchers discovered challenges in collecting various items and gathering a large enough sample to support HRA. In general, testing differences generated by design factors in the human-system interface can cause difficulties for the full-scope simulator, since changing a design already configured and programmed into the simulator is no easy task. Furthermore, projecting a full-scope study is

P. M. Arezes and R. L. Boring (Eds.): AHFE 2020, AISC 1204, pp. 362–368, 2020.
https://doi.org/10.1007/978-3-030-50946-0_48

relatively resource-intensive and time-consuming, and presupposes utilities' cooperation in partially releasing collected data. Therefore, it is strictly limited to those few organizations that can satisfy such conditions.

As a complementary approach to a full-scope study, Idaho National Laboratory (INL) developed Rancor Microworld, a simplified simulator, to generate HRA data from student subjects. The simulator was developed to reproduce the important characteristics of real nuclear operations and evaluate theoretical and practical design concepts [5]. This paper aims to identify collectible items within Rancor Microworld versus those in full-scope simulators. In the paper, collectible items and their varying analysis levels within Rancor Microworld are identified in comparison with full-scope simulators. The approach for treating the items is then suggested, along with experiment directions for the future.

2 Characteristics of Different Analysis Levels in Rancor Microworld

Rancor Microworld is a simplified simulation environment that reproduces the important characteristics of real operations at nuclear power plants (NPPs) [5]. Figure 1 shows the interface for Rancor Microworld. As opposed to a full-scope simulator, it is equipped with alarms, indicators, a piping and instrumentation diagram, and controllable components (i.e., pumps and valves) representing at least the minimum components required to implement the system.

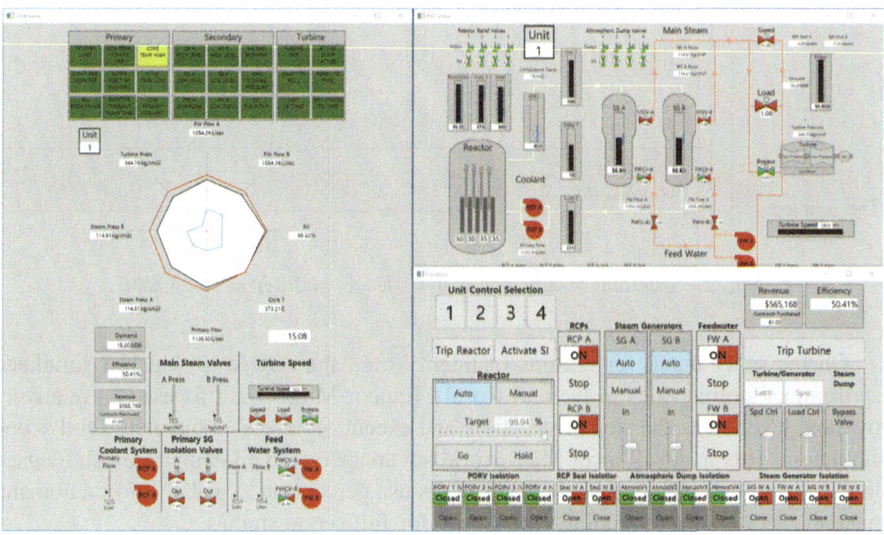

Fig. 1. Rancor Microworld interface.

This section compares the different analysis levels between Rancor Microworld and existing full-scope studies. The four levels considered in this comparison are 1) task level, 2) step level, 3) instruction level, and 4) execution level. The task level corresponds to a strategy such as "feed and bleed," an important long-term cooling strategy for maintaining core safety in emergency situations at NPPs. This level normally consists of several steps within NPP operational procedures. The step level is a procedure step level composed of several actions. Controlling reactor coolant temperature, such as by opening the atmospheric dump valve, is an example of a procedure step level. The instruction level matches the action level included in each procedure step. Lastly, the execution level, the simplest task unit, consists of actions such as looking at or reaching for an object.

Figure 2 summarizes the various analysis levels in different simulators. Full-scope studies collect data in the task, step, and instruction levels. In the recent studies mentioned in the Introduction, the SACADA database [3] focuses on the task and step levels, whereas the HuREX framework [2] concentrates on the instruction level. The execution level is rarely considered in full-scope studies, since they already focus on so many items in the larger task units; therefore, information from items in small task units may be missed or ignored.

Real world ⇕ The simplest simulator

	Task level	Step level	Instruction level	Execution level
Full-scope simulator	○	○	○	△
⋮	⋮	⋮	⋮	⋮
Rancor Microworld	△	○	○	○

Fig. 2. Spectrum of different analysis levels in different simulators.

On the other hand, for reasons of fidelity (i.e., the degree to which a simulated environment corresponds to the real world), Rancor Microworld focuses on relatively low item levels (i.e., the step, instruction, and execution levels) compared to full-scope simulators. This makes Rancor Microworld advantageous in allowing research focused on the execution level, such as how operators use perception to gather information and make decisions of confounding complexity for full-scope studies.

3 Collectible Items in Rancor Versus Full-Scope Simulators

To investigate collectible items in Rancor Microworld, this study defines a relationship for collectible items between Rancor Microworld and full-scope simulators. The Venn diagram in Fig. 3 indicates this relationship and encompasses four sections: 1) Section A (items only collectible in full-scope simulators), 2) Section B (items collectible in both Rancor Microworld and full-scope simulators), 3) Section C (items only collectible in Rancor Microworld), and 4) Section D (items not collectible in either Rancor Microworld or full-scope simulators). The study described in this paper focuses on Sections B and C.

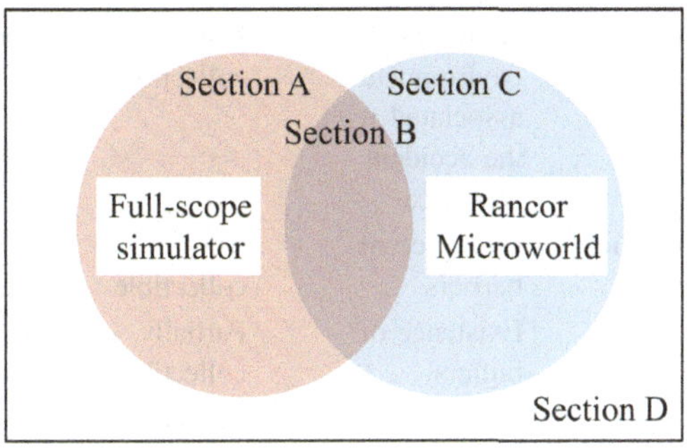

Fig. 3. A relationship of collectible items between Rancor Microworld and full-scope simulators.

A technical report from KAERI [6] suggests a catalog of generic HRA data items collected from documents issued from the perspective of HRA practitioners and human factors engineers. It includes a total of 89 generic HRA data items in seven different categories: 1) environment, 2) human-system interface, 3) organization, 4) procedure, 5) task, 6) evaluation/success criteria, 7) and performance context. The details are shown in [6], including measurable instances for each data item.

For this paper, the catalog of generic HRA data items was used to evaluate collectible items in the Rancor Microworld and full-scope simulators, then categorize the ones that correspond to Sections B or C. Table 1 gives example evaluation results for collectible items in full-scope simulators and Rancor Microworld. Certain data items—for example, "adverse environment associated with the accident sequence"—are collectible using Rancor Microworld, but not collectible using full-scope simulators. In fact, Rancor Microworld is portable and can be installed on a laptop; therefore, the data-collecting experiment can be performed wherever a desk, chair, and power source are available. For this reason, data can be collected in a facility equipped with a specific adverse environment, such as seismic experimentation equipment.

In Table 1, most HRA data items fall under the human-system interface category and are evaluated as collectible via Rancor Microworld. For full-scope simulators, it may not be easy to change a design already configured and programmed into the simulator. On the other hand, the Rancor Microworld design can be easily changed,

Table 1. Example evaluation results for collectible items in a full scope simulator and Rancor Microworld

Category	Subcategory	Data item	Evaluation		Classification
			Full-scope simulator	Rancor Microworld	
Environment	Workplace	Appropriateness of ingress and egress paths	Not collectible	Collectible	Section C
		Appropriateness of workspace envelope	Partially collectible	Partially collectible	Section B
	Work environment	Adverse environment associated with the accident sequence	Not collectible	Collectible	Section C
Human-system interface	Ergonomics	Existence of barriers	Not collectible	Partially collectible	Section C
		Existence of buffers	Partially collectible	Partially collectible	Section B
		Provision of memory aids	Partially collectible	Partially collectible	Section B
	Panel design	Conformity of standards, conventions, and nomenclature	Not collectible	Collectible	Section C
		Availability of indications	Partially collectible	Collectible	Section B
		Availability of controls	Partially collectible	Collectible	Section B
	Status indication	Existence of wrong or inadequate information	Not collectible	Collectible	Section C
		Appropriateness of task feedback	Partially collectible	Collectible	Section C
		Provision of clear decision criteria	Partially collectible	Collectible	Section C

making it favorable for evaluating aspects of human factors engineering, such as information coding and visual representation.

4 Direction for Designing Empirical Studies Using Rancor

INL posited several empirical studies using Rancor Microworld. The goal of this research is to 1) infer actual-operator data collected from a full-scope simulator by using Microworld data from student subjects and 2) collect additional data potentially missed in full-scope research. The following sections detail these goals.

4.1 Treatment of Items Collectible in Both Rancor Microworld and Full-Scope Simulators

Items collectible in both Rancor Microworld and full-scope simulators align with the first of the aforementioned goals. Figure 4 outlines a method for inferring actual-operator data collected from a full-scope simulator by using student data from Rancor Microworld. To achieve this, the differences between actual operators and students first need investigated. In the author's previous study [7], an investigation into the differences in human performance between actual operators and students using Rancor Microworld was initiated. Secondly, the differences that occur when operators use different simulators with varying levels of complexity need to be understood. Lastly, based on the results of the first and second steps, actual-operator data from full-scope simulators can be approximately inferred through the student data from Rancor Microworld.

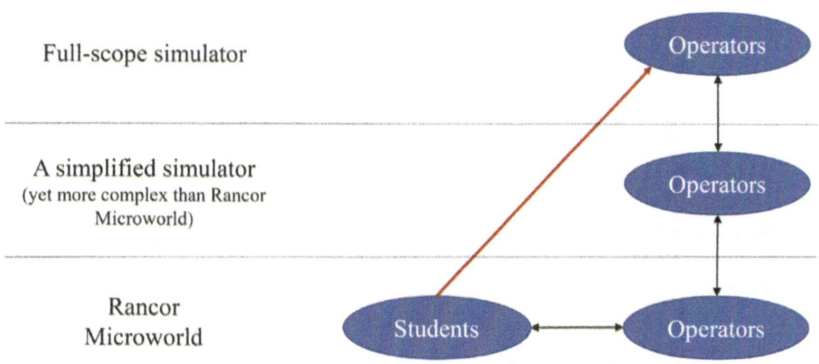

Fig. 4. Method for inferring actual-operator data from a full-scope simulator by using student data from Rancor Microworld.

4.2 Treatment of Items Only Collectible in Rancor Microworld

Items only collectible in Rancor Microworld would contribute to existing efforts to collect additional HRA data that are difficult to gather using a full-scope simulator. In HRA, the concept of a performance-shaping factor (i.e., any factor influencing human performance) has been used to highlight error contributors and adjust nominal human error probabilities [8]. Through empirical studies using Rancor Microworld and student subjects, it is possible to collect a variety of performance-shaping factor data not collected in the full-scope studies. Furthermore, when designing an interface, the results of this study could serve as a reference for selecting the best graphical design among various human factors engineering principles.

5 Conclusion

This paper suggests collectible items within Rancor Microworld compared to full-scope simulators. Collectible items and their levels in Rancor Microworld are identified in comparison with full-scope simulators. A method for treating the items is suggested with proposed experiment directions to be treated in the future. This study represents an ongoing effort to experimentally collect data using Rancor Microworld. Experiments introduced in this paper will be conducted at a future date. Results of experiments using Rancor Microworld are expected to foster a variety of HRA data items, with sufficient sample sizes to inform HRA.

Acknowledgements. This work of authorship was prepared as an account of work sponsored by Idaho National Laboratory, an agency of the U.S. Government. Neither the U.S. Government, nor any agency thereof, nor any of their employees, makes any warranty, express or implied, or assumes any legal liability or responsibility for the accuracy, completeness, or usefulness of any information, apparatus, product, or process disclosed, or represents that its use would not infringe on privately-owned rights. Idaho National Laboratory is a multi-program laboratory operated by Battelle Energy Alliance, LLC, for the U.S. Department of Energy under Contract DE-AC07-05ID14517.

References

1. Hannaman, G., Spurgin, A., Lukic, Y.: Human cognitive reliability model for PRA analysis. NUS-4531 (1984)
2. Jung, W., et al.: HuREX–A framework of HRA data collection from simulators in nuclear power plants. Reliab. Eng. Syst. Saf. **194**, 106235 (2020)
3. Chang, Y.J., et al.: The SACADA database for human reliability and human performance. Reliab. Eng. Syst. Saf. **125**, 117–133 (2014)
4. Bye, A., et al.: International HRA Empirical Study—Phase 2 Report: Results from Comparing HRA Method Predictions to Simulator Data from SGTR Scenarios. U.S. Nuclear Regulatory Commission (2011)
5. Ulrich, T.A.: The Development and Evaluation of Attention and Situation Awareness Measures in Nuclear Process Control Using the Rancor Microworld Environment. University of Idaho (2017)
6. Park, J., et al.: A guideline to collect HRA data in the simulator of nuclear power plants. KAERI/TR-5206/2013 (2013)
7. Kim, J., et al.: An experimental design on the use of rancor microworld simulator: a comparison of human performances between actual operators and students. In: Transactions of the Korean Nuclear Society Autumn Meeting, Goyang, Korea, 24–25 October 2019
8. Park, J., et al.: An experimental investigation on relationship between PSFs and operator performances in the digital main control room. Ann. Nuclear Energy **101**, 58–68 (2017)

Research on Prevention and Control Countermeasures of Team Situation Awareness Errors in Digital Nuclear Power Plants

Pengcheng Li[1,2], Xiaofang Li[1], and Licao Dai[1(✉)]

[1] Human Factor Institute, University of South China, Hengyang 421001, Hunan, China
dailpc@sina.com
[2] Department of Industrial Engineering, Tsinghua University, Beijing 100084, China

Abstract. To identify the main situation awareness issues of digital nuclear power plants. Based on the established team situation awareness error analysis method, the collected human factor event reports of digital nuclear power plants were analyzed to identify the main team situation awareness errors and its cause factors. Specific systematic prevention and control measures were proposed according to the defective cause factors for the prevention and reduction the occurrence of team situation awareness errors.

Keywords: Team situation awareness error · Performance influencing factors · Prevention and control countermeasure · Digital nuclear power plants

1 Introduction

In complex social-technical systems such as nuclear power plants (NPPs), due to the complexity and diversity of tasks, an operator does not have sufficient knowledge, experience and time to complete such complex tasks. Therefore, the monitoring and operation of abnormal conditions of the system are completed by a team. The performance of a team is essential to the safety of NPPs. Many studies had shown that team performance level is positively related to team situation awareness (TSA) level [1–3]. Therefore, the reduction of TSA errors is an important way for improving team's performance and the safety of NPPs. At present, there are a number of studies on human error prevention. For example, Park K.S. [4] analyzed human error prevention countermeasures, including improving the situation level, changing the personnel, and reducing the system impact. Kontogiannis T. [5] suggested to prevent and control human error by error recovery means. Reason thought that human error should be prevented from the perspective of a organization. U.S. Department of Energy [6] provided some human error prevention tools for prevention of human error. However, these researchers focus on human error prevention, not on TSA errors. At the same time, these prevention countermeasures are not systematic. Therefore, this article

P. M. Arezes and R. L. Boring (Eds.): AHFE 2020, AISC 1204, pp. 369–376, 2020.
https://doi.org/10.1007/978-3-030-50946-0_49

proposes prevention countermeasures of TSA errors from the perspective of the entire process and defense-in-depth.

2 Analysis Model of TSA Errors

Based on analysis of cognitive process, and combined with relevant theories such as system theory and information processing theory, a TSA error analysis framework of digital NPPs was shown in Fig. 1. Individual factors, team factors, situation factors, and organizational factors affect team cognitive process, thus leads to TSA errors, including team perception errors, team understanding errors, and team assessment errors, each aspect can be broken down into more specific errors (see Fig. 1). Because human factor incident reports from digital NPPs do not involve the external factors of the organization of NPPs, the external factors of organization are not considered here such as the supervision department and government factors, and only the internal factors of the organization of NPPs are considered. The definition and classification of specific factors and errors refer to the reference [7].

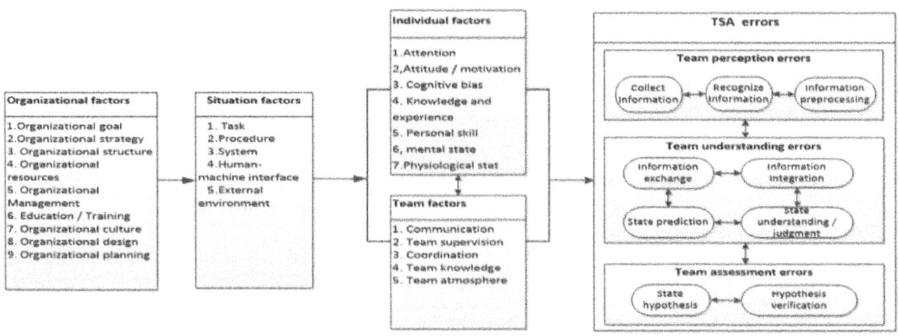

Fig. 1. The analysis framework of TSA errors in digital NPPs

3 Identification of the Main TSA Errors and Influencing Factors

3.1 Main TSA Errors

In order to identify the main TSA errors and influencing factors in digital NPPs, a total of 184 incident reports from 2010 to 2017 were collected from digital NPPs, of which 145 were related to human errors (human errors includes TSA errors), the percentage is 78.8%, the remaining incident reports are about equipment or technology failure. Based on the established TSA error analysis framework and its classification system, human factor event reports analysis was performed to identify the main TSA error modes and performance influencing factors (PSFs) in digital NPPs. In the process of analyzing human factor incidents, a human factor event may contain 1, 2 or even more TSA errors. The same type of TSA errors are merged into 1 without repeated calculations.

The relevant PSFs influencing TSA errors are also statistically analyzed. The statistical results are shown in Fig. 2.

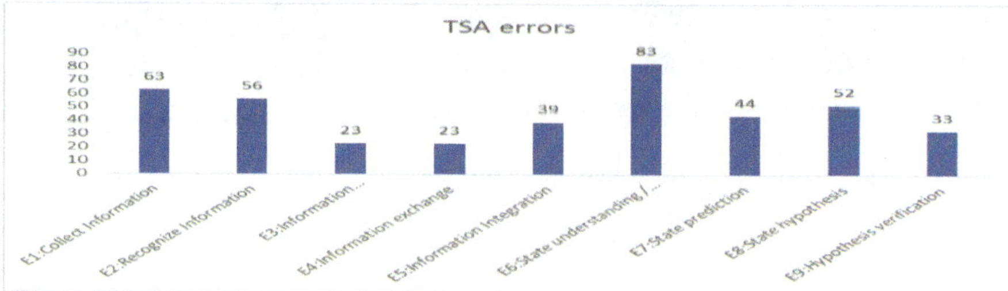

Fig. 2. The results of TSA errors

It can be seen from Fig. 2 that the understanding errors (including information exchange errors, information integration errors, state judgment/understanding errors, and state prediction errors) have the highest occurrence frequency, and are the main TSA errors. The next is team perception error and team assessment error. Among team understandings error, the state judgment/interpretation errors (frequency: 83) accounted for 57.24% of the total number of human error event reports, which may be due to incomplete information collection and insufficient information understanding etc. The second highest percentage is information collection errors (frequency: 63), accounting for 43.45% of the total number of human error events, which may be due to lack of knowledge and experience of team members, poor work attitudes or inattention, poor human-machine interface design and system design etc. Information recognition errors (frequency 56) accounted for 38.62% of the total number of human error events, which may be caused by insufficient risk awareness or insufficient understanding of information etc.

3.2 Main PSFs

Among the 145 sample data of human error events, the analysis results of PSFs that triggered TSA errors are shown in Figs. 3, 4, 5 and 6, including individual factors, team factors, situation factors and organizational factors.

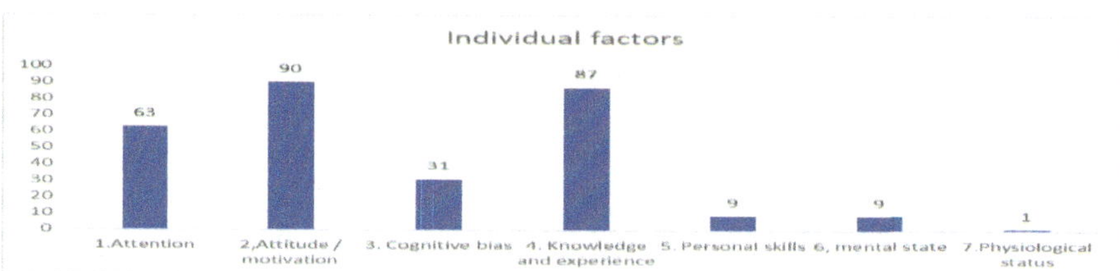

Fig. 3. Individual factors

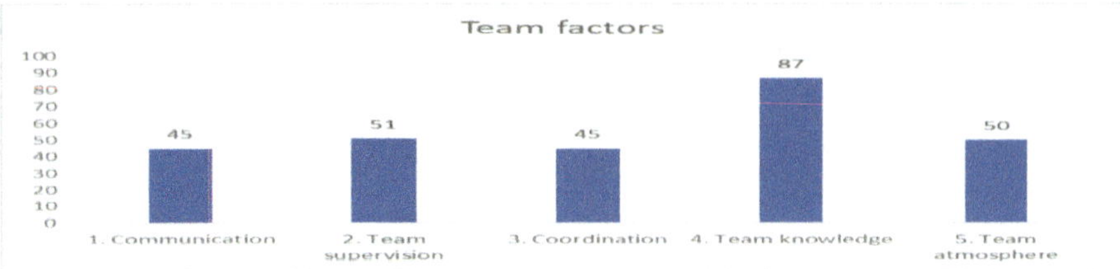

Fig. 4. Team factors

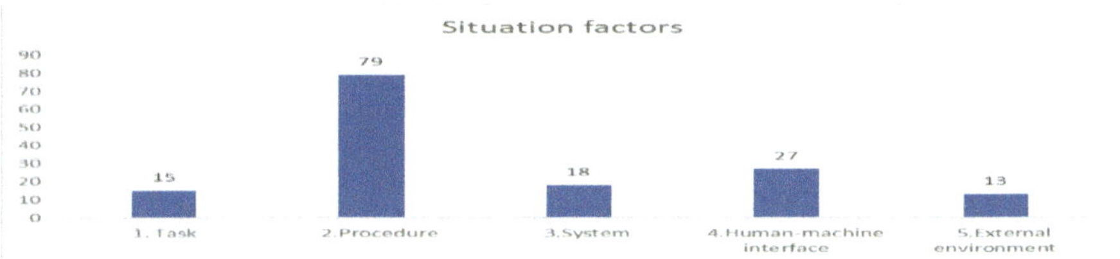

Fig. 5. Situation factors

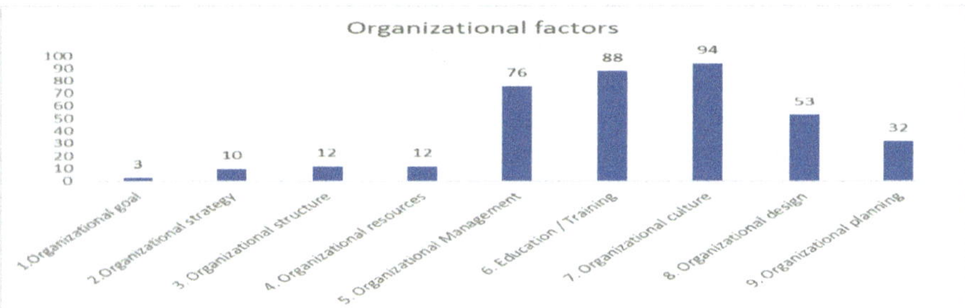

Fig. 6. Organizational factors

In the aspect of individual factors, according to Fig. 3, motivation/attitude (frequency 90) is the main performance influencing factor, which may be due to the lack of questioning attitude and risk awareness, overconfidence, procedure violation, or over simplification of operation and shortcut during the task implement; secondly, PSF is inadequate knowledge and experience factor (frequency 87), it due to insufficient communication or education/training. In the aspect of team factors, as shown in Fig. 4, the main factor influencing TSA error is team knowledge (frequency 87). The individual is a part of the team. If the individual knowledge and experience are insufficient, then the team knowledge and experience are insufficient. The next factor is team supervision (frequency 51), which may be due to lack of supervision, inadequate supervision, interference from external environment or the organization has not formed a good supervision mechanism etc. Furthermore, in the aspect of situation factors, as shown in Fig. 5, procedure factor (frequency 79) is the main factor triggering error, which has some problems such as insufficient or no description of precautions, incomplete operation steps, and no or incomplete relevant warnings, poor interface design of electronic procedure etc. Next, it is the human-machine interface or human-system interface factor (frequency 27), it due to there are some issues in the interface design such as inadequate information display, inadequate interface layout and complex interface management tasks etc. Finally, It can be seen from Fig. 6 related to organizational factors that the main PSFs are organizational culture (frequency 94), and education/training (frequency 88) etc. Based on the above analysis results, we should focus on the main PSFs causing TSA errors and put forward preventive measures to reduce the occurrence of TSA errors.

4 Prevention and Control Strategies of TSA Errors

Generally speaking, people themselves are unstable and changeable. A person's behavior is always affected by many aspects, including physical condition, psychological pressure, environmental impact, etc. Therefore, when a person's behavior is disturbed by these factors, it is easy to produce human error. Therefore, we should control and reduce the influence of these factors. In order to prevent the TSA errors, from the point of task process, prevention and control countermeasures can be provided form three aspects, namely before, during and after the work. In addition, from the cause chain point of view, prevention and control can be carried out from the government and regulatory departments (not considered here), organizational factors, situation factors, team factors and personal factors. The combination of the two views forms a systematic TSA error prevention and control system as shown in Table 1 for specific explanation.

Table 1. TSA multi process and multi-level comprehensive prevention system

Process cause	Beforehand	Halfway	Afterward
Individual	Task preview, to improve the level of knowledge and experience	Questioning attitude	Self-examination
Team	Pre-employment meeting, Simulation drill	Three stage communication etc.	To strengthen supervision etc.
Situation	To improve the state level of situation factors etc.	To reduce interference	System response such as alarm
Organization	To improve organization and management level, etc.	To provide support, such as emergency support	Timely feedback, learning and correction

4.1 Prevention and Control in Advance

Prevention and control in advance is to identify and prevent the potential defects that induce TSA errors, including organizational defects, situation defects, team and individual defects. This includes many measures, for example, in the aspect of individual factors, personnel selection, evaluation, etc. Team part should have clear role positioning, task allocation, personnel allocation, communication and cooperation rule, etc. In the aspect of situation factor, it is mainly to identify the potential design questions issues about human-computer interface, digital procedures, technical system, working environment, task etc. As much as possible to improve these PSFs level to reduce operators' working load and pressure level. In terms of organization, it is the root cause of human error (including TSA errors). Therefore, the prevention and control of human error should make the state level of various organizational factors affecting TSA error in the best state. Especially organizational culture, training, organizational management and organizational design. Secondly, we should do a good job in preparing for the task, including task preview, pre-employment meeting, simulation drill etc.

4.2 Prevention and Control During Work

The control during work is a process or behavior that makes a good plan, arrangement, implementation and control of the upcoming tasks. Due to various objective reasons, it may be difficult to completely eliminate all kinds of errors in advance prevention. Therefore, in order to make the task complete smoothly, in-process control can generally be controlled by using the "human error prevention tool". For example, in terms of individuals, to reduce their own errors and team errors in the process of work, it is necessary to use the tools such as questioning attitude, self-check of stars, abide by operating procedures, etc. In terms of teams, it is necessary to use three-stage communication, team monitoring and other tools. In terms of situation factors, it is necessary to reduce the number of simultaneous tasks, reduce the complexity and urgency of tasks (through reasonably arrange the number of personnel, automatic control and other means), reduce the interference caused by the working environment (eliminate noise, reduce the interference of irrelevant external personnel), continuously optimize the human-computer interface and procedures in practice, improve the working efficiency. In terms of organization, in the emergency handling process of abnormal incidents or accidents, the organization should provide advanced management methods, tools and means, and provide corresponding personnel staff, resource support, etc. to coordinate the work of the team.

4.3 Prevention and Control After Work

It should be noted that TSA errors are cognitive errors, but they can be identified by the discovery of active human error. Human error prevention strategy is to intervene between behavior and error, while error recovery is to intervene between error and consequence. Error recovery system includes (1) error detection; (2) error explanation or plan; (3) error correction; (4) prevention and minimization of error consequences [5]. Based on previous research results [8], the strategies to support error detection include:

(1) communication or supervision; (2) feedback; (3) restrictive function; (4) standardized check; (5) contingency plans.

Based on the above strategies, there are three effective ways to detect errors, one is to conduct independent inspection and evaluation by others; the other is to provide information feedback or limit and stop function to detect system failure clues to detect errors; the third is operators conduct routine inspection on their behaviors or suspect the occurrence of errors to carry out self-monitoring or self-check to detect mistakes. The first strategy is based on effective team communication and supervision to detect errors in the plan. The second strategy is based on the design of technical system and human-computer interface to provide feedback, limit and stop functions. The third strategy is based on correct and effective operator training plan to improve the level of knowledge and experience and form good operation habits and error detection ability. In addition, the emergency plan is also a way of self-monitoring. Operators can clearly describe their decision-making standards through the emergency plan. When they find that their processing strategies cannot meet the decision-making standards, they can find errors in the plan, which is very important for dynamic emergency events. The error explanation (cause identification) belongs to the cognitive stage, and there are difficulties to detect brain explanation errors for individual operator, so it is mainly detected through self-monitoring, questioning attitude, communication and supervision, and automatic cause diagnosis etc. The error correction stage (behavior consequence stage) is the same as the behavior implementation stage. The recovery ways to prevent and minimize the consequences of errors includes error-tolerant design through system design, delay the propagation of errors, and reduce the impact of errors through safety barriers and limiting functions. It should be noted that there is partial repetition between error recovery ways and human error prevention tools. For example, work monitoring means to control the occurrence of errors and detect and recover the errors.

5 Conclusion

TSA is very important for complex high-risk systems such as NPPs. Once a TSA error occurs, a major accident may occur. Therefore, it is necessary to prevent and control TSA errors. The main TSA errors and PSFs are identified by analyzing event reports in digital NPPs. The main TSA errors are state understanding errors and information collection errors etc. Main PSFs are: motivation/attitude, knowledge and experience etc. from individual factors, team knowledge and team supervision etc. from team factors, procedure factor and human-system interface factors etc. from situation factors, and organizational culture and education/training factors etc. from organizational factors. Finally, a systematic TSA error prevention and control system is provided, including prevention and control in advance, during work and after work, and provided the special countermeasures.

Acknowledgments. The financial support by the National Natural Science Foundation of China (No. 51674145), Natural Science Foundation of Hunan Province (No. 2017JJ2222) are gratefully acknowledged.

References

1. Kim, S.K., Park, J.Y., Byun, S.N.: Crew resource management training for improving team performance of operators in Korean advanced nuclear power plant. In: IEEE International Conference on Industrial Engineering and Engineering Management, Hong Kong, China, pp. 2055–2059 (2009)
2. Lin, C.J., Yenn, T.C., Yang, C.W.: Evaluation of operators' performance for automation design in the fully digital control room of nuclear power plants. Hum. Factors Ergon. Manuf. Serv. Ind. **20**,10–23 (2010)
3. Lin, C.J., Hsieh, T.L., Yang, C.W., Huang, R.J.: The impact of computer-based procedures on team performance, communication, and situation awareness. Int. J. Ind. Ergon. **51**, 21–29 (2016)
4. Park, K.S.: Human Reliability Analysis, Prediction, and Prevention of Human Errors. Elsevier, New York (1987)
5. Kontogiannis, T.A.: Framework for the analysis of cognitive reliability in complex systems: a recovery centred approach. Reliab. Eng. Syst. Saf. **58**, 233–248 (1997)
6. U.S. Department of Energy.: Human performance improvement handbook. Volume 2: human performance tools for individuals, work teams, and management. DOE-HDBK-1028-2009, Washington, D.C. (2009)
7. Li, X.F.: Team situation awareness errors analysis in digital nuclear power plants. Master Dissertation, University of South China, Hengyang, Hunan, China (2019)
8. Li, P.C., Zhang, L., Dai, L.C., et al.: Study on operator's situation awareness error and prevention and control in nuclear power plants. Atomic Energy Sci. Technol. **50**(2), 323–331 (2016)

A Study of Crew Error on the Interface Between Passive Side Stick and Electronic Flight Control System in Cockpit of Civil Aircraft

Zhang Yinbo(✉), Zhu Yao, Zhou Yang, Lu Shasha, and Meng Hua

COMAC ShangHai Aircraft Design and Research Institute,
No. 5188 Jinke Road, Pudong, New Area, Shanghai, China
zhangyinbo@comac.cc

Abstract. CS 25.1302 requires that, to the extent practicable, installed equipment must enable the flight crew to manage errors resulting from the kinds of flight crew interactions with the equipment that can be reasonably expected in service, assuming the flight crew is acting in good faith. According to this requirement, civil aircraft design should consider human factor influence on the human-machine interface between systems. The purpose of crew error analysis in aircraft design is to identify the possible mistakes which are made by flight crew during their operation on equipment in the cockpit and to find the root cause of any possible incidents, in order to take proper measures for mistake prevention or fault-tolerance. This thesis proposed a new method for a dedicated design of a civil aircraft, using this method, this thesis identified the crew errors related to human factors, analyzed the cause and consequences of the crew errors, and proposed proper measure that can be taken to prevent the mistakes.

Keywords: Human factor · Crew error · Side stick · Electronic flight control system

1 Introduction

AMC 25.1302 [1] states that "reasonably expected errors in service" related to cockpit human-machine interface design should be identified, that is, the potential crew error.

According to the integration, comprehensive and complexity analysis for a civil aircraft, passive side stick and electronic flight control system of this aircraft are two complicated systems which are newly designed and highly integrated design features of the aircraft.

Based on the analysis result of intended flight crew tasks, this thesis identified and analyzed potential crew errors in the human-machine interface between passive side stick and electronic flight control system in the cockpit of civil aircraft.

P. M. Arezes and R. L. Boring (Eds.): AHFE 2020, AISC 1204, pp. 377–384, 2020.
https://doi.org/10.1007/978-3-030-50946-0_50

2 Background

2.1 Design Feature Description

The electronic flight control system of the civil aircraft uses a digital Fly by Wire with a passive side stick as the pilot's main control input device. With redundant flight control digital computer, actuator electronic control unit and remote control electronic unit as core processing components, with electromechanical and electro-hydraulic servo actuators as executing components, the corresponding flight control surfaces can be controlled through the calculation of the control law, so as to realize full-time and full-authorization Fly by Wire. Compared with the traditional side stick, the passive side stick has the advantages of reducing the weight of the control system, optimizing the cockpit display-control layout, improving the quality of aircraft control, reducing the flight crew workload and facilitating maintenance.

According to the characteristics of the civil aircraft, the passive side stick and electronic flight control system not only contains the passive side stick control unit intended for the pilot and co-pilot input and the primary flight control mode, it also includes all the control components related to the electronic flight control system. In addition to the control equipment above, there is also a rudder pedal brake assembly, a speed brake lever, a trim control panel, there are also related indications which include a sidestick authority indicator, a Synoptic page of flight control system diagram (on MFD), trim information display on Engine Indication and Crew Alerting System and Crew Alerting System messages related to the primary flight control system.

2.2 Analysis Method of Flight Crew Errors

Crew errors are defined as situations that deviate from the intention of the crew or the consequence due to the actions or omissions of the flight crew. There are four commonly used human error analysis methods: Systematic Human Error Reduction Approach (SHERPA) [2], Human Error Template (HET) [3], Hazard and Operability Study (HAZOP) [4], and Human Error Identification in System Tool (HEIST) [5]. The advantages and disadvantages of the four methods are shown in Table 1.

Table 1. Advantages and disadvantages of the methods for human error analysis

Method	Advantages	Disadvantages
SHERPA	1) Data supporting credibility and validation 2) Versatile, short learning time, easy to use	1) costs too much time for large complex tasks 2) Need to perform corresponding hierarchical task analysis
HET	1) Short learning time and less training required 2) Error checklist is generated based on pilot error type and frequency analysis, and has the highest accuracy for aircraft cockpit evaluation	1) costs too much time for large complex tasks 2) Need to perform corresponding hierarchical task analysis

(*continued*)

Table 1. (*continued*)

Method	Advantages	Disadvantages
HAZOP	1) Short learning time and less training required 2) Using HAZOP correctly can identify all possible errors in the system as much as possible	1) costs too much time 2) the guidance tends to be limited to nuclear energy, chemical industry 3) the guidance is not specific enough
HEIST	1) Use classification tables to help analysts identify potential errors 2) provides relevant results for each error issue and management measures for existing designed error	1) Need to use a lot of raw data 2) It is used for a single purpose and has the lowest accuracy compared to other analysis methods when analyzing flight missions

3 New Flight Crew Error Analysis Method

The purpose of crew error analysis is to identify errors that may occur during the interaction between the crew and the cockpit human-machine interface during product development phase, and analyze the causes of the errors in order to add error prevention or fault tolerance measures in the aircraft design. However, none of the above four methods alone can achieve the stated purpose. Therefore, based on the above four methods, this thesis has developed a more suitable error analysis method for the cockpit of the civil aircraft, that is, Human Error Template+ (HET+). The analysis steps of HET + error analysis method is shown in Table 2 (Fig. 1).

Table 2. Steps of HET+

Steps	Activities needed	Output
1. Build up a team	Build up an analysis team, which members include the pilot, human factor specialist, system design engineer, cockpit design engineer and safety analysis specialist. The pilot directly take part in the preliminary identification of crew error, likelihood analysis and hazards level analysis. The human factor specialist. System design engineer, cockpit design engineer take part in the whole process. The safety analysis specialist mainly take part in the hazards level analysis	Crew error analysis team

(*continued*)

Table 2. (*continued*)

Steps	Activities needed	Output
2. Hierarchical task analysis	Analyze the related tasks intended for the analysis objects (such as side stick and EFCS), in order to obtain the bottom behavior in every normal and abnormal task	Bottom-level task behavior
3. Based on the error check-list, preliminary identification of possible errors in the crew when performing tasks	Using the error checklist [6] to perform error identification on the underlying task behaviors analyzed in step two, and preliminary identification of design-related errors that may occur when performing the current task, including: 1) Potential or possible mistakes identified by pilots based on experience when performing similar tasks on other aircraft types; 2) Mistakes that may occur in combination with the civil aircraft human-machine interface design features	Preliminary identification of design-related potential crew error
4. Causes analysis of crew error	With regard to the crew errors in this step, analyze the reasons that cause the errors. The causes of errors may include: interface design issues (potential non-compliance with AMC25.1302) and other issues	Analysis of the causes of crew errors based on the existing human-machine interface design characteristics
5. Hazards analysis	Based on subjective judgment, analyze the likelihood of errors occurring during the execution of the current task, and determine if they are H (High), M (Medium), or L (Low). Analyze the possible consequences of errors, analyze their impact on aircraft safety margins or functional	Hazard level of crew error

(*continued*)

Table 2. *(continued)*

Steps	Activities needed	Output
	capabilities, workload or physical discomfort to the flight crew, and injuries to members other than the flight crew. Determine the level of hazard corresponding to potential errors, I (No safety impact), II (minor), III (major), IV (hazardous), V (catastrophic). Crew errors with a likelihood of H and a hazard of III, IV, V or errors with a likelihood of H and a hazard of IV, V will be analyzed furthermore in the next step	
6. Analysis of error management measures	Analyze the management measures for potential human errors with hazard levels III, IV	Error management measures corresponding to potential human errors

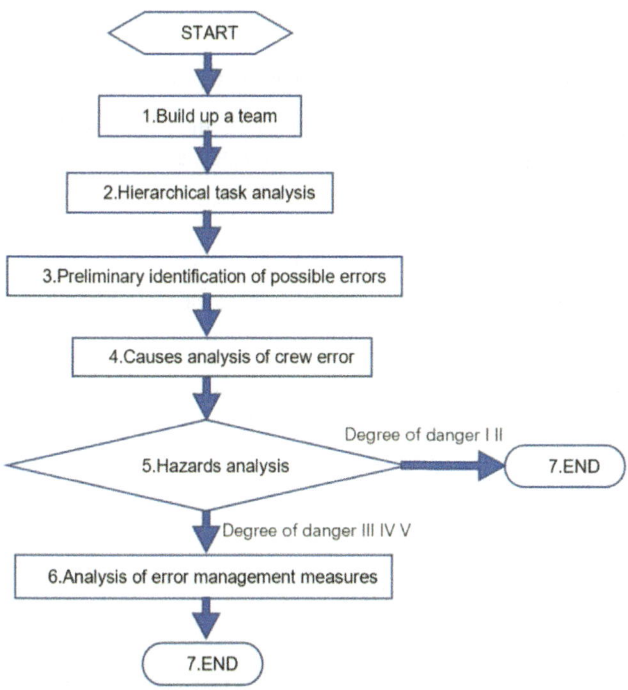

Fig. 1. HET + Crew error analysis method flow

4 Analysis and Discussion

Based on the tasks intended for flight crew, which is the input of step 3 of HET+. This thesis analyzed the crew error on the interface between passive side stick and electronic flight control system.

4.1 Preliminary Identification of Potential Crew Errors (Step 3)

Based on the error analysis checklist, this thesis analyzed each underlying behavior obtained from the task analysis (such as operations manual) [7], and preliminarily identified potential design-related potential errors.

Table 3. Preliminary identification of potential crew errors based on stall warning task

Number	Abnormal/emergency crew task	Bottom flight crew task	Error type	Potential error
1	Low energy warning (Voice warning "SPEED, SPEED, SPEED")	1.1- Increase thrust until alert disappears note: when the speed is less than VLS-TBD kt, ASP (Auto-throttle Speed Protection) will initiate the auto throttle, which can get the aircraft back to preselected speed	A2	More likely to be too late when increasing thrust
			A3	Forgot to increase thrust
			A...	......

4.2 Cause Analysis of Potential Unit Errors (Step 4)

For the identified design-related potential unit errors, the causes of their occurrence are analyzed, and the results obtained are shown in Table 4.

Table 4. Cause analysis of potential errors

Number	Task name	Bottom flight crew task	Error type	Potential error	Cause of the error
1	Abnormal crew task-Stall Warning	1.1-Lower nose to decrease angle of attack	A2	Pitch down too late	The crew did not operate in accordance with the operating procedures, which is not a problem of interface design
......	......	......	......	......	......

4.3 Analysis of Potential Error Dangerous Degree (Step 5)

Based on the identified design-related potential crew errors, combined with the functional risk assessment of the aircraft's main flight control system, the equivalent functional failure after the error occurred was analyzed, and the degree of danger after the error was analyzed. The results obtained are shown in Table 5.

Table 5. Preliminary analysis of the degree of potential error risk analysis

Number	Task name	Bottom flight crew task	Error type	Potential error	Flight control system FHA failure status	Degree of danger
1	Abnormal crew task-Stall Warning	1.1-Lower nose to decrease angle of attack	A2	Pitch down too late	Loss of pitch attitude protection	III
……	……	……	……	……	……	……

4.4 Error Management Measures (Step 6)

This section analyzes the error management measures for potential human errors with danger levels III, IV, and V. The analysis results are shown in Table 6.

Table 6. Analysis of potential error management measures

Number	Task name	Bottom flight crew task	Error type	Potential error	Degree of danger	Analysis of error management measures
1	Abnormal crew task-Stall Warning	1.1-Lower nose to decrease angle of attack	A2	Pitch down too late	III	The angle-of-attack protection function is activated when the aircraft's angle of attack is higher than a pre-set threshold of the aircraft type
……	……	……	……	……	……	……

5 Conclusion

The conclusion can be made that the new crew error analysis method (HET+) is helpful for identifying possible errors in flight crew's flight task. The HET+ flowchart can be used to perform crew error assessment on any system on the aircraft. This thesis uses the abnormal crew task-stall warning as an example to briefly explain how to perform crew error analysis and evaluation. Through the level-by-level analysis of this method, the error cause and hazard level of the bottom level task can be found, and then error management measures can be given to prevent crew errors.

References

1. EASA, "Certification Specifications for Large Aeroplanes CS25 AMC", AMC25.1309, EASA, 2009/017/R, Cologne, Germany (2009)
2. Hannaman, G.W., Spurgin, A.J.: Systematic Human Action Reliability Model for PRA Analysis EPRI-NP-3585. Electric Power Research Institute, Palo Alto (1983)
3. Marshall, A., Stanton, N., Young, M., Salmon, P., Harris, D., Demagalski, J., Waldmann, T., Dekker, S.: Development of the human error template- a new methodology for assessing design induced errors on aircraft flight decks. ERRORPRED Final ReportE!1970 (2003)
4. Kenndy, R., Jones, H., Shorrock, S., Kirwan, B.A.: HAZOP analysis of a future ATM system. Contemporary Ergonomics (2000)
5. Kirwan, B.: A Guide to Practical Human Reliability Assessment. Taylor and Francis, London (1994)
6. Stanton, N.A., Salmon, P.M., Walker, G.H., Baber, C., Jenkins, D.P.: Human Factor Methods-A Practical Guide for Engineering and Design, Human Factor Integration Defence Technology Center. Ashgate Publishing Limited (2005)
7. Boeing. Boeing 737NG Operations Manual Normal Procedures Amplified Procedures, Flight Deck Preparation. Boeing Published, Seattle, Washington (2000)

Author Index

P. M. Arezes and R. L. Boring (Eds.): AHFE 2020, AISC 1204, pp. 385–387, 2020.
https://doi.org/10.1007/978-3-030-50946-0

The manufacturer's authorised representative in the EU is Springer Nature Customer Service Centre GmbH, Europaplatz 3, 69115 Heidelberg, Germany. If you have any concerns regarding our products, please contact ProductSafety@springernature.com

Printed and bound by CPI Group (UK) Ltd, Croydon, CR0 4YY
30/09/2026
02225509-0013